Dying on the Vine

Dying on the Vine

How Phylloxera Transformed Wine

George Gale

UNIVERSITY OF CALIFORNIA PRESS
Berkeley · London · Los Angeles

University of California Press, one of the most distinguished university presses in the United States, enriches
lives around the world by advancing scholarship in the
humanities, social sciences, and natural sciences. Its
activities are supported by the UC Press Foundation and
by philanthropic contributions from individuals and
institutions. For more information, visit www.ucpress.edu.

University of California Press
Berkeley and Los Angeles, California

University of California Press, Ltd.
London, England

Library of Congress Cataloging-in-Publication Data

Gale, George, 1943–.
  Dying on the vine : how phylloxera transformed
wine / George Gale.
    p. cm.
  Includes bibliographical references and index.
  ISBN 978-0-520-26548-6 (cloth, alk. paper)
    1. Grapes— Diseases and pests—History—19th
century. 2. Grapes—Diseases and pests—20th
century. 3. Phylloxera. I. Title. II. Title: How
phylloxera transformed wine.

SB608.G7G33 2011
634.8'2752—dc22   2010031697

Manufactured in the United States of America

19 18 17 16 15 14 13 12 11
10 9 8 7 6 5 4 3 2 1

This book is printed on Cascades Enviro 100, a
100% post consumer waste, recycled, de-inked fiber.
FSC recycled certified and processed chlorine free. It
is acid free, Ecologo certified, and manufactured by
BioGas energy.

Contents

List of Illustrations   *vii*

Acknowledgments   *ix*

Introduction   *1*

1. Disaster Strikes: "All your vines are fatally condemned to disappear, Monsieur"   *13*

2. La Défense: Sand, Submersion, and Sulfiding   *51*

3. La Reconstitution   *79*

4. The Underground Battle: Grafting on American Rootstock   *120*

5. Phylloxera Makes the European Grand Tour   *163*

6. The Bug Goes South: New Venues, Same Story   *184*

7. The Old Americans, or How the Fox Conquered Europe   *201*

8. Phylloxera Breaks Out (Twice) in California   *211*

Conclusion   *247*

Appendix A. Life Cycle of Phylloxera   *251*

Appendix B. American Wild Grape Species   *253*

Appendix C. Old American Varieties   *257*

Notes   *259*

Glossary   *285*

Bibliography   *287*

Index   *303*

Illustrations

Figures

1. J.-É. Planchon / *17*
2. Phylloxera / *20*
3. F.E. Guérin-Méneville / *21*
4. Victor Signoret / *22*
5. C.V. Riley / *34*
6. Phylloxera life cycle / *36*
7. Inspecting the vines / *44*
8. Steam-powered pump flooding vineyard / *66*
9. Sulfide treatment team / *73*
10. Pal injector / *74*
11. Statue of Alexis Millardet / *92*
12. Gaston Bazille / *101*
13. Marguerite, la Duchesse de Fitz-James, 1867 / *109*
14. T.V. Munson / *113*
15. Field graft / *125*
16. AxR1 leaves / *137*
17. Pierre Viala / *144*
18. Planchon monument / *158*
19. George C. Husmann / *216*
20. Professor Bioletti pruning a vine / *217*
21. Jeffrey Granett in his lab / *233*

Tables

1. Proposed uses of American species / 111
2. The scale of the grafting job, 1880 / 127
3. Chauzit's recommendations for rootstack varieties based on active lime tolerance / 155

Acknowledgments

A huge number of people helped me along the way as I wrote this book—none of whom, of course, is the least bit responsible for any of my mistakes. The UMKC Office of Research Administration funded my visits to the École Nationale Supérieure d'Agronomie de Montpellier and the University of California, Davis. At Montpellier I was welcomed by J.-P. Legros and hosted wonderfully by J. Michaux, and in the *bibliothèque* I was particularly well advised by Mme. B. Bye and M.J. Argelles; the Department of Viticulture chair A. Carbonneau was also especially helpful and hospitable. At Davis, J. Skarstad in Special Collections, J. Granett in Entomology, and P. Teller and K. Whelan in History and Philosophy of Science were very helpful. Pittsburgh's Center for Philosophy of Science supported me famously for a semester; in particular, R. Olby, J. Lennox, and P. Machamer gave me exciting things to think about. J. Whiting, D. Downie, A. Walker, and J. Granett (especially!) have been very generous with their time and expertise all along the way. N. Shanks was always there to help, too. Earlier versions of this research were presented at HSS-Atlanta, HSS-San Diego, and HSS-Vancouver; at colloquia in Davis and at Virginia Tech and East Tennessee State; and at the 4th International Pitt Center Fellows Conference, Bariloche, Argentina, June 2000. Finally, I must thank Ann Mylott, who originally pointed me in the direction of medical models of disease.

Introduction

In the mid-1860s, a near-microscopic yellow insect, the grape phyllox-era,[1] invaded the South of France and began killing the native vines, the Cabernets, Chardonnays, Syrahs, and all their kin. Within fifty years the invasion had spread throughout Europe and had jumped the oceans to Africa, South America, Australia, and California, laying waste to vineyards wherever it landed. It was a biological disaster of worldwide proportions, a disaster that ruined national economies, destroyed agri-cultural systems, and destabilized cultures, causing massive migrations of peoples to fan out over the face of the globe, bringing social and politi-cal change wherever they went. Although bits and pieces of this tale have been told,[2] much remains unchronicled and many lessons unlearned. In the story that follows, I attempt to describe in detail several crucial as-pects of the phylloxera disaster for the first time, and along the way to reveal—often in a near day-to-day fashion—the lives of some of the major players who successfully fought the battle against the invading devastator of vines.

Three major foci emerge from the story. First, the phylloxera case is the original and best model for studying the dynamics of the interaction between human beings and an invasive species as it colonizes a new habitat. Careful attention to the events of the phylloxera invasion can direct our attention to currently unfolding invasive phenomena. Second, the French response to the phylloxera invasion provides the original model for the genesis and development of one important type of Big Science, that powerful synergistic force combining government, industry, and research institutions. Study of the mechanisms of the birth, growth, and development of Big Science in phylloxerated nineteenth-century France can teach us much about the nature of Big Science in our own century. Finally, the phylloxera case exemplifies important issues in the history, philosophy, and sociology of science, providing data for controversies that grip these disciplines in their conflicting claims over how science actually works. As we shall see, our battle with phylloxera today is as alive and important as it was more than a century and a half ago.

Yet overarching these three issues is one ultimate, sobering truth: phylloxera is not gone. It is still there—in the soil and in the air—everywhere that it has traveled over the last century and a half. The bug is not—indeed, it cannot be—vanquished. Our struggle with this devastator of vines is never ending, since powering the struggle are the unstoppable forces of evolution. For every defense we muster, the bug ultimately will evolve a successful offense. There is no victory in this struggle, neither for us and our vines, nor for the bug. We must always be alert, especially to the lessons that will be revealed by this book's examination of the past battles. Never again can we let ourselves be lulled into complacency by temporary defensive successes. California's phylloxera disaster of the 1980s and '90s occurred precisely because the defenders got cocky and relaxed their vigilance. If nothing else, this book should show the disastrous consequences of drifting off to sleep in our unending struggle with the bug.

INVASIVE SPECIES

Our biological world is globalizing as rapidly as, and more dangerously than, our economic world. Invasive species, imported life forms as varied as animals, plants, bacteria, and viruses, today wreak havoc on indigenous systems at all levels of organization, from wetland communities to agricultural production systems to human social webs. Introduced into new ecologies where they can "make a living,"[3] foreign life forms flourish without natural enemies, overwhelming native species, degrading the local ecology, and frequently spreading widely, in the process wiping out whole biological economies. At the moment North America faces invasive threats on many fronts. Examples are ready at hand.

Zebra mussels *(Dreissena polymorpha),* bivalves from Russian freshwater lakes, first appeared in the Great Lakes in 1988. Since that time they have spread into watersheds of the Ohio and Mississippi rivers, and they are expected to continue to infest the eastern United States and Canada. Lacking natural enemies, they have increased in population, threatening not only local wildlife but also many human enterprises, including water filtration systems, power generation, and food sources, as they clog waterways and machinery.

Purple loosestrife *(Lythrum salicaria),* a European perennial flowering plant, has invaded wetlands throughout the Northeast, destroying native plants such as cattails and degrading the entire wetland ecological system. Since they have no competition, an end to loosestrife's spread is not in sight.

Dutch elm disease *(Ophistoma ulmi),* an Asian fungus spread by a beetle, has decimated populations of native American elms throughout the eastern United States, changing the look and feel of the entire countryside and cityscape. Although resistant hybrid elms have recently appeared on the market, there is no telling whether this will ultimately remedy the horrible devastation caused by this invasive species.[4]

Here we have only three examples of the destruction caused by invasive species. Unfortunately, these examples could be multiplied many times over, and, what is worse, the dangers are accelerating rapidly.[5] Although many factors are involved in the rise of invasive species, three are particularly important.

Global climate change affords new environments for invasive species to colonize. Although past climates might have been hostile to the existence of certain life forms, general warming trends allow the gradual spread northward of warm-climate pests. This is believed to be the mechanism allowing Pierce's disease, a bacterial infection spread by sucking insects such as the glassy-winged sharpshooter, to infect vineyards at ever-higher latitudes. Since the disease is sensitive to cold winter temperatures, as average temperatures rise, the bacterium survives further north. Other examples of temperature-sensitive invasive pests abound.

Increasing global transport and travel spread life-forms more quickly and more frequently than ever before. Zebra mussels are thought to have arrived in North America in the ballast—or perhaps on the anchor chains—of cargo ships arriving from Russian waters. It is suspected that commerce brought phylloxera to France. But commerce is not the only vehicle. Tourists today go everywhere, and in the process they transport foreign life-forms both wittingly and unwittingly. An amazing number of critters can live in the bit of mud on the bottom of a trekker's shoe.

The relentless focus on monoculture—the planting, raising, and growing of one variety or species of produce or livestock—provides an incredibly soft target for an invasive disease species as well. Current North American corn-growing practices provide a vast "lawn" just waiting for the next invasive blight, against which all those millions of acres of genetically identical corn plants haven't the slightest resistance.[6]

In each of these cases, and in myriad others, the social, political, economic, and scientific establishments are called upon to respond. Although all invasive species have the potential to lead to negative effects, some are downright disastrous, as the recent corn blight invasion and the ongoing Dutch elm epidemic illustrate.

But none of this is news. A century and a half ago the worst of all known invasive species disasters began in France and, in the end, spread to most of the civilized world.[7] This was the grapevine phylloxera catastrophe, the result of an invasion of a species of sucking insect from North America, which, in dreadful progression, wiped out the cultivated grapevine in Europe, Asia, Africa, South America, and Australia. Millions of people were displaced, national economies were ruined, agricultural systems were destroyed, and scientists, growers, politicians, and ordinary people from all levels of society were thrust into frantic action against the unknown and vicious killer of vines.

Phylloxera's invasion of the European vineyard heartland is the first and, until now, the worst instance of intercontinental invasive species. Before the advent of safe, reliable, and speedy transport between North America and Europe, the phylloxera bug could not survive the journey on the many vines and cuttings sent eastward across the Atlantic (Ordish 1972). But once steamships began plying the trade routes, the bug could survive the trip, alight in Europe, and, soon enough, find a marvelously undefended environment in which to make a living at the foot of the European grapevine, destroying untold thousands of kilometers across the continent.

The phylloxera case serves as a model for the story of all invasive species battles. Moreover, the phylloxera case didn't happen just once, between 1867 and 1900, in Europe; it happened again, in California, between 1980 and 2000, with a reprise of many of the themes from its first appearance. But the phylloxera story isn't unique. Three stages of the phylloxera story are repeated time and again in the stories of other invasive species. First, denial—"It can't happen here"—is the universal initial stage of invasive species debacles. Watching the growers of Burgundy come up with reasons why the disaster that was ruining the Midi wouldn't—indeed, couldn't—happen to them is no different from watching the denial expressed by growers faced with corn blight, or foresters faced with Japanese beetles, or communities faced with new diseases such as influenza A (H1N1). Denial certainly was the immediate response of Northern California winegrowers when phylloxera was found on some of their vines in 1980.

Denial is invariably followed by control efforts such as physical or administrative barriers, cordons sanitaires, or quarantines, which inevitably fail, as they did in every country battling the phylloxera bug.[8] Yet this is the first choice of the authorities in every invasive species case, inevitably to little or no avail, and typically at great cost to some interested community, whether farmers, hunters, foresters, or fishers. Careful examination of the phylloxera dynamic shows that physical barriers are most correctly understood as, at best, buyers of precious time, defensive holding actions to be taken while the real defense is conceived, tested, and deployed.

Eradication efforts, such as the use of industrial chemical agents, usually join the efforts at control. Unfortunately, they also inevitably fail.[9] These agents cost too much, require expert application, and have insidious side effects. European experience with CS_2, an industrial solvent effective as a pesticide against the phylloxera, conclusively demonstrates the limitations of this approach. We shall see why the industrial chemical approach failed to stop phylloxera, a valuable cautionary tale against such approaches against today's invaders.

Each of these three stages appears to a greater or lesser degree whenever invasive species events turn into crises or threaten to become disasters. But the invasive species model provided by phylloxera is not its only relevance. It also reveals to us the origin of an important form of Big Science.

PHYLLOXERA AND THE BIRTH OF BIG SCIENCE

A second but equally significant model is provided by the response to the phylloxera crisis: the clearly recognizable birth of Big Science, that powerful amalgam of government, industry, and research universities dedicated to conquering a massive threat to the common wealth.[10] France's effort to beat back and triumph over the devastating bug foreshadows other Big Science efforts such as the Manhattan Project and the Human Genome Project. The methods, structure, and organization that evolved during France's national counterattack on the insect find their counterparts in today's massive efforts.

At first the counterattack was a strictly local effort, spearheaded by quasi-autonomous nongovernmental organizations (QUANGOs) such as the Société Centrale d'Agriculture de l'Hérault (SCAH), a group of private citizens—growers, merchants, professionals from medicine, law, and the university, and other interested parties—who had come together years earlier because of their interest in the region's agriculture. The University of Montpellier and the nearby École Agronomique were rapidly drawn into the fray, as were QUANGOs such as the local and regional agricultural expositions.[11]

Two reasons for the essentially regional initial response to the danger were the political and cultural remoteness of the Midi from Paris and the total preoccupation of Paris with the disastrous war against Prussia between 1870 and 1872. Only after the war did Paris realize the scale of the calamity unfolding in the south and begin to mobilize in response to the crisis. First, of course, it formed a national committee, based in Paris, composed of scientists, industrialists, and a few private citizens. For the most part the members were highly visible actors from in and around Paris. J.-É. Planchon, the Montpellier botanist and co-discoverer of the bug, plus one or two others from the affected regions were the only exceptions to this general rule.

In addition to this administrative action taken by the Ministry of Agriculture, the legislators slowly awakened into action as well, first by offering a large cash prize to anyone who discovered a method of defeating the invader. The amount of the reward was radically increased almost immediately, but the money was never awarded. Soon thereafter the legislators began to pass laws instituting cordons sanitaires, quarantines, and other restrictions on the transport of vine material. Centralized teams were organized to inspect, evaluate, and enforce the restrictions. Local opposition to these teams' activities was widespread and sometimes violent.

On the positive side, the government focused on Big Science solutions, especially funding the research and development of insecticidal systems, and then, once a useful product had been found, organizing public-private production and distribution systems,[12] changing laws to allow growers to form syndicates, and then, finally, subsidizing the syndicates to purchase the product. A parallel development was the institution in each of the afflicted *départements* of a phylloxera commission, or, more formally, the Central Committee for Study and Vigilance of Phylloxera, which met monthly and published or otherwise publicly recorded the record of its meetings. By the time the next invasive pest, American black rot fungus, hit the vineyards, the institutional structure of the commissions, both national and departmental, were onto the problem in very short order.

In the end, granting funds to the research institutions, particularly the University of Montpellier and the École Agronomique, provided the most practical—and democratic—solutions. From these and other research centers knowledge about grafting—and, indeed, matériel for the process—were deployed via a widespread teaching effort, concentrated mostly in the workshop model.

By 1900 France had an agricultural public-private research structure that was the envy of the industrialized world.[13] Big Science had arrived, and been proved, in the phylloxera crisis.

PHYLLOXERA AND THE HISTORIANS, PHILOSOPHERS, AND SOCIOLOGISTS OF SCIENCE

There is much for the historians, philosophers, and sociologists of science to learn from the phylloxera disaster. Three topics stand out: the major role of argument in science, the relation between theory and practice, and the degree of social construction in science, a major bone of contention between philosophers and sociologists. Let us look at these issues in order.

Philosophers since Aristotle have talked about the role of logic in science, and Aristotle in particular focused on the relationship between demonstration and knowledge. Modern symbolic logic became the interpretive tool of choice for the logical positivist school of philosophy of science during the middle of the twentieth century. But these discussions, with their narrow focus on the analysis of the role of single—or select— arguments, miss a crucial, quotidian aspect of argument in science. Scientists live and breathe—indeed, their intellectual lives are constantly bathed in—argument. While it goes too far, the claim by some rhetoricians that the practice of science is itself a rhetorical activity gets at something essential about science (Harris 1997, xv).[14] Unfortunately, the constant and essential argumentative activity of scientists has not seen much analysis by philosophers until now.[15] But such analysis is called for, since, in the end, according to the philosopher, the relation between scientists and their beliefs ought not be based on arbitrary or whimsical choices but upon principled, argued reasons, whether the scientists are adhering to a theory (Franklin 1990, 487) or supplanting one theory with another (Burian 1990, 167). Exhibiting the central place of argument in scientific belief is particularly important in light of claims by sociologists of science, especially members of the "strong" program such as David Bloor and Barry Barnes, that "change in the content of scientific knowledge is to be explained or understood in terms of the social and/or cognitive interests of the scientists involved" (Franklin 1990, 487). Only close analysis of actual scientific argument will reveal whether or not the sociologists are correct.

Here three intense extended arguments are examined in detail: the seven-year controversy over whether phylloxera was the cause of the vine disease or was only its effect; Prosper Gervais' four-year campaign against the *américanistes*, those scientists who believed that saving France from the phylloxera would necessarily involve the use of American grapevines; and, finally, the several-year-long dispute during which Alexis Millardet tried to persuade his fellow scientists, vignerons, and landholders of the efficacy of *Vitis berlandieri* grapevines from the alkali regions of Texas as rootstocks for use in the limestone regions of France. Each of these extended periods of argumentation is explored in minute detail, at some points on a day-by-day basis, revealing the exchanges between members of the various factions in the debate and evaluating their contributions to the argument. As we shall see, contrary to what the sociologists claim, argument based on evidence—observation, experiment, practice—overwhelmingly marks the conclusions of the examined controversies.

A second point of interest to historians, philosophers, and sociologists of science in the phylloxera case is the role of the interaction between theory and practice. This modality of the scientific enterprise is not well known and not studied sufficiently. Studying the phylloxera case will reveal some crucial features of the theory/practice interaction. At some moments in the ongoing battle between humans and bug, theory led practice; at other times, theory misled practice. At still other times, low-level generalizations and even generalized observations led practice, which in turn created and/or corrected theory. Experiment, an element of practice, demonstrates similarly complicated interaction with theory. As Alan Franklin notes, "Typically, the interaction seems to be rather complex.

Experiment may confirm or refute a theory or call for a new theory. Theory may suggest new experiments, offer a new interpretation of known results, or help to validate an experimental result. These are, of course, not the only possibilities" (Franklin 1986, 163). But even though there are no easy generalizations (other than that the interaction between theory and practice is complex), some of the features revealed will be important. For example, as some theorists of science have maintained, it is always possible for a scientist to support a pet theory, no matter the evidence against it, as did Guérin-Méneville, who held the phylloxera-effect theory until his death (de Ceris 1873, 674).

To these interactions, two more elements must be added. First is the key role played in the phylloxera case by what can only be termed serendipity. One notable example was Faucon noticing in 1868 that when his previously flooded land emerged into the sun, the bugs were in remission. It was chance that his vineyard sector was flooded, but skill that allowed Faucon to make this observation. A similar case occurred in 1869, when the Phylloxera Commission was investigating the situation in Bordeaux and made the salient observation that Grenache vines—the variety most susceptible to phylloxera—growing in very sandy soil were healthy while those around them were sick and dying. It was good fortune that the commission decided at the last minute to go that way, but good science that resulted in the observation being made. In another example, how lucky was it that Laliman's American vines were interplanted with *vinifera,* allowing him to compare the different states of health in the same location? An important element of science, this: as Pasteur noted, chance favors the prepared mind.

Second, the fitful progress attained during the interaction between theory and practice differs from the fairly smooth, uninterrupted progress in the technological realm. Theories, on the one hand, "are revised, rejected, overthrown," and "the process is not cumulative, since former beliefs may be readily abandoned in the light of new evidence or better theories" (Pearce and Pearce 1989, 407). Certainly this is what happened when the Montpellierian phylloxera-cause theory (the view that the phylloxera was the cause of the disease) overthrew the established phylloxera-effect theory (the view that the phylloxera arrived only after an unknown disease had sickened the vines). Once the prevailing theory shifted, beliefs and practices entrained by the previous theory were abandoned.

This is not what generally happens with technological progress. As Pearce and Peace (1989) write, "By contrast, technological knowledge is basically cumulative in the sense that successful technologies are not subsequently open to 'refutation' as being 'mistaken,' though they may well become superseded through innovative change" (407). Response to the phylloxera crisis required the development of a number of parallel technologies. For example, methods of deep fertilization developed under the phylloxera-effect theory during early stages of the debate were retained after the triumph of the phylloxera-cause theory. Even though their origin was in a now-discredited theory, these methods worked, and were thus retained. Other examples abound. For example, techniques of immersion and sand planting resulted in constant improvement during the thirty years following their initial discovery. In another example, the growth and training requirements of American varieties were unknown at the beginning of the crisis, and, as all interested parties discovered, were quite different from the requirements of traditional varieties. Solving these problems was essential before efficient use could be made of the Americans vines either as direct producers or as rootstocks. In many cases, cooperation with American wine people assisted development, but some problems, such as the propagation of *V. aestivalis* varieties for wine production, or *V. berlandieri* varieties for rootstocks, proved intractable. Similarly, in another area, techniques, processes, and equipment for grafting needed development; in this case, aside from some minor borrowing of technique from early experience in grafting fruit trees, technological knowledge about grafting started at zero. On another front, once CS_2 had been isolated as a useful insecticide, an entire industry—from raw inputs to nationwide distribution to end-user equipment and techniques—had to be developed from the ground up. And, finally, traditional viticultural practices underwent massive changes as vignerons had to learn how to plant, manage, and, most importantly, protect by spraying the new grafted-type vines, thereby developing what came to be called the "new viticulture."

What is evident everywhere in all this is the extreme fluidity of the territory between theory, practice, and technology. Over the thirty years of war against the bug, a rich record of genuine science was graven in the minds of the participants, and, luckily for us, printed in all manner of texts. We will examine this record in detail, thereby providing a record of our own about how science in a crisis is done, paying particular attention to interactions among theory, practice (including experiment), and technology.

This brings us to the last of our three main topics. Philosophers and sociologists of science have wrangled many times over the question of how much of science is "constructed," that is, invented by the scientist's imagination rather than discovered by the scientist's observation. Sociologists have long claimed that much of theoretical science is socially constructed, in other words, invented by scientists according to their interests, theoretical commitments, institutional constraints, and other factors. The best-known sociological case studies of science focus upon examples in which two or more theoretical possibilities conflict with one another, with the sociologists arguing that, ultimately, scientists' choices among competing theories are made according to what is socially determining, not what is actually present in the world. Most of these case studies concern subatomic particles. The poster child of constructionist theory is Andrew Pickering's (1984) study of the quark; a similarly admired work is Trevor Pinch's (1986) case study of solar neutrino detection.[16]

The phylloxera case provides an excellent example for philosophers and sociologists to study. Right at the beginning, the phylloxera disease entered a scientific world split quite unevenly between an established theory (the phylloxera-effect view, held by a very powerful scientific elite) and a novel theory (the phylloxera-cause view, held by a very small and unempowered minority community). From the moment the bug was discovered on the roots of some grapevines in an unimportant Rhône Valley vineyard, these two communities battled over whose version of the bug's role was appropriate. Things could not have been set up better for the social constructionist view of science.

Admittedly, the situation is remarkable. It is a genuine case of two groups of theorists looking at exactly the same phenomenon and seeing two completely different things. The phenomenon itself is nearly microscopic, so it requires careful scrutiny to see what is going on, namely, the bugs eating the vine's roots. But, this having been observed, the question remains: what is actually going on? Are the bugs causing the disease? Or are they only responding parasitically to sick and dying vines?

According to social constructionist theory, the dominant community, with its stronger interests, more powerful connections, and tighter institutions, should be able to impose its theoretical world upon the much weaker minority community. According to philosophers, however, the decision by scientists about which of the two theories to accept should depend not upon social forces, but rather upon what the scientists actually found to be the case in the world.

Who is right, sociologist or philosopher? We shall see! This marvelous case is now ready to be laid out for the examination of all.

CHAPTER I

Disaster Strikes

"All your vines are fatally condemned
to disappear, Monsieur"

THE DISASTER BEGINS

In summer 1866 a few grapevines in an obscure vineyard along the Rhône, in the South of France, withered and died. Others around them began to show signs of the same progression. Over the next thirty years the withering disease would spread throughout France and Europe and into North and South America, Africa, and Australia, destroying traditional wine growing and wine making everywhere it invaded. What was the nature and origin of this dreadful disease? More importantly, what could be done to stop it? Answering these questions generated enormous debate among scientists, much heat, and, at first, not much light. Just getting to the point where the cause of disease was understood and agreed upon by most involved took nearly seven years from the time those dying grapevines were first noticed.

In the end it was not the scientists but the practical and resourceful people on the ground—grape growers, winemakers, landowners—who led the way in resolving the scientific controversy, thereby making possible the renaissance of wine growing, and wine itself, around the world. The story is long and complicated, so let us begin at the beginning.

It is not precisely known when the *la nouvelle maladie de la vigne*— "the new disease of the vine"—first appeared.[1] According to J.-É. Planchon, Montpellier's famous botany professor and one of the central figures in the story, in the late 1860s "an unknown disease menaced certain vignobles,[2] on both slopes of the Bas-Rhône; at Pujault, in the Gard, there were perhaps vague glimpses of this disease in 1863" (Planchon 1874, 546).[3] But by the summer of 1866 there were clear signs that something was dreadfully wrong in the vineyards of St-Martin-de-Crau, near Arles. Toward the end of July leaves on a large number of plants had lost their healthy green color and become dark red. The problem had spread from a small number of originally affected vines to those nearby, and from those to others, radiating in circles like "the gradual spreading out of a spot of oil" (Planchon 1874, 553).[4] The first published report on the problem was provided a year later by M. Delorme, a veterinarian from Arles, in a letter to the president of the agricultural exposition at Aix-en-Provence. Dr. Delorme notes that, after the reddening of the leaves, the disease "withers the bunches of grapes, dries them out, as well as drying out the tips of the roots" (Cazalis 1869a, 225). By the spring of 1867 all the affected vines were dead, and by end of the 1867 growing season, "vines dead or sick occupied an extent of about five hectares,[5] which produced next to no crop" (quoted in Pouget 1990, 10).[6]

Planchon's description cannot be bettered: "Everywhere the gradual invasion presented the same phases: after a latent period, some isolated points of attack appeared; during the course of the year, these local points enlarged themselves. . . . At the same time multiplication by new foci—advance colonies thrown to distances of several leagues around the centers developed the preceding years; in a word, the radiating aggravation of an already confirmed evil" (Planchon 1874, 553). Although *la nouvelle maladie de la vigne* spread rapidly enough, until 1869 its effects were confined to six departments of the lower Rhône, where grapes destined for ordinary table wine were grown in mass production vineyards. But, "apparently after an independent introduction," the disease spread to the Médoc region near Bordeaux, one of the most important areas for producing high-quality wine in all of France (Stevenson 1980, 47).

At first the growers did not realize either the seriousness of *la nouvelle maladie de la vigne* or its genuine novelty, attempting to see in the new disease the shape of a familiar one. "Always disposed to connect new facts to ones already known, the peasants of the Vaucluse called the disease *le blanquet,* or root rot" scoffs Planchon, "even though the conditions—of soil, of weather, etc.—for the latter were clearly not met by the actual situation of the new disease" (Planchon 1874, 546). As Pouget remarks, "The presence, in some cases, of rot on the dead roots led some to think for a while of the damage from *pourridié* (*Armillaria* root disease). But very quickly it was realized that the attacks of the disease could not be explained by this cause alone, most notably because of the great speed of the symptoms' expansion" (Pouget 1990, 10–11). Yet even given the speed of the expansion, the disease's effects were at first swallowed up in the immense tracts of land given over to growing vines: in 1865, the peak year of French wine growing, almost 2.5 million hectares were planted in wine grapes. Moreover, "phylloxera did not appear everywhere at once, and its impact was variable in time and space" (Pouget 1990, 50). For example, even by the mid-1870s in the Hérault, the most intensively planted department, "some communes possessed not a single producing vine, while others, often quite nearby, registered a record harvest" (Pouget 1990, 50).

Reactions to the new disease varied enormously from vignoble to vignoble. In those most affected, the unease of the winegrowers, who were "seeing their vines grow enfeebled and then irremediably perish," grew rapidly, especially because "the cause of their perishing was not known, and a way of combating it was scarcely envisaged" (Pouget 1990, 11). But in those vignobles that were as yet unaffected, people appeared unconcerned. Attitudes ranged from straightforward denial to wistful hope against hope. But once the plague hit, the reaction was ever the same: "The French winegrower . . . passed from indifference to incredulity, then to inquietude, and finally to despair" (Garrier 1989, 45). In the Hérault, belief came slowly. Wrote F. Cazalis in an editorial in *Le Messager Agricole du Midi* of Montpellier in 1871: "Despite these repeated warnings, the population of this rich *département* does not want to believe that the scourge which menaces their vineyards could lead . . . to the almost complete destruction of their vines" (quoted in Stevenson 1980, 59).

Regardless of human interests, the spread of the *nouvelle maladie* was rapid and destructive. In the three major vignobles in the Midi, the 1870s saw precipitous drops in vine-growing area: le Gard, which had 88,000 hectares of vines in 1871, had only 15,000 in 1879; during a similar period, the Hérault plummeted from 220,000 to 90,000 hectares; and the Vauclause went from 20,000 to 9,000 (Lachiver 1988, 416).

Though these statistics refer only to the southern growing regions, wine growing was one of the most significant national institutions. As Vialla noted in 1876, "The ravages of the Phylloxera do not cease extending; the Minister of Agriculture recently told the Chamber of Deputies that, of fifty-five[7] *départements* in France that cultivate the vine, already there are twenty-three of them under attack" (Vialla and Planchon 1877, 3). In 1870, eight million people in France lived directly off the vine (Millardet 1877, 82); 17 percent of the French work force was involved in wine production, which amounted to 25 percent of the farm economy.

For the individual small landholder, a hectare of wine grapes provided 400 percent more value than any other crop. But by 1880, nearly half of overall French wine production had ceased, ruining enormous numbers of smallholders, throwing large numbers of people out of work, and, in the end, causing vast numbers of local economic dislocations, all of which made it entirely clear that the *nouvelle maladie* had achieved disastrous proportions (Pouget 1990, 5).

But what must not be lost in all this is the human dimension. While the plague was certainly a disaster in national and economic terms, it was at every juncture an individual and personal disaster as well. Joseph de Pesquidoux, a grower from the Gard, provides a gripping account of what the times were like:

> One downsizes equipment and material, lets people go, reduces expenses. One retreats into oneself as in a depression. The beast wins everywhere. In its wake solitude invades all the land. And the horizon takes on an unfamiliar aspect, made up of empty and desolate space. As a palpable sign of the plague, one sees all along the roads huge carts overladen with dead vines, leading one to the funeral pyre. (quoted in Garrier 1989, 50)

A tragedy of grand proportions was unfolding, and the nation was forced into mobilization against *la nouvelle maladie de la vigne,* whatever it was.

LOOKING FOR THE CAUSE

Discovering the nature of the malady was the first job. On 6 July 1868, at the request of the Société d'agriculture de Vaucluse and M. Gautier, the mayor of Saint-Rémy, a three-member commission was appointed by the Société centrale d'agriculture de l'Hérault (SCAH) to inspect stricken vineyards on the west bank of the Rhône.[8] Members of the commission included Gaston Bazille, president of the SCAH; Jules-Émile Planchon, a physician and professor of pharmacy and botany at Montpellier; and Felix Sahut, a well-regarded winegrower. The commission began its work on 15 July, spent three days at the task, and immediately reported their results in all available media.

At first they focused upon dead vines, examining the shoots and their remaining foliage, and then they dug them up to inspect the roots. Nothing of interest was found. After some time at this fruitless activity, someone—it remains unclear who it was—suggested that a neighboring healthy vine be pulled up, over the protestations of the grower. Magni-

FIGURE 1. J.-É. Planchon, University of Montpellier botanist, co-discoverer (and describer) of *Phylloxera vastatrix*. Courtesy Institute of Botany of Montpellier.

fying glasses in hand, the committee turned their attention to the living roots. Their results were so striking that they deserve quoting at length:

> Suddenly under the magnifying lens of the instrument appeared an insect, a plant louse of yellowish color, tight on the wood, sucking the sap. One looked more attentively; it is not one, it is not ten, but hundreds, thousands of *pucerons* that one perceived,[9] all in various stages of development. They are everywhere: on the deepest roots as well on as the shallow, on the thick underground parts as well as on the most delicate rootlets. . . . During three days we found—at every place the malady had attacked—these innumerable insects. At Saint-Rémy, at Gravaison, in the Crau, at Châteauneuf-du-Pape, at Orange, everywhere on the roots of the suffering vines, the loupe showed us thousands of lice sucking the sap. . . . Great and small proprietors, simple workers, each grabbed the loupe in order to see for himself the enemy that had been discovered. . . . We cannot forget the nearly joyous exclamations of two or three peasants from the neighborhood of Orange, who cried "Ah! There's the enemy. It's good, we will make it perish! Courage, our vines can be reborn, our ruin is no longer certain, at last we can defend ourselves." (Cazalis 1869a, 237)

The effect of this discovery upon Planchon and the other members of the commission was immediate: "It is useless to look elsewhere for the cause, unhappily too evident, of the malady and the deaths. . . . What is now necessary to find is no longer the cause of the malady, it is its remedy" (Cazalis 1869a, 238). Unfortunately, it was not to be nearly that straightforward.

OPPOSITION ARISES

Face to face with the sucking bugs, which Planchon soon dubbed *Phylloxera vastatrix* (devastator of vines), the committee took their observations to be conclusive: they were watching the vines die before their very eyes from the depredations of the insects. And knowing what we know today about the disaster, it is all too easy to agree with the Montpellierians that, of course, the bug was the cause of the dead and dying vines. Yet we— and Planchon and the others—would be in the distinct minority here. Most of the rest of France, and especially the scientific establishment, immediately rejected the southerners' theory about the cause of the disease and advanced in its stead a contrary theory, one based in the long-standing dominant medical model. It was this contrary phylloxera-effect theory, in which the appearance of the bug was taken to be only an incidental consequence of an earlier disease, which was to stand in opposition to the committee's view throughout the next seven years of intellectual conflict. Planchon's description of the contrary theory is trenchant:

> Fundamentally, the diverse theories about the role of phylloxera in the new malady of the vine may be reduced to two: the *phylloxera effect* and the *phylloxera cause*. . . . According to the first theory, phylloxera would be the result of an enfeeblement, a previous alteration in the health of the vine, due, according to some theorists, to long-term monoculture, or wrong training of the vine (too short, too long, too severe pruning); according to other theorists, intemperate weather, or, as in other fruit-tree culture, because of asexual reproduction rather than via seeds. (Planchon 1874, 554; emphasis added)

According to Planchon's litany here, the original disease, the "enfeeblement," might be due to five different causes: monoculture, training that was too long or too short, intemperate weather, or asexual reproduction. Although these putative causes vary widely in their nature, the effect is always the same: a stricken plant, open to invasion by the phylloxera.

In opposition to this is the phylloxera-cause theory, Planchon's own, which maintains that it is the bug's sucking per se that debilitates and eventually kills the vine. It is this theory that the scientific establishment, centered principally in Paris, rejected out of hand based on their long-term allegiance to the officially mandated medical model of disease. As Legros remarks, "The experts of the capital, without ever leaving their offices, uttered their advice on the malady hitting the vines of the Midi. For them, the bug was not the cause of the malady" (Legros 1997, 33). As members of the establishment, the experts in Paris were believers in the official theory, the phylloxera-effect theory.

The differences between the two groups of scientists may be put quite straightforwardly: When Planchon and his colleagues from Montpellier observed the bugs sucking on the roots, they saw the event as *the phylloxera causing the vine's decline*. But when Guérin-Méneville or Signoret observed the bugs sucking on the roots, they saw the event as *the phylloxera as an effect of the vine's poor health*.

Part of the difficulty facing Planchon and the committee was professional and disciplinary. Although the three committee members were experienced in matters of the vine and botany in general, they were not entomologists. This counted against them: while the discovery of phylloxera made a sensation in the viticultural communities of the south, "it was not the same in the scientific world of the time since some, putting in doubt the role of Phylloxera in the withering of the vines, ironically called Planchon, Bazille, and Sahut, 'the entomologists of l'Hérault' " (Pouget 1990, 12). In response to the critics' irony, the ever-outspoken Duchesse de Fitz-James, owner of a large wine estate, defended Planchon, Bazille, and Sahut: while it was true that "not one of them is an entomologist by profession," in the end they were right, and "the number of ignoramuses who think they know more defies arithmetic" (Fitz-James 1881, 889). Since none of the three were insect scientists, it was easy to ridicule their theory that insects were the cause of the new vine disease.

But not only entomologists opposed Planchon and his colleagues. Among those arrayed against the Montpellierians, the four most significant were:

F.-E. Guérin-Méneville: an entomologist who solved the silkworm pébrine problem, thereby saving the silk industry;

Charles Naudin: botanist and director at the Jardin des Plantes, the national botanical garden in Paris;

Victor Signoret: president of the Entomological Society of France;

A.-H. Trimoulet: leading entomologist in Bordeaux and secretary for the Linnaean Society of Bordeaux.

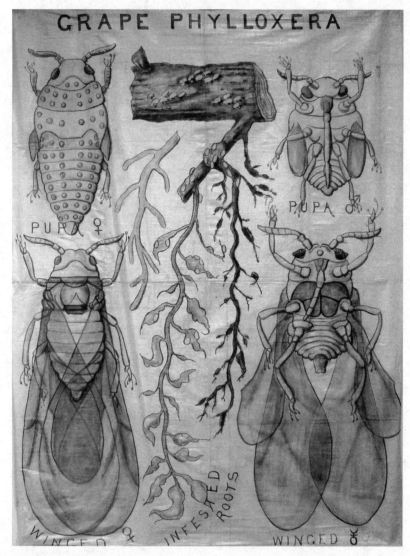

FIGURE 2. Phylloxera, by C.V. Riley, the American entomologist who played a crucial role in the identification of the bug. Courtesy Charles Valentine Riley Charts, Morse Department of Special Collections, Hale Library, Kansas State University.

FIGURE 3. F. E. Guérin-Méneville. Courtesy
Senckenberg Deutsches Entomologisches Institut
(SDEI), Image Archive.

Each of these men had a well-deserved reputation as a serious scientist in a field applicable to the dispute. Certainly their backgrounds explain some of their hostility to the ideas of the southerners.

Yet professional jealousy and disciplinary protection are not the most important forces in the controversy between the scientists. Underlying the dispute is a significant *theoretical* difference between the two sides. In the end, what sustains the seven years of intense debate between the Montpellierians and the scientific establishment is a fundamental intellectual collision between two opposed sets of scientific beliefs. These dynamics deserve full discussion.

THEORY VERSUS THEORY: "SEEING AS"

Our theories, suspicions, and beliefs strongly condition what we perceive the world to be like. As Louis Pasteur put it, we observe what we

FIGURE 4. Victor Signoret. Courtesy Collection
de la Société entomologique de France, Paris.

are prepared to observe (Pasteur 1854). Under ordinary conditions, when
we hear the house creak at night, we think it's normal, only the floor-
boards settling or the joists cooling off. But let there be a series of night-
time break-ins around the neighborhood and that very same sound
suddenly becomes something different, something ominous. What we
think shapes what we perceive. This interaction between thinking and
perceiving extends from ordinary life into scientific investigations—to
the role of phylloxera in the new vine disease, for example.

The American philosopher N.R. Hanson, who for years studied the
interaction between scientific theorizing and scientific observation, captured his findings in a famous slogan: "All seeing is seeing as" (Hanson
1965, 19).[10] Although Hanson's saying sounds cryptic, its meaning is
actually fairly straightforward. When we see, observe, or otherwise perceive something, we never perceive it simply as an unorganized set of
sensations; rather, we perceive it as some *kind* of something, some type

of object or event. And, more important, what we perceive that something to be, what object or event we take it to be, is constrained by our theories and beliefs about what the world is like.

Hanson argues his case with an example that he invents from the history of science. He asks us to consider a short conversation between two famous scientists, Tycho Brahe and Johannes Kepler (Hanson 1965, 18). This conversation actually could have happened: the two men worked together for years. But there was a particularly crucial difference (among many) between the two men. Tycho believed that the sun orbited the earth (the geocentric theory) while Kepler believed that the earth orbited the sun (the heliocentric theory). Hanson imagined the two men seated on a hillside on a clear morning at dawn, talking about what they were observing. The conversation might have gone something like this:

Tycho: The sun is coming up.

Kepler: Actually, it's not coming up, the earth is turning under us.

Tycho: What? You don't believe your eyes? Just look at the sun!

Kepler: I am looking at the sun, and it's not rising, the earth is turning.

Tycho: No it's not.

Kepler: Yes it is.

How is this discussion, this fundamental disagreement, possible? In some important sense, the two men are "seeing" the same thing: there's a big yellow ball in front of them, with a dark line underneath it, and both call the yellow ball "the sun" and the dark line "the horizon." But beyond these simple points of agreement, they are "seeing" two quite different things. One sees the sun as moving, while the other sees the sun as stationary. And, as far as science is concerned, it is this second sense of seeing, the *seeing as,* that is important to their discussion; it is the *seeing as* as an inference of some kind that directs their scientific behavior, constrains how they disagree with one another, and informs how they frame their dispute. Each man views the dawn exactly as his theoretical beliefs prepare him to see it.

Disagreements like this are common in science, including the case of phylloxera. Planchon and his colleagues see the sucking bug as the cause of the disease, but when Guérin-Méneville and Signoret observe this same event, they see the bug as feeding on a previously sick vine. Just like Tycho and Kepler, our two sides have precisely equal and opposite *seeings as,* due to the precisely equal and opposite theories in which they believe. We need to look carefully at the nature of these two theories.

TWO THEORIES OF DISEASE: INSIDE AND OUTSIDE

European plant and animal scientists of the mid-nineteenth century were inevitably trained in medical schools before they began their more specialized studies of botany and zoology. For this very straightforward reason, the theories of plant and animal disease used by botanists and zoologists closely resembled the theories of human disease they learned in medical school.

There are a great many theories of disease. Given the antiquity of the subject, this is only to be expected. Yet this vast number of theories is associated with only two general perspectives: disease caused by an entity external to the body (the exogenous theory) and disease as an internal bodily process gone wrong (the endogenous theory). As Lord Cohen notes, "The two notions varying a little in content and occasionally overlapping have persisted" since antiquity, "the dominance of the one or the other at different epochs reflecting either the philosophy of the time or the influence and teaching of outstanding personalities" (Cohen 1961, 160). In general, the external-causal model of disease has been called *ontological* by medical historians, in reference to the disease entity's independent existence. At various times the entity acquired different names; during the early years of the period we will be focusing on, ontological disease theories typically featured a "contagion" or, slightly later, a "germ" as the causal entity. Obviously, the phylloxera-cause view held by Planchon and his colleagues from Montpellier is an ontological theory. The opposing model featured bodily processes—within the sickened person, animal, or plant—and their interrelationships as the causal entity. These sorts of models are usually called *physiological*. Just as with the ontological model, different epochs viewed different processes as pivotal. During the Galenic era, for example, the balance among the bodily humors was the focal point of explanation and theorizing about disease. Guérin-Méneville and Signoret's phylloxera-effect position is a classic example of physiological thinking.

By their very nature, physiological explanations are multifactorial—the body, after all, contains a very complex set of systems. Indeed, physiological practitioners and theorists alike believed that "causes could act singly or in concert in any given instance of disease" (Pelling

1993, 312). This makes physiological explanations very difficult to investigate: since there are many factors involved, and they can interact in many different ways, no single determinative test can ever give a definitive answer to the question "What is causing the problem?" Moreover, and most important for the phylloxera crisis, physiological thinking does not neglect the local environment as an essential context for the body's internal processes. Climate (seasonal change, for example) and topography (such as swamps, streams, and soils) were significant factors, and, as noted medical historian E.H. Ackerknecht (1948, 592) points out, there was particular interest in the relation between disease and its geography. The French took this line of investigation farther than any other nation and were the most systematic and successful at it. In the years just before the French Revolution, around 1770, the Royal Society of Medicine of Paris "organized an extensive network of observers in every province ... whose task it was to make thrice-daily recordings of weather conditions and keep detailed lists of current diseases" (Hannaway 1993, 299). As we shall see, recourse to climate and other aspects of local environment were a main feature of physiologist thinking during the phylloxera crisis.

In addition to environmental factors, two other processes, *diatheses* and *degeneration,* dominated physiological thinking. The first of these referred to some weakness or susceptibility to a particular disease, usually heritable, found in particular individuals and families. Our closest analogy today would be something like allergies, which render individuals prone to suffer from particular environmental agents. Degeneration, described by such thinkers as Charles Lyell and Charles Darwin, is a process hypothesized in species that is akin to aging in individuals: just as individuals weaken and ultimately die, species over the span of generations weaken, degenerate, and die as well. This is particularly true of plant species that are reproduced asexually, from cuttings, as are grapevines.

Each of these three processes figures at one point or another in the various attacks made upon the phylloxera-cause arguments of Planchon and his colleagues. Indeed, each is clearly noted in his characterization of his opponents' position that we saw earlier: "According to the first theory, phylloxera would be the result of an *enfeeblement, a previous alteration* in the health of the vine, due, according to some theorists, to long-term monoculture, or *wrong training* of the vine (too short, too long, too-severe pruning); according to other theorists, *intemperate weather,* or, as in other fruit-tree culture, because of *asexual reproduc-tion* rather than via seeds" (Planchon 1874, 554; emphasis added).

"Enfeeblement" and "previous alteration" are diatheses; "wrong train-ing" and "intemperate weather" are environmental factors; and "asex-ual reproduction" is a degenerative process. Thus here is the whole list of physiological causes of the vine's weakening, something that happened *prior* to the bugs' arrival and, in fact, *caused* the bugs' arrival. Clearly, then, the phylloxera-effect theorists are physiologists. Moreover, they were the dominant scientific population in France at the time, and, fi-nally, they were all medically trained. So two further points need to be explained: first, how did physiological thinking come to dominate French medical thinking, and, second, how is it that Planchon—himself medically trained—and his colleagues deviated from this dominant the-oretical model?

MEDICINE IN FRANCE, 1800–1875

France had three national medical schools in this period: in Paris, Stras-bourg, and Montpellier. The curriculum at each school was generally specified by the National Academy of Medicine, although individual professors were allowed to make some local variations to their courses. Around the turn of the century, French medical school education had turned from ontological theorizing to embrace physiological thinking squarely and forthrightly. Reasons for the decline of the ontological perspective in the early nineteenth century are justifiable: it had failed to produce useful predictions or practice during the long series of epi-demics of yellow fever, cholera, and plague. It had failed especially mis-erably during the 1793 Philadelphia yellow fever outbreak, which marked the beginning of the end for this perspective on disease. In 1828 the French Academy outlawed the teaching of ontological theories of disease and required the teaching of physiological theories. As Ackerknecht notes, "It is almost impossible to exaggerate the importance of this step of the French Academy, the leading medical corporation in the leading medical country of the period. It set the pattern for the Western world" (1948, 573). This directive was not officially repudiated until late in the century, long past the phylloxera controversy and its resolution.

Thus, the medical education received by all the principals in the con-troversy over the role of phylloxera was rigorously physiological. Ac-cording to this education, the phylloxera were not the prime factor in the disease; indeed, they were merely late arrivals at the funeral of the vines.

So argued Guérin-Méneville, Signoret, and all the other French scientists—all except Planchon. Why did Planchon deviate from the standard model of thinking about disease? What made his case different?

The crucial factor here is Louis Pasteur (1922), whose 1859 paper on fermentations had made a radical proposal about the nature and causes of fermentations. Fermentations, Pasteur said, are inevitably caused by an independently living microorganism that, from its place in the environment, invades the fermentable body and starts the process we call fermentation. A clear ontological theory, this, and it is terribly important for theories of disease.

Analogies between fermentation and disease have been invoked since Aristotle. With the theories of René Descartes and Robert Boyle the connection was made even stronger. By the nineteenth century the conceptual relationship had become so strong that fermentation theory was routinely invoked to explain diseases. One result was Justus Liebig's theory of fermentation, a specifically physiological theory that dominated the science of the time and was used to explain disease around 1850. But, when Pasteur's theory of fermentation threatened to supplant Liebig's in the 1860s, its counterpart disease theory was threatened as well:

> Pasteur's explanation of varieties of fermentation offered a model for current preoccupation about disease in suggesting the presence (varying with locality) of micro-organisms in the air; a specific, one-to-one relationship between the micro-organism and the type of fermentation; a similar specific dependency between the organism, its food, and other chemical features of the environment in which it multiplied. . . . The germ theory was established as an issue of immediate clinical relevance to medical practitioners by the antiseptic methods of surgery advocated by Lister from the 1860s. (Pelling 1993, 327–28; cf. also Gale 1979)

Conceptual change happens slowly in France: after all, Joseph Lister, the first to attempt to apply Pasteur's findings in the operating room, was in Scotland! However, even though Pasteur's influence began small, its importance during the initial stages of the phylloxera crisis cannot be discounted. This is particularly true because Camille Saintpierre, a young, intelligent, and energetic scientist who dedicated his 1860 thesis on putrefaction and fermentation to Pasteur, came that same year to the Montpellier faculty, bringing his Pasteurianism with him. Writing at the end of his thesis about the debate between Liebigian physiologists and Pasteurian ontologists, Saintpierre exclaims, "Experience will judge the debate, by marching from now on in the direction laid out by recent discoveries. M. Pasteur announces new researches; the scientific world waits with impatience. This École will totally follow the path he has so surely traced" (1860, 125). Saintpierre could not have known how soon the claim made in this battle cry would be put to the most extreme test.

Saintpierre and Planchon served on the same faculty, where they became fast friends and close colleagues. In his classes Saintpierre taught the Pasteurian ontological theory of fermentation. Quite naturally, respect for the analogous theory of disease—the germ theory—grew at the same time. By 1867, when the first undertones of the phylloxera malady had begun to be heard, Saintpierre's theories had become established at the medical faculty in Montpellier and at the nearby agricultural college. Ontological theory was in the air there.

By the time Planchon first saw the bugs eating the roots, his theoretical expectations had switched: he *expected* to see some active agent in the environment causing the malady in the vines. And, true to Pasteur's dictum, his expectations prepared him to see phylloxera as the cause of the malady. Signoret and the others, however, did not know Saintpierre, nor had they spent seven years following the brilliant young scientist as he worked out Pasteurian theory in the lab and in the clinic. Consequently, for them it was only possible to see the phylloxera as the effect of some previous enfeeblement, climatic process, or degeneration, precisely as their education had taught them to see.

Disputes such these are extremely difficult to resolve. Right from the start the two sides differ about what they are seeing, and hence talk about seemingly different basic phenomena. They talk past one another, as Thomas Kuhn famously said, "passing like ships in the night" (Kuhn 1993, 3). The flavor of the controversy is revealing.

LIKE SHIPS IN THE NIGHT

The dispute went public almost immediately. The ontological side of the controversy first surfaced on 15 July 1868, when the three Montpellierians concluded that "it is useless to look elsewhere for the cause, unhappily too evident, of the malady and the deaths. . . . What is now necessary to find is no longer the cause of the malady, it is its remedy" (Planchon, quoted in Cazalis 1869a, 238).

Physiologists became formally active two days earlier, on the thirteenth, during a special session held at the agricultural exposition in Montpellier. The focus of the session was a report by Henri Marès, permanent secretary of the SCAH, who had made his own investigation of the destruction.[11] Marès' conclusion about the cause of the malady could not have differed more from that of the commission. According to Marès, the "great freeze of last winter, severe without interruption from 30 December to 11 January, is the determinate cause of the deplorable state of the attacked vineyards, and there is no particular malady of the vine" (Cazalis 1869a, 234). Marès' claim that there is no "particular" malady is revealing: physiological explanations, as noted earlier, are typically multifactorial, with a combination of general processes or events presented as causes. Within days Marès' view was seconded by J. Barral, editor of the influential *Journal d'agriculture*: "These new lice, to which M. Planchon has definitively given the name of *philloxera [sic] vastatrix,* appear primarily to be only the consequence of the disease due to meteorological circumstances whose effects have been augmented by the condition of the soil of the attacked vines" (Barral 1868, 20 September, 725). Barral concluded his review of the situation with the claim that "the plant lice that one now finds on the sick vines are therefore only parasites that have come after the event. They will constitute later on a special problem, now that they are born in such great numbers" (Barral 1868, 20 September, 727).

Barral here refers to weather and soil as causing the disease, which then attracts the lice to the sick vine. Such richness of explanatory models is precisely what should be expected given the overall dominance of physiological thinking among the participants. The thinking processes of Marès and Barral reveal many of the multicausal features inherent to the physiological mode of explanation. Marès, for example, refers to weather and geography as the general causes of the decline—and denies the existence of any particular (especially ontological) cause. Barral clearly opts for a diatheses: the plants have been weakened by climate and soil type, the insects have invaded, and now they will constitute their own "special problem." Moreover, it is clear not only that the explanations offered are not mutually exclusive, but also that they would be extremely difficult to test by any kind of evidence since none is unique to any one of them.

In contrast, the proposal by the Montpellierians is simple and straightforward: the insect is the unique cause. In terms of what is to follow, the simplicity and straightforwardness of the Montpellier hypothesis has a distinct methodological advantage over the multicausal explanation(s) offered by the other side: it can be tested and/or falsified in a fairly direct fashion. But first the *phylloxeristes* had to work out replies to the physiologists.

The response to Marès, Barral, and others by the Montpellierian pathogenetists was immediate and strong. Gaston Bazille, a member of the original team of discoverers, reacted against "Editor Barral's condemnation of our views, in sanctioning those that we call erroneous" (Barral 1868, 20 November, 521). It is worth looking at Bazille's reply at length:

> I take particular exception to Barral's statement "the plant lice that one now finds on the sick vines are therefore only parasites that have come after the event." I ask permission to prove what I am quite sure of, that the perishing of the vine in Provence is due almost entirely to the plant louse on the roots, to the phylloxera, as M. Planchon has named it. . . . It is possible that the original lice did choose in preference a sick vine; however, once established, they have become in and of themselves, and independent of all other facts— cold, drought, impoverished soil, excesses of humidity—a cause, and unhappily a cause very actively withering and killing the remaining vines. The opinions of our adversaries don't explain at all the advancement of the disease, or the present state of the vineyards of Provence. (Barral 1868, 20 November, 522)

Most importantly, Bazille in his response does not directly address one of the main questions posed by Barral and the others: what was the immediate and precipitate cause of the initial stage of the phylloxera? Indeed, Bazille leaves just enough room for a diatheses type of explanation, allowing, say, the weather, to slip in. But unlike the strict physiologists, he does not settle on any particular event as the immediate cause of the phylloxera. In his mind, the power of the phylloxera itself—as an ontological entity—to cause mischief completely overwhelmed any interest in the proximate cause of the coming of the bug. By 1875 the issue of the original proximate cause would have lessened considerably in importance. At the beginning, however, the exact precipitating cause was precisely the issue. This is what may be expected in a wildly multicausal explanatory environment.

The earliest phase of the controversy concluded three weeks later with a vigorous debate among the main parties conducted during a session on agriculture at the 7 December meeting of the Congrès scientifique de France. Dr. Félix Cazalis reported on the meeting for *L'Insectologie agricole,* a new entomological journal. In his report he noted that "the present malady of the vine has been attributed to general causes from climatic influences. It could be that certain circumstances have favored the development of the louse; but these general circumstances are still obscure and can not be determined a priori" (Cazalis 1869b, 29).

Henri Marès, a climatic-physiologist thinker, played a prominent role at the meeting. He got into a debate with growers Ripert and Laval, both of whom argued, along with Cazalis himself, that it was difficult to understand how the very same climatic conditions could favor the development of the louse in Provence, yet, in the adjoining Languedoc, development was not favored. Perhaps, Cazalis wonders,

> aren't there, in the valley of the Rhône, particular causes to naturally explain this apparent contradiction? Causes inherent in the soil, in the subsoil, in the vegetation, in the grape variety cultivated, the style of culture, the mode of planting, etc. etc. . . . Yet M. Marès believes therefore in an external influence, an atmospheric influence that has favored the invasion of the insect. . . . Thus, if the weather had been better the vine would have budded more vigorously; it would have restrengthened its wood and roots and the louse would not have been the stronger. (Cazalis 1869b, 31)

Here Cazalis imputes to Marès the classic physiologist view: had it not been for certain environmental factors, the physiology of the vine would have been robust, in better balance, and thus better able to resist the depredations of the insect.

Planchon was also in attendance at the meeting. He waited until the end and then, in what Cazalis reports to have been a clear and concise summing up, picks Marès' position apart. Planchon's unavoidable sticking point was that general climatic causes extend over large areas, yet the same cause apparently has opposed effects in adjoining areas, areas that are otherwise similar except for the ravages of the insect. How could this be possible, Planchon asks. Cazalis reports no answer from Marès.

The meeting then ended on a most interesting remark by M. Hortolès, who observed that there was one simple thing to be done to settle the issue. "Tobacco water, he said, is sovereign for destroying lice. Why not apply it to a certain number of seriously invaded and ill vines in Provence, destroying the lice? If the vines regain their health and vigor, then no more doubt could remain that the insects were not the true cause of the evil" (Cazalis 1869b, 31). With this, the meeting ended.

Hortolès' point went straight to the methodological issue mentioned earlier. One advantage of the *phylloxeriste* position is that it is testable, and directly so. Hortolès recommended precisely such a test. At this point, however, there was no follow-up on the suggestion. But within two years

Baron Thénard began large-scale field tests of insecticidal remedies, exactly à la Hortolès.

The controversy reported above stretched from 13 July to 7 December 1868, the period from the initial discovery of the phylloxera bug until the first national scientific meeting to consider the disease. It was not a long period of time, yet the positions taken, the arguments made, and the criticisms exchanged during this short period essentially covered the entire range of ideas that would recur during the next seven years, until 1875, when the cause of the disease would be settled. Moreover, underlying the range of positions and arguments were the welter of disease models extant at the time and available for use by the participants in this debate.

The saddest point is that the two sides weren't making any progress talking to each other. No consensus position was even envisaged, let alone on the table. Both sides went their own way in the theoretical dispute. Practical matters were suspended for the moment while the theorists dueled it out. Yet, theoretical disputes notwithstanding, some scientific progress was about to be made, and soon.

A YEAR OF DISCOVERIES

Controversy over the role of phylloxera in the new malady peaked in 1869–70. As 1869 began, districts to the north of the infestation began to take notice and to begin their own investigations. Isère's initial call was typical:

> The prefect of the department of Isère has named a commission to study the malady *pourri des* racines [root rot],[12] directed by M. Sequin, dean of the Faculty of Sciences at Grenoble and director of Grenoble's Botanical Garden. Questions to determine: First, is the trouble produced by a plant louse? Second, on the contrary, does the plant louse appear only on plant stocks already altered by some pre-existent morbid cause, and hence it is only auxiliary in completing or hastening the death of the plant? Third, is the plant louse everywhere the same? Fourth, the plant louse: its habits and means of destruction. Fifth, if an original cause, still unknown, attracts the plant louse, study and find methods to repel its appearance and propagation. Sixth, what influences are most important for development of the disease: Climate? Soil? Fertilizer? (Barral 1869, 45)

As this list of questions makes clear, all the relevant options are in play, from (first) the straight ontological view of Planchon and the Montpellierians, to (second) the diatheses view, and on to (fifth) the miasmatic and (sixth) the physiological views. Clearly the situation in Isère was wide open to begin with, and very little that happened during the ensuing year appeared—at least on the surface—to offer much help in settling the issues. Strong attacks on Planchon and his colleagues appeared from both the Linnaean Society of Bordeaux and Signoret, whose paper *Phylloxéra vastatrix cause prétendue de la maladie actuelle de la vigne* (*Phylloxera vastatrix*, alleged cause of the present malady of the vine) caused quite a stir, and managed to rally the phylloxera-effect scientists to ever greater levels of polemic.[13] Opposing scientists of course responded; the scientific debate grew more and more bitter. However, even in the midst of this learned combat, some results were beginning to accumulate.

Most notable was the progress Planchon and his allies made, both in discovering new details about the insect and its life and behavior, and in gaining new converts to their view. At the time of the insect's discovery, Planchon thought that it was "a type from a new genus," and he promised to provide more details very soon (Cazalis 1869a, 238). He provisionally named it *rhizaphis*, or "root aphid." But he knew that fuller identification would need to await examination of a winged stage of the bug. To that end, he took several specimens whose carapaces showed incipient wing buds and kept them in his lab at the university. On 28 August the wings emerged and Planchon observed "an elegant little midge, or, better, a cicada in miniature, displaying four spread-out, transparent wings. Thus my *rhizaphis* became a *phylloxera*, because, except for some differences of detail, it was difficult to distinguish from the *phylloxera quèrcus*, an insect that lives on the leaves of white oaks and whose presence is marked by a yellowing around the point it punctures" (Planchon 1874, 547). The obvious scientific questions were thus raised: "Where did it come from? Had it been described? What were its closest relatives?" (Planchon 1874, 547). Planchon vigorously set out to answer these questions. His project was aided by a number of people, several of whom need to be mentioned.

Planchon's brother-in-law, Jules Lichtenstein, worked most closely with him. Lichtenstein was a botanist who also was highly regarded as an amateur entomologist—indeed, he was a member of the Entomological Society of France. Over the next eighteen months the pair would succeed in answering some of the most troubling questions about the origin and life history of the insect.

Strangely enough, given his already-noted outspoken affirmation of the phylloxera-effect position, Victor Signoret, then France's leading

FIGURE 5. Portrait of C. V. Riley (1876), American entomologist. Courtesy Charles Valentine Riley Collection, Special Collections, National Agricultural Library.

entomologist, cooperated willingly with Planchon and Lichtenstein in the identification of the insect and in discovering the intricacies of its life cycle. Although Signoret never left his lab in Paris—remaining even during the chaos of the Paris Commune, sending correspondence via balloon post (Morrow 1961, 26)—he sent specimens to and received them from all the major parties in the dispute, even while continuing his polemic against many of them. One of his most important correspondents was Charles Valentine Riley.

C. V. Riley was born in Britain, educated at excellent schools and colleges in England, France, and Germany, and then orphaned at seventeen. He emigrated to America, where, after nearly starving to death, he picked up a job illustrating, and then writing about, his first love, insects. After working for a while at Chicago's *Prairie Farmer*, one of the Midwest's leading journals, and Civil War service in 1864, Riley became, in rapid succession, the state entomologist of Illinois and then of

Missouri. His *Annual Reports on the Noxious, Beneficial and Other Insects of the State of Missouri* became almost instantly famous for their unique blend of humor, quotes from classical literature, technical advice, and, above all, enormous practical wisdom. All seven editions sold out immediately upon publication.

Riley became aware of the phylloxera problem as soon as it was published by Planchon. His first hunch was that Planchon and Lichtenstein's insect was the same as the so-called *pemphigus* aphid identified in 1854 by New York's state entomologist Asa Fitch, which Riley himself had redescribed in the *Prairie Farmer* in August 1866 (Morrow 1961, 25). The major issue, however, was that Fitch's insect lived strictly on the leaves of American grapevines, while Planchon's lived strictly on the roots of European grapevines. Riley's hunch was generated by his idea that the different circumstances of the American and European environments resulted in different lifestyles.[14] In the late summer of 1869, Riley wrote both Signoret and Lichtenstein about his suspicions. The timing was propitious: on 11 July, while inspecting vines around Sorgues, Planchon discovered "on two vines of a variety called *tinto*[15] numerous leaf galls [wartlike growths] similar to those of the American *pemphigus*" (Planchon 1874, 548). Several days later Bordeaux experimenter Leo Laliman—who had imported a great number of American vines and planted them in a trial block to test their resistance to powdery mildew—sent Planchon some American vine leaves with galls on them. Planchon and Lichtenstein immediately hypothesized that the leaf insect and the root insect must be two life stages of the same bug: "Suspecting that these two insects, so different in appearance, were forms of the same animal modified by the milieu, one to the underground life (the root type), the other to aerial life (the leaf gall type), M. Lichtenstein and I came to the idea that the *pemphigus vitifoliae* of Fitch was nothing other than our *phylloxera vastatrix*" (Planchon 1874, 548).

Planchon and Lichtenstein immediately tested one aspect of their suspicion: they put American types on the roots of an uninfected European plant. Within two days the insects began to feed on the roots, and their behavior soon became characteristic. As the two noted, "*So we have a fact clearly established through experiment:* the phylloxera of the leaves, or the gall-forming and aerial type, can [behaviorally] become the phylloxera of the roots, i.e., the root-louse and underground form of the same insect. Nevertheless it remains to be discovered how, in the process of nature, the affiliation between one form and another

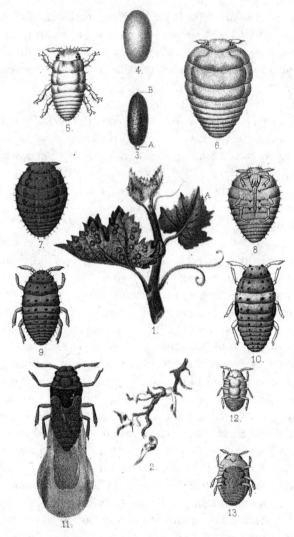

FIGURE 6. By 1890 many of the phases of phylloxera's very complex life cycle had been observed and described. This illustration (from Mayet 1890) depicts thirteen distinct phases. See Appendix A for a fuller discussion.

takes place" (Lichtenstein and Planchon 1870, 181). But, of course, much remained to be discovered: What was the life history of this obviously very complicated insect? And, aside from the scientific interest—which some complained was the main motivation for Planchon, Lichtenstein, and Signoret[16]—what steps might be taken against it? Most important, how was wine, the lifeblood of La Belle France, to be saved from this scourge?

SOME PRACTICAL PROGRESS

Although the battle is well and truly joined between the researchers on the two sides, more practical men—growers, owners, and agricultural workers—were unhappy with the emphasis upon theory and research; they wanted practical solutions. Louis Faucon, a well-known and respected grower from the Provence side of the Rhône, was particularly outspoken. In a 14 April *Journal d'agriculture pratique* article entitled "The New Malady of the Vine," Faucon sharply criticized the learned establishment (with the notable exception of Planchon):

> There have been many learned dissertations. *But the learned doctors don't pay enough attention to the simple factor of curing the sick vines.* Planchon is an exception, and his patient practical work will find a cure and win for him a noble recompense. The eminent *agronomes* are too critical of cures found to date. They should attack the problem at hand and give us new solutions to try, if not better ones. *The eminent savants even travesty practical empirical work being done in the field by amateurs,* who, at great sacrifice and expense, are showing some results. (Faucon 1870, 512; emphasis in original)

Faucon ends his complaint by remarking, "I bring to this great question merely *the modest observations* of a simple practitioner whose works have only the feeble merit of being based on *experiment.*" Faucon is too modest by half. Starting in 1868 he had devised a practical method not just to stop the progress of the new malady, but to also completely reverse its damage. The concept was simple enough: he drowned the insects by flooding his vineyards.[17]

Faucon was a very savvy and well-educated grower. As the plague raged around him, finally enveloping his own sixteen hectares, he read everything he could in the agricultural press. He tried all the remedies recommended by the anti-*phylloxeristes:* careful cultivation, lots of fertilizer, lots of irrigation, repeated sulfurings. Nothing worked. Faced with the complete failure of cultural solutions, Faucon converted to the *phylloxeriste* camp (Vialla 1869, 358). It was almost too late. By 1869

Faucon's vineyard was showing how desperate his situation had become: production that year was only 927 gallons, down from a peak, two years earlier, of 24,500 gallons. Then occurred one of those fateful happy accidents. In the late spring flood, a small parcel of Faucon's land along the river was submerged for over a month. When he investigated the vine roots as the land dried out, he was surprised to find that the phylloxera had disappeared. The now-dry vines immediately rebounded, setting both a good crop and Faucon diligently to work, preparing dikes, canals, and other structures necessary to flood his entire vineyard. He also wrote Planchon to ask for advice on the survival duration of phylloxera submerged in water. Planchon, fascinated, did the experiments. The champion *puceron* survived submersion for twenty-two days before expiring (Morrow 1961, 7). A national inspection committee, chaired by the École's Vialla, arrived at Faucon's place, near Arles, in mid-July; they were cautiously hopeful about, and certainly impressed by, the evident good health of the previously inundated parcel (Vialla 1869, 358).

Four months later Faucon attended the national Viticultural Congress at Beaune, 8–10 November 1869. There, to a very interested audience, he reported on his success, estimated how much water and how long the flooding would take based upon Planchon's work, and enthusiastically advocated flooding to everyone within hearing (Barral 1870, 666–67). Although not many vineyards in France were level enough for flooding to be applicable, nor was the water supply everywhere sufficient, there was still an appreciable acreage eligible for the treatment.[18] Although in later years Faucon frequently complained that the response to his good news wasn't anywhere near fast enough, in the long run flooding saved a significant percentage of the French vignoble (Faucon 1872). His technique is still in use on a small number of estates, 130 years after phylloxera was first discovered.[19]

Faucon's technique had two immediate effects, one practical and one theoretical. On the practical side, it was suddenly evident that at least some of the French vignoble—that capable of being flooded—would not perish before the plague. Faucon's success made a telling theoretical point as well: not only had the remedies of the physiologist-botanists failed, but a technique directly attributable to *phylloxeriste* thinking had saved him. Drown the beasts, save your vineyard. This point was not lost on the vignerons in the region.

Another presentation at the Beaune Congress had enormous import not just for the theoretical controversy, but for the very future of French quality viticulture. Leo Laliman had shown up at the congress, bearing luxurious foliage, replete with lush crops, from several varieties of American vines. In his presentation at the meeting, he reported that his European vines had all perished that season, but his several neighboring stands of American vines were all as healthy as they could be. Indeed, anyone with eyes could see what he was holding in his hands and would be forced to agree (Barral 1869, 666–67). Obviously, the American vines were somehow resistant to the attacks of the insects; although, as Laliman remarked loudly and in no uncertain terms, wine from the surviving American grapes was undrinkable, but perhaps suitable varieties for winemaking could be found.

Many eager ears heard Laliman's testimony. Almost immediately the search was on for American varieties suitable for direct production of wine in areas where the French vines had succumbed to the insect. By 1873 more than half a million rooted American vines were imported into France (Morrow 1961, 22–23). Although it most certainly could not have been known at the time, Laliman had pointed his peers toward the ultimate solution for the phylloxera problem.

Two other *phylloxeriste* ideas were tried. Having observed that sandier soils seemed somehow to preserve the vine's health,[20] some growers planted test vineyards in the sand dunes southwest of Montpellier, near the eastern bank of the Rhône's mouth. No right-thinking vigneron would have planted vines in these dunes before,[21] but these were desperate times. Although the vines in the nearest soil-based vineyard were completely phylloxerated, the vines in the sand flourished, remaining free from the infestation.[22] This is another example of the slogan just then gaining currency in the south: *"Sublata causa, tollitur effectus"* (Suppress the cause, take away the effect).

Trials of another method to suppress the insect took place in Bordeaux. In Floriac, Baron Thénard, the well-known chemist, was trying a risky experiment. The baron believed that it should be possible to find some substance that would have an insecticidal effect upon the *puceron*. After trying a long list of things, he settled upon carbon disulfide, a volatile chemical solvent just then becoming available in French industry. Thénard and his colleagues worked out a mechanical system to inject the chemical around the plant, and then proceeded to treat a large parcel. Sure enough, the pests died. Unfortunately, so did the vines (Laliman 1872, 20). Thénard retreated to reconsider. He reformulated his method, and his next trial a year later had more favorable results. After treatment the insects perished, but the vines did not; indeed, they slowly began to recover from the bug's effects. Again, *Sublata causa, tollitur effectus.*

Successes of *phylloxeriste* thinking began to be noticed in the agricultural press. For example, in the 11 August issue of the *Journal d'agriculture,* editor M.A. de Ceris's "Chronique agricole" column contained the following passage: "The phylloxera *cause or effect* is no longer a debatable issue. M. de Serres at Orange has put the insect on healthy vines and it has killed them. M. Faucon at Graveson, on the contrary, has cleared away all the aphids from vines severely attacked by means of a prolonged submersion, and it has saved them. Destroy the phylloxera and you save the vines, it is an established fact" (de Ceris 1870, 210).

Unfortunately, Planchon himself did not share in de Ceris's optimism that the debate was indeed over. In a manuscript dated just one month earlier, he decried "the subtle distinctions between the phylloxera cause and the phylloxera consequence" with which "several distinguished persons have lulled the public of our viticultural region into a perilous illusion. Today the malady of the vine has made its first entrance into our department [the Hérault]; the extension of the phylloxera as a destructive parasite is evident for all eyes to see, but two years have been lost in vain discussion[23] without acting resolutely against a visible and direct enemy" (Planchon 1870–71, 5). Planchon's bitterness toward the physiologist-botanists is evident here; it was never to lessen. Two years later, in his renowned article in the leading political journal *Revue des Deux Mondes,* he derided them scathingly:

> If these hypotheses are given the honor of being entertained at all, it is only in small part due to a spirit of impartiality, but mostly it is because the scientific method remains so foreign to our French public that the truths of common sense, and especially facts of natural history, need to be demonstrated against the hollow subtleties, reasonings empty and outside of facts, needing only a bit of rhetoric to seduce not only the ignorant but as well those who are educated and otherwise intelligent. (Planchon 1874, 554)

Planchon's view here was never much modified. He believed that "distinguished persons" like Signoret, Trimoulet, and Guérin-Méneville were ignorant of "the truths of common sense," and thus unable to see what he could see so clearly himself: that the bug, and only the bug, was uniquely and directly responsible for the disease. Any other belief he found incredible. But then that is the nature of seeing-as disputes: literally, neither side can see the other side. And so the controversy raged on among the various groups in the scientific community. Paris, meanwhile, finally awakened to the disaster in the southern vineyards in late 1870.

BUGS AND BUREAUCRATS

Until that point, all energies in the capital were focused on the war with Prussia; it was only when the Ministry of Agriculture began to realize how huge the potential revenue—and tax—losses would be come harvest time that action was hastily initiated. First and foremost was the creation of the Commission supérieure du phylloxéra, brought into being by the minister himself. Appointed to the commission was a diverse range of members, including "senators, deputies, savants [Planchon], presidents of *Agricultural Societies* [Bazille, Marès], and functionaries from the Ministry of Agriculture" (Pouget 1990, 30). The commission was presided over by the celebrated chemist J.B. Dumas, permanent secretary of the Academy of Sciences. Dumas took an active role in the commission, but his influence had both good and bad effects in the eventual outcome of the crisis.

The commission took two immediate actions. First, they instituted a prize of 20,000 francs to be awarded to anyone "who finds a practical and efficacious procedure capable of combating the new malady of the vine characterized by the Phylloxera" (quoted in Pouget 1990, 31).[24] The commission's carefully noncommital employment of the term "characterized by the Phylloxera" should be noted: at this point, it was taking no stand on the controversy over "phylloxera, cause or effect?" Dumas, however, within the next two years was to strongly commit the commission, and its funds, in favor of Planchon's phylloxera-cause analysis.

The second action of the commission was to recommend to official government bodies in all wine-growing departments the institution of local *comités d'etude et de vigilance départementaux* (departmental committees for study and vigilance). Local committees were to be modeled upon the departmental phylloxera commission set up independently two years earlier in Montpellier by the Société centrale d'agriculture de l'Hérault. But, true to form, most departments were quite slow in following the recommendation of the superior commission, assuming, as always, "it can't happen here." Côte d'Or, the most important viticultural region in Burgundy, didn't set up its study and vigilance committee until July 1874. Indeed, it was not until the terrible year of 1876, when the bug broke out from the south almost simultaneously into nearly every other wine-growing region, that the affected departments formed their committees (Morrow 1961, 31). But once started, the committees were, on the whole, quite active. Many met monthly and immediately published the results of the meeting. From the time of each committee's founding, it inevitably came to assume the role of clearinghouse for information going both ways, up and down the chain of command. Many of the committees sponsored workshops, expositions, and lectures by principals in research and practice. Planchon, for example, was a regular on the circuit. When, in the early 1880s, the phylloxera threat began to lessen, the committees were perfectly placed as the next threat from America—black rot—exploded on the scene. In the end, the network of vigilance committees served to link together for the first time what previously had only ever been a noncommunicating group of independent viticultural states and fiefdoms.

During the winter of 1870 and spring of 1871, research continued at Montpellier and in various other places. The bug continued to expand its infected area, and the partisans of either side in the "phylloxera, cause or effect" controversy continued to snipe at one another. But, slowly and quietly, the phylloxera-cause side was gaining the upper hand.

One development in 1871 was particularly instrumental. Early in the year Riley published his *Third Annual Report on the Noxious, Beneficial and Other Insects of the State of Missouri,* in which he reported correspondence with Lichtenstein and Signoret,[25] both of whom agreed with him that the leaf-dwelling specimens sent from America appeared to be very similar, if not identical, to the root-dwelling version in France. Riley's report also commented upon the divergence between Signoret's phylloxera-effect views, and the phylloxera-cause views of Lichtenstein. Obviously referring to Signoret's *Phylloxéra vastatrix cause prétendue,* Riley notes that Signoret "claimed that [the disease] had a botanical rather than an entomological cause, that it was principally due to drought, bad soil and poor culture, and that the *Phylloxera* was therefore incidental" (Riley 1871, 86). Riley argued strongly against Signoret's views. After remarking that "this insect should be found only on such roots as are already diseased is highly improbable," he argues that "there can be no reasonable doubt that M. Lichtenstein is right in attributing the disease directly to the *Phylloxera.*" Moreover, he concludes, "The history of our louse, which I shall now proceed to give, corroborates M. Lichtenstein's views" (Riley 1871, 87).

That so important a figure as Riley conclusively adopted the phylloxera-cause view did not go unnoticed in France. Laliman, in a long review of the controversy published soon thereafter in *Le Messager Agricole,* quotes Riley's argument in complete detail, using it to back his own attack upon the phylloxera-effect partisans (Laliman 1872, 21).[26]

Later that year Riley took part in one of the first examples of international scientific cooperation. Knowing that Riley was visiting family in London, Planchon and Lichtenstein approached the SCAH president, Louis Vialla, and suggested that the Missouri entomologist should be invited to visit Montpellier in order to observe firsthand the ravages of the insect. Riley accepted the invitation. He came to Montpellier, where he was greeted and hosted by Planchon, who took him out on field trips throughout the region, where he was treated royally. Eventually he found specimens of the insect from all life stages, which he took with him back to Missouri, preserved in acetic acid (Legros and Argeles 1994, 214).[27] Next spring, in his *Fourth Report,* Riley announced that, without doubt, the insects were the same:

> That the two are identical there can no longer be any shadow of a doubt. I have critically examined the living lice in the fields of France, and brought with me, from that country, both winged male and female specimens, preserved in acetic acid. I find that the insect has exactly the same habits here as there, and that winged specimens which I bred last fall from the roots of our vines, accord perfectly with those brought over with me. (Riley 1872a, 2)

Thus, the identity of the insect was settled. But there was still the question of its origin: where had it come from?

At the start there was no thought that the bug had come from America. When the 1868 commission traveled around on its inspection tour, it did not spend any time looking at the plantings of American vines, either in the lower Rhône or in Bordeaux. After all, the American vines were totally healthy, so why inspect them? They certainly had nothing to do with the bug. Planchon held this opinion (Pouget 1990, 16). Only in 1869, when leaf galls were found in both the south and Bordeaux and the connection was made between Asa Fitch's 1854 observation of the bug (which he had named *Pemphigus vitifolii*) and the French insect did it "become logical to suppose that the parasite had been introduced into France from America" (Pouget 1990, 16). When it was later revealed that the phylloxera found on European vines in English greenhouses in 1863 had definitely been brought in on American vines, this was taken to be "an irrefutable proof" (Pouget 1990, 16).

But, of course, it was no such thing. Leo Laliman, among others, always denied it; but this, perhaps, was because many thought that it was his American vines that had started the infection in Bordeaux. Others argued that the bug had been hidden, inert, in Europe for decades or even centuries (Ordish 1972, 48–49). Yet, beginning in "1870, the American

FIGURE 7. C.V. Riley (in Stetson), J.-É. Planchon, and Jules Lichtenstein inspecting the vines.

origin of the phylloxera, escaped from imported vines, was a common-place in all the scientific and viticultural communities, even though it lacked being demonstrated in an irrefutable manner" (Pouget 1990, 17). Thus, by mutual agreement, the bug was American.

The way was thus opened for the ultimate consequence of phylloxera-cause thinking: finding vines that had natural resistance to the Ameri-can bug. Given the practical observations of Laliman and others, cou-pled with the Darwinian theoretical explanation provided by Riley (Riley 1871, 91; 1872c, 624), the solution to the phylloxera crisis would be found in its very origin: vines from America. This led to what many French authors, beginning with Planchon, have called the great phylloxera paradox: the source of the disease and the cure for the disease are the same![28] This, of course, had been presciently proposed by Laliman when he showed up at the national exposition in 1869 bearing lush crops of American grapes grown in his phylloxerated Bordeaux vineyards.

THE TIDE TURNS IN THE SOUTH

Laliman's initial suggestion had been that foreign vines could be used as *direct producers*, that is, the American vines could be planted directly in French soil, and their fruit used to make wine, without the interposition of French vines at all. Although many growers worried about the quality of the wine from American vines, as did Laliman, the suggestion was taken up with a vengeance. In winter 1872–73 Missouri's Bush and Sons and Meissner received orders for more than 400,000 cuttings to be sent to Montpellier and environs. As we will see later, enormous difficulties attended these desperate efforts.

Pouget quite rightly describes this period as *"l'expérimentation anarchique des vignes américaines"* (Pouget 1990, 51). The later disaster of this period—the widespread dying of ill-conceived widespread plantings of untested American vines—only solidified the already-sizable animosity against the American vines, which, after all, were seen as the carriers of the insect scourge in the first place. Animosity reached all the way to the top. The Academy of Sciences, which had been charged by the government to direct the battle against the insect, "at first was not at all favorable to use of the American vines; its prejudices were prolonged for a long time afterward" (Convert 1900, 513). One major question that the academy had was whether any of the American vines truly had resistance to the pest, and, if so, which ones were they?[29] This latter question was difficult and involved. No systematic taxonomy of the American vines had been developed. It simply was not known how many species there were, what was their geographical distribution, and what were their adaptations to soil and terrain. Moreover, nomenclature was slipshod—a single variety could have more than one name, and one name could be used for more than one variety. It would take a real specialist in grape botany, an ampelographer, to begin the job of sorting out identification, and quickly, before the importation anarchy in France got completely out of hand. Pasteur and the academy made two moves to begin settling these issues, one immediate and short-term, the other more deliberate and long-range. In early summer of 1873, the academy approached the Sociéte centrale d'agriculture de l'Hérault and asked it to commission Planchon to make an immediate tour of the wine-growing areas of the United States, looking especially at the questions of resistance, identification, and adaptation (Convert 1900, 513). It was expected that Riley would host the French scientist and see that his energy was most efficiently used. Planchon visited during August, September, and October. His soon-published results were both impressive and extremely useful (Planchon 1875).[30]

Second, in early 1874 Pasteur and the academy directed Millardet, then professor at Nancy, to bring logic and order to the classification of the American vines. Millardet was the best choice for the job: not only was he arguably the finest botanist in France at the time, but he was also supported by the initial holdings of what soon became (and remains) the finest grape varietal vineyard in the world, at the École in Montpellier.

Montpellier was also involved in the other effort deriving from the phylloxera-cause theory, namely, insecticides. From the very start, phylloxera-causists had hoped to find something, some chemical or biological entity, that would kill the insect. As we saw earlier, Baron Thénard utilized carbon disulfide in the first large-scale field trials of a candidate insecticide. Although he was initially unsuccessful, he persisted in his efforts, even as others began serious experimentation. At Montpellier, Professors Durand and Jeannemot performed their first field trials from 8 May to 8 June 1872, using vines in a phylloxerated vineyard called Mas de las Sorres, just south of town (Ordish 1972, 69). Their results were inconclusive. Looking forward to the probable passage of a new law raising the prize for a solution to the phylloxera crisis from the 20,000 francs (in 1870) to as high as 300,000 francs (in 1873 or 1874),[31] they rented the majority of the vineyard from its owner, M. Fermaud.[32] During the summer of 1873, they increased the size of each trial block from five vines to a five-by-five plot. Two rows of vines down the middle of the vineyard served as untreated controls. In all, some 16,000 vines were used over the four years of the experiment.

During the next three years, suggestions for methods to try were forwarded from the academy to the Montpellier professors; out of the 696 proposals forwarded, 317 were actually tried (Commission Départmentale 1877). Given the eccentricity of some of the suggestions deemed sensible enough to try (goat's urine, shrimp bouillon, garlic peels), one is left bemused, wondering about the suggestions the academy deemed unworthy of forwarding on to Montpellier. Planchon was scathing in

his denunciation of the whole prize-seeking, public-suggestion process: "The remains of this dossier of stupidities reveal a sad day for the
state of mind of the great public in terms of its scientific instruction"
(quoted in Pouget 1990, 31). In the end, nothing worked, and the vines
all died.

Meanwhile, the growing success of the phylloxera-cause hypothesis
in the south did not silence the botanists and others on the phylloxera-
effect side. They continued to argue for their views, frequently in high
places and before high personages. Guérin-Méneville, for example, communicated a paper to the Academy of Sciences, which was duly presented
at the meeting of 20 October 1873. Nothing new appeared therein. Guérin-
Méneville argued once more that "*this parasite is not the cause, but a consequence of the malady of the vines*" (de Ceris 1873; emphasis in original). His argument proposed that the insect had always been around,
but because of "its smallness, its hidden life, and its insignificance as a
zoological species," it had not been noticed by researchers and remained
"undistinguished among the innumerable species of the groups of parasites within which it belonged." Phylloxera's incredible population explosion, its "exaggerated multiplication . . . is only a phenomenon consequent
upon a vegetal [botanical] malady." This latter malady, of course, was
caused by the "grand meteorological perturbations" with which everyone
is familiar (de Ceris 1873, 674).

But Guérin-Méneville concludes with a proposal of sorts, more of a
wish really, that, in both tone and content, shows that he knows the game
is up. It is worth quoting in full.

> I believe that it would be useful to instigate some practical research on this
> [changing the physiological culture of the vine], and that the rewards offered
> by the government and the agricultural societies for research into scientific
> methods for destruction of the parasite could better be provided to *agri-
> culteurs* who succeed, by practical and cultural methods, to save the vine
> from the malady that ameliorates the extraordinary and prodigious develop-
> ment of *Phylloxera*. (de Ceris 1873, 674)

Clearly, Guérin-Méneville sees in the government's and agricultural societies' shift toward funding phylloxera-cause programs the demise of
his own position's viability. Almost wistfully, it would seem, he believes
that at least some of the encouragement should be directed toward
more research into the proposals from the phylloxera-effect camp. But
it was not to be. As editor de Ceris remarks in his comments on this

paper, Guérin-Méneville remains ever faithful to his doctrine, but nonetheless the experimenters are most likely to be productive by continuing to focus on destroying the insect.

Again it must be noted how significantly the single-cause theory of the *phylloxeristes* differs from the multiple-cause theory of the *antiphylloxeristes,* most evidently because of the ease of testing the former and the difficulty of testing the latter. Guérin-Méneville wants research money to go into testing the physiological models, but he knows that it will not happen, not only because there have been some *phylloxeriste* successes, but also because finding more successes (or even failures) is much more feasible on the ontological model than on the physiological one.

The old botanist died not long thereafter, and, with the death of their strongest spokesman, the physiologist/phylloxera-effect side lost most of its public visibility. Indeed, Guérin-Méneville's passing marked the first official recognition of the ultimate victory of Planchon and the phylloxera-cause scientists, which came in the foreword of a review by Falières.

MONTPELLIER'S OFFICIAL VICTORY

M. E. Falières was a leading viticulturalist in the Gironde, with a well-known vineyard in the top-of-the-line Saint-Émilion region. In early 1874, Falières published a review of the state of the debate between the *phylloxeristes* and the physiologists, including among the latter Guérin-Méneville, Naudin, Signoret, and Trimoulet. His analysis was quite careful, thorough, and, in the end, committed completely to the *phylloxeriste* position. Falières sent a reprint of the article to J.B. Dumas, president of the National Phylloxera Commission, in Paris. Dumas liked the article very much, indeed, so much that he wrote a letter to Falières praising his article and resolved to present it personally to the Academy of Science the following day. Then, after expressing gratitude that so important an area as Saint-Émilion had committed itself to the struggle, Dumas went on to say, "You have been too indulgent to the promoters of the dangerous idea of the *Phylloxera-effect;* this idea has caused the ruin of a great number of vines and reduced to powerlessness the most authoritative among those who wish to devote themselves to this so grave question" (Falières 1874, 1). With this comment, Dumas not only reveals that official opinion had swung solidly in favor of the phylloxera-cause theory, but also that the promoters of the opposing theory were viewed as holding an opinion that was dangerous and ruinous.

A second piece of evidence that official judgment on the "Montpellier entomologists" had undergone a significant reversal is contained in Maurice Girard's 1874 monograph *Le Phylloxéra de la vigne: son organisation, ses moeurs*. Girard was certainly one of the most respected life scientists in France, a *docteur es sciences*, a delegate of the academy, and former president of the Entomological Society of France, who had spent two years working with the bug in that difficult chalky soil of the Charentais—Cognac country. His monograph quickly established itself as the most reliable and most thorough review of the phylloxera situation to date.

Girard minces no words. On page 14, the chapter begins with the title "The Phylloxera Is the Direct Cause of the Malady." He glosses this title in the following clear fashion:

> As I began my studies on the current malady, from the very start I have followed the true experimental method; I have backgrounded in my mind all agronomic or botanical theories, all the elements of meteorology, I have searched solely to observe the facts and to deduce from them all their consequences according to the inflexible laws of logic. If new facts trump my views, I will quite simply just change my opinion . . . but I must say, with the most sincere conviction, that until now, all that I have seen roots ever more deeply in my mind the proposition that I have given as the title of this chapter. I know that I will have many contradictors; but they must demonstrate by the facts that they have good reasons. (Girard 1874, 14–15)

The malady is caused by the phylloxera; indeed, the malady *is* the phylloxera. But Girard isn't content simply to affirm this proposition; rather, he throws down the gauntlet before the botanists, daring them to "demonstrate by the facts" that their position is reasonable. But, of course, that is precisely what the anti-*phylloxeristes* cannot do. They have no clear-cut practical successes to bring forth as evidence. Unlike the *phylloxeristes*, who can point to submersion, sand, and American vines, they are left with no successes of their own to point to.

Thus, the Montpellierians, who seven years before were scoffed at by the scientific establishment in Paris, have now recruited to their side the most influential members of that same establishment. So far as official opinion is concerned, so far as *agronomes*, viticulturalists, and freeholders in the south are concerned, the phylloxera is the cause of the malady, and salvation lies in thwarting its voracious appetite for the roots of the vine.

Unfortunately, winning the battle for the hearts and minds of officialdom and the south was only the beginning of a thirty-year struggle.

CHAPTER 2

La Défense

Sand, Submersion, and Sulfiding

THE BUG BREAKS OUT

Official recognition of the bug's "unique responsibility" for the disease at first did little to abet the growing chaos. Between 1875 and 1881 the bug swept northward up the Rhône from the Hérault into southern Beaujolais and beyond. In the Gironde, the bug jumped the river and rapidly proceeded north, devastating the Charentais' Cognac grapes. And, in a grand two-pronged pincers movement, the north arm of the westerly invasion from Montpellier met up with the eastward flow from Bordeaux in the region just east of Figeac; and the great southern arc of the vineyards of Montpellier, Carcassonne, Auch, and Bordeaux were simultaneously destroyed by the pincer's lower prong. As expected, in areas newly under attack by the bug, disbelief, denial, and bravado were the attitudes of the day. But in the areas already devastated by the insect, any and all courses of action offering even the slightest hint of relief were seized upon immediately. For the most part, remediation consisted of accepting the American vines in one role or another. The Hérault, first to lose its vines, was first to make efforts to recover. These efforts were led by the energetic members of SCAH and the equally resourceful faculty of the university and the École at Montpellier.

Phase differences over space and time—different regions of the country were in different phases of encounter with the plague at different times—contributed mightily to the chaos. There was considerable difference in behavior among regions still relatively free of the bug, those where the struggle was fully engaged, and those where the vignoble was totally lost. Mancey, a village of the Saône-et-Loire, in lower Burgundy, offers a doleful example. In 1876 the first cases of "the spot" were observed in several of the vignobles around town. Yet M. Millot, the mayor, was still full of hope that some vagary of climate in the region would save them (Millot 1876, 10). Three years later, all was lost. But still there were doubters. A reporter for *La vigne française* described the responses of a group of regional viticulturalists taking a guided tour of Mancey. At first the visitors saw only a healthy vineyard, full of vines,

> apparently green and vigorous. Already those visitors who believe that the evil has been exaggerated triumphantly say that the phylloxera is far from causing at Mancey all those ravages so often talked about. "Patience!" respond those among us who have already made a previous visit to Mancey. . . . We follow our guides and turn right. . . . As soon as we arrive at mid-slope, a dolorous spectacle, lamentable, offers itself to our eyes. As far as our view extends, we see only yellow patches, desiccated leaves, and dead or dying vines. We are face to face with what is conventionally called "the patch" *[la tache]*, but what one more justly calls "the disaster of Mancey." ("Une visite au vignoble" 1879, 43–44)

Several months later, the attitude in the whole department had finally shifted:

> The more considerable extent that the phylloxera invasion takes in the department the more the indifference of the vignerons disappears to make way for—unhappily—a very justified uneasiness. Indeed, the number of reported invasions in the department goes up rapidly, not because the bug is moving any more quickly, but because the vignerons, now seriously alarmed, have finally decided to call to the attention of the committees of vigilance the state of certain parcels of land that, already some time ago, they saw dry up. ("Rapports des Comités" 1880, 110)

Finally taking the threat seriously, the vignerons reported the infestations, but it was far too late. Had they not ridiculed the vignerons of the Hérault for bad care of their vines and told themselves "It can't happen here," the Burgundians might have been able to avoid some of the disasters that befell them.

Other regions far from the devastation of the Midi showed the same patterns of behavior. In Champagne, the *comice agricole* (similar to the administration of the county fair) of Épernay scheduled a meeting in early March 1877 at which Monsieur JL spoke about the troubles in the Midi.[1] JL's talk was obviously intended to raise "a cry of alarm" against the dangers represented by "those who can have the security in which seems to sleep a very great number of proprietors" (Comice Agricole d'Épernay 1877, 51). In his talk, JL claimed that once a vine has been attacked, it is a dead vine. The *comice* vice-president countered that things most likely were not as worrisome as all that. Moreover, there had been no sight of the bug yet in Champagne; it would seem to be limited to the Midi. After all, Mancey was still hopeful, he noted, and they were much closer to the problem.[2]

Three years later Georges Vimont had ascended to the vice-presidency of the *comice*. His attitude was, if anything, even more militant than JL's.

> Conclusion: phylloxera will come to Champagne. During its incessant march it has encountered the most diverse terrains, some analogous to ours; drier climates and more humid climates, hotter climates and colder ones than ours; but no circumstance of soil or of climate has been able to stop it. Conclusion: phylloxera will come to Champagne and it will develop here just as it has everywhere else! . . . Always a vine attacked by it, and not cleansed of it, is dead! Once attacked no vine left to itself returns to health. (Vimont 1880, 13)

But Vimont's words for the most part fell on deaf ears, ears that would remain deaf for a surprisingly long time. Thus ten years later, in Champagne's district of Mardeuil, handwritten notes from vignerons collected at the *comice* still frequently expressed such sentiments as "the existence of the philoxera [sic] is not sufficiently proved"; another wrote, "the philoxera [sic] doesn't exist, it's an imaginary malady" (Guy 1997, 24). And the growers of the Midi—already nearly a dozen years into reconstituting their vineyards with the American vines—were still being roundly castigated for taking bad care of their vines, thereby bringing the disaster down upon themselves!

Thus, what one generally finds during the period 1875–90, and more particularly 1875–81, is a vast countryside—something roughly the size of Missouri, Iowa, and Kansas taken together—undergoing the differential sequence of phases of a disaster that, region-by-region, was relentlessly ruining its economy, dislocating its inhabitants, and threatening the communal future. Against this backdrop of disaster, made even more disruptive because of its phase differences, some sort of a coordinated campaign needed to be initiated and sustained. Remarkable as it might seem, this is exactly what happened.

LA DÉFENSE

Once the bug had been accepted as the unique cause of the malady, thought immediately turned to stopping it, preferably by stopping it dead in its tracks. Especially in official circles in Paris, the explicit goal was to preserve the traditional wine varieties, traditional growing practices, and, wherever possible, traditional vignobles. Cornu and Dumas, speaking for the Academy of Science's national phylloxera commission, put it this way: "The commission considers the phylloxera as the cause of the malady of the vine. It proposes for a precise goal the conservation of the French vines; since their principal types are the product of secular practice, it is important above all to save them" (Cornu et al. 1876, 1). This sentiment is completely understandable. France's worldwide oenological fame was based upon the traditional varieties, grown and vinified in the traditional ways. At least this held for the most famous of French wine regions, Bordeaux, Burgundy, and Champagne, whose export brought in sizable profits for the state in terms of taxes, export duties, and the like. For the mass-production vignobles in areas such as the Midi's Languedoc, the Hérault, and Gard, however, it was not a question of fame and exports. Rather, the far simpler issue was the huge production of sound, ordinary drinking wine and the equally huge state income from the taxes thereupon. In the decade prior to the phylloxera invasion, production of both high-quality and ordinary wines had become highly developed, highly specialized, and highly efficient. So far as Paris was concerned, there was not the slightest thought of moving away from this tried and true "secular practice." Thus the central focus of the battle against phylloxera became the defense of prevailing viticultural practice from the invading American disease. Because of this focus, the first organized phase of the battle against phylloxera soon acquired the quasi-official title "La Défense."

But phylloxera was not the only invading foreigner that troubled official France. For reasons that have never been completely clarified, the reigning members of the national phylloxera commission, including Dumas, M. Cornu, and F. Mouillefert, were dead set against any and all uses of the American vines, either as grafting material, direct producers, or hybridizing parents.[3] From their point of view, the ultimate goal of La Défense was French viticulture, full stop. In an official publication from the academy, the commission wrote that they conceived of American vines as "a menace to French vines," which must be protected since

American vines were "permanent focuses of phylloxera" (Commission du Phylloxéra 1877, 4).

One possible reason for the official anti-Americanism is the apparent paradox mentioned above: how was it conceivable that the very same entity that brought the plague simultaneously would bring the solution?

> The future would reveal that the remedy was introduced with the American vines at the same time as the "evil." But what appears to us today as evident was considered, at the debut of phylloxera's appearance, to be inconceivable and contrary to the most elementary laws of logic. . . . In the minds of many, the American vines, carriers of the "evil," could in no way contribute to vanquishing it. This was the origin of the tenacious and systematic mistrust of the American vines manifested by certain adversaries who ignored the notions of resistance, of tolerance, and of sensitivity of the various varieties to the phylloxera. (Pouget 1990, 51)

It is quite clear here that by "certain adversaries" Pouget means Dumas and the other savant officials, who, after all, should have known better than to "ignore" basic biological concepts such as variability in resistance, tolerance, and sensitivity.

Whether or not the apparent paradox is sufficient to explain the anti-Americanism of the commission, "its prejudices were for a long time prolonged" (Convert 1990, 513); moreover, they "incontestably held back" the eventual solution involving American vines (Pouget 1990, 89). As R. Dezeimeris, who was not only a serious experimentalist-viticulturalist, but also the consul general of the Gironde, put it in a widely disseminated public denouncement of the national commission, "It is the sheer obstinacy of the higher administration that makes them able to see only one thing: the defense" (quoted in Pouget 1990, 42). Even Planchon, always level-headed, open-minded, and cautiously optimistic, found the atmosphere at the national level impossible to take. As G. Foëx later reported, Planchon

> arrived in this assembly [the National Phylloxera Commission] hoping to do some useful work, but, faced with the systematic hostility toward everything that he and the majority of his colleagues could contribute to encouraging employment of American vines, he realized that the only useful role he could play would be to find ways, with his friend G. Bazille and some other persons with a clear view on these matters, to oppose the adoption [by the commission] of overly rigorous restrictions on the use of the [American] vines. (Foëx 1900, 40)

Unfortunately, Planchon was not able to forestall the most draconian defensive move of the national commission: the laws of 15 July 1878 and 2 August 1879. With these, as Garrier notes, "the American predator confronts the French bureaucracy" (Garrier 1989, 64).

Provisions of the laws established three zones: "uninjured," "partially phylloxerated," and "totally phylloxerated." Depending upon its zone of residence, a vignoble was prescribed various activities and proscribed others. For example, vignobles in uninjured zones were completely restricted from any traffic in American vine material; perimeters were established as boundaries of cordons sanitaires; annual inspection of all vineyards was required; and nurseries and gardens were patrolled. On the positive side, the law provided for subventions to pay for insecticides, submersion, and other authorized treatments, and it authorized communal syndicates to buy, distribute, and apply the treatments. Only in totally phylloxerated zones was traffic in—and thus experimentation with—American vines allowed. This prohibition insured that, by administrative fiat alone, the eventual solution "[would] be, incontestably, retarded" (Garrier 1989, 64). Planchon's distress at the prohibition was evident:

> Therefore, at the very moment when one starts along the path of reconstituting the vignobles by means of resistant varieties, what good is it for Europe to ask for help that it refuses, and that America can give us? Mercy of God! The soil belongs to all mankind, and it would be a puerile self-love that makes the choice of varieties destined to serve as graft stock a question of national jealousy! (Planchon 1877, 259–60)

Restrictions on traffic in American vines amounted to a quarantine. But quarantines are useful only if they manage to buy time for research into long-term solutions. By this standard, the quarantines set up by the laws of 1878 and 1879 were intrinsically flawed. Because they confined experimentation with American vines to those regions that were already totally contaminated, research into American vine–related solutions that might offer protection to uncontaminated regions were prohibited in those very regions. According to the thinking of those managing La Défense, the quarantines were an end in themselves, not a means to an end. Thus the quarantines could not and did not serve their central purpose: buying time for research.

But in terms of the human misery involved, the quarantines were insignificant compared to what followed from the policies of the cordons sanitaires, which caused hardship in the extreme. From the start, vignerons were required to report any suspicious symptoms in their vineyards. Following the report, inspectors would arrive to observe the quality and quantity of damage. Were the damage found to be from phylloxera, the vineyard would be condemned to uprooting and burning on the spot. Local mayors would keep statistics, and once a triggering proportion of vineyards was found to be phylloxerated, the entirety of the relevant vignoble would be officially declared partially or totally phylloxerated, and forthwith uprooted and burned. Only with this "officialization of the total contamination of the arrondissement, therefore ending the interdiction of American vines," could progress toward reconstituting the region begin (Garrier 1989, 64). American vines could save a vignoble, but only if the vignoble were destroyed first.

Needless to say, the vignerons resisted. In the Var, for example, vignerons forcibly prevented experts from the departmental phylloxera commission from inspecting vineyards, and laws that proscribed transporting and burning dead vines as fuel were not respected. And when the viticultural society of Lyons asked for a strict application of the rules for uprooting and burning, the minister cautioned that "it would be more prudent to act by means of persuasion and counsel" than by force of law (Garrier 1989, 67).

Most onerous of all was the *traitement d'extinction*. In regions that were only lightly infected, or whose infection was localized in isolated patches, extremely strong doses of carbon disulfide were used to kill everything in the soil, including the vine itself, leaving swaths of scorched earth through which, it was hoped, the bugs could not pass (Pouget 1990, 39). The technique had been developed, and used successfully, against the bug in the Swiss vignoble around Geneva. But the method was not so successful in France, perhaps because it took "the more methodical and more disciplined Swiss and German growers" to use it "successfully" (Garrier 1989, 67). In fact, at best the method could only be "a means of retarding the progression of the phylloxera"; since winged forms of the bug existed, scorched earth could provide "only a prophylactic method with a temporary and hazardous efficacy that rapidly showed its limits" (Pouget 1990, 40–41). As far as the bug and its progress were concerned, treatments for extinction were inconsequential. But the human consequences were truly awful.

A young son of a vigneron in Gers described the situation with the following tragic words:

Misery hit us quickly. It began by our living on our last resources. Then began a kind of rationing. To give you an idea of how it was, we cooked, as in the olden days, cornbread, along with which one ate a small morsel of true bread, bread from wheat.[4] For drinking, we returned to water. The little bit of wine spared from the plague was saved for use in sickness or during the heavy toil of the summer. . . . Wood was the only thing that we didn't lack. All the countryside uprooted its vines. They were piled before the doors like stacks of hay. (Garrier 1989, 50)

The estimate is that from 1871 to 1873 Narbonne satisfied nearly 40 percent of its fuel needs by burning uprooted vines. Everywhere in the Midi the empty fields looked like a desert; in areas treated to extinction, the soil was sterilized—nothing grew there at all. Year by year, "the phylloxera pursued its work of devastation, wasting away everything in its passage," until, in the end, there was nothing but "a desolated countryside" (*Compte-rendu* 1882, 101). Everywhere "the French vignerons, region by region, passed from indifference to incredulity, then to unease, and finally to despair" (Garrier 1989, 45). In many regions depopulation followed immediately upon the declaration of total phylloxeration. People moved to the cities, or emigrated to Tunisia and Algeria, hoping to start over as vignerons in a new, uncontaminated vignoble.[5]

Clearly, official anti-Americanism and the policies it engendered via the laws of 1878 and 1879 were in every sense disastrous. Yet, almost in spite of the laws, some aspects of La Défense worked, and worked well. Let us now turn to these limited but genuine successes—sand and submersion.

LES VIGNES DES SABLES: GROWING VINES ON BEACHES

Observations about the potential for growing vines in the sand were made very early in the phylloxera disaster. In summer of 1869 the Vialla commission noted that "some vines of the most susceptible variety [Grenache] seem to have survived on a band of very sandy soil" (Vialla 1869, 347). Although these vines ultimately died, the idea itself found fertile soil (Azam 1882, 498).

Aigues-Mortes is a medieval town, replete with city walls and one old tower. It is situated on a level stretch of sand, about four kilometers from a lagoon on the Mediterranean. In 1865 around eighty hectares of vines were planted in the environs of the old town, mostly in the sand

(Gachassin-Lafite 1882, 138). By 1869–70, the insect had wiped out most of the surrounding vineyards, those situated on the normal alluvial soils of the region. But the vines planted in the sandy plain were different: one saw there "the luxuriance of the vegetation, which formed an oasis of green in the middle of vast, uncultivated extents" (Pouget 1990, 42). Something about the sand had stopped the bug: "Today one can see in the middle of a countryside whose fields have been completely devastated, veritable islands, alas! Not terribly extensive, spared as if by miracle. The sands truly possess this precious power" (Gachassin-Lafite 1882, 137). As we shall soon see, the source of this precious power remained in doubt for some time.

A vigneron from Aigues-Mortes, M. C. Bayle, made some of the first public remarks about the success of the sand-planted vines. His observations were very quickly verified by several authorities. In 1874 an official announcement appeared: soils with more than 75 percent of fine sand permitted neither the wanderings nor the survival of the phylloxera (Garrier 1989, 72). As the neighboring vines planted upon alluvial soil continued to perish, Bayle and several other locals bought up land—at that point priced dirt cheap at 150 francs per hectare—and immediately began planting vines. By the time of the 1881 conference in Bordeaux, Bayle's planted land was going for 5,000 to 6,000 francs per hectare (Gachassin-Lafite 1882, 138). Very soon after Bayle started planting, entrepreneurs from Sète, from the Camargue, indeed, all the way west to the Aude, were engaged in planting vines in the sand: "The first efforts having succeeded completely, huge plantings began; all the sands that are susceptible are covered with vines" (Gachassin-Lafite 1882, 139).

One of the largest players was the Compagnie des Salins du Midi, a huge concern engaged in producing salt from the marshes and lagoons along the coast to the southwest and southeast of Montpellier. By 1890 the firm had converted seven hundred hectares from salt production to vines (Garrier 1989, 74). They were extremely careful about setting up the operation. First, the dunes were thoroughly leveled and fertilized generously, and rigorously selected vines were planted. The machinery for both vine growing and wine making was also the most modern and expensive available. But their expenses were well justified: because of their typically above-average harvest (more than one hundred hectoliters per hectare, a goodly amount in any vignoble), and the high returns due to elevated prices and efficient marketing, revenue averaged nearly two thousand francs per hectare, double that of the surrounding Languedoc (Garrier 1989, 73).

Yet planting in the sand wasn't easy: "The strong coastal breezes and especially the mistral are the great enemies; they uplift the sand and uncover the vine roots" (Gachassin-Lafite 1882, 139). To prevent this, a mulch of reeds was planted between the rows. Once the reeds were established, the sand was considerably stabilized. M. Vernette, of Béziers, designed and developed a special machine that disked open four furrows into which small rooted pieces of reed could be dropped as the machine was dragged over the dune (Ordish 1972, 95).

Fertility was frequently a problem. Sand is essentially inert as far as biochemical activity is concerned. Left to themselves growing in unameloriated sand, most plants soon perish from undernourishment. With the exception of those dunes where the madder root had been previously farmed, the Mediterranean sands were pretty much sterile.[6] Fertilization was required in order for the vines to perform well. But therein lay a problem. Most typically, in the vignobles with adjacent lagoons, marshes, or swamps, bottom mud was initially the fertilizer of choice. It was rich in nutrients swept down into the delta region from the mountains and plains further north, and it also had the benefit of being wet. When sluiced upon the dune vineyards, mud produced exactly the jump in fertility expected. Unfortunately, the mud also turned the soil from sand into something else: a sandy alluvial clay, to which the phylloxera returned in droves, immediately killing the carefully planted vines (Azam 1882, 501).

The solution was chemical fertilizers—phosphates, superphosphates, and nitrates—whose use by then was fairly well understood. Chemical fertilizers increased the fertility of the dunes without changing the soil type at all.

Other problems involved the salt content of water from both the top (overflowing seawater) and the bottom (the subterranean water table). If the dunes were subject to frequent flooding from surrounding marine sources, salt buildup in the soil could seriously affect the health of the vine, even to the point of killing it. (Gachassin-Lafite 1882, 140). Similarly, a very shallow water table containing salt water instead of fresh water could kill the vine.

Not all types of sand were appropriate for planting vines. Although there was some controversy over the exact numbers, official advice published in 1879 argued that soils with more than 60 percent siliceous sand and less than 12 percent chalk were safe for vines (Vène 1882, 430). But these numbers were, to say the least, inexact. A. Vène, secretary of the Commission on Vines of the Société d'agriculture de la Gironde notes that some soils with more than 82 percent siliceous sand did not protect the vines from the bug, while others with chalk contents varying from 1 to 35 percent were in fact secure. Based on these observations, Vène was led to conclude "that the composition of the soils in vignobles in which the phylloxera cannot exercise its ravages is so variable, and the number of analyses carried out on these same soils is so small, that it makes it at the moment impossible to affirm, according to a physical and chemical analysis of a given soil, whether the vine will or won't be invaded by the phylloxera there" (Vène 1882, 430).

Similar inconclusiveness, not to mention controversy, attended attempts to explain how and why sand protected against the dread insect. One of the first attempts to explain the protection came from Professor A. Marion, from Marseille, who argued that "the sand had an indeterminate insecticidal action" (Pouget 1990, 43). This theory was rejected by the work of others, particularly Professor V. Vannuccini, who was on loan from the Italian government to Montpellier. Vannuccini argued that water invaded the interstices between the sand grains and thus asphyxiated the bugs (Ordish 1972, 95; Pouget 1990, 43).[7] Others argued that the small grain size of the sand allowed for a compacting not only around the rootlets themselves, but all throughout the body of sand. This compactness blocked any phylloxerian movement from rootlet to rootlet, let alone from vine to vine (Azam 1882, 497). Citing the well-known dictum "build on the sand," Professor Azam argued that "pure sand is the least compressible" of all soils, and this because of its lack of interstices. From this, Azam hoped "to deduce . . . information of use to all viticulture" (Azam 1882, 497).

Azam's "deduction" is quite interesting. He analyzed the soil from the Rhône delta, which had been successfully planted, with the hope of comparing it to the sands on the Atlantic estuaries in his own home region in the Gironde. Unfortunately, Azam remarked, the sand grains from the east were considerably smaller than those of the Atlantic region. This did not bode well. He writes, "Please permit me here a vulgar but striking comparison: can a mouse traverse a pile of beans? This appears to me difficult or impossible. Can it traverse a pile of walnuts? With less difficulty. A pile of oranges? Without pain, but still less easily than it would cross a pile of melons" (Azam 1882, 500). Azam's point is well taken, but perhaps not correct: certainly the larger spheres of the melons are easier for the mouse to traverse than are the smaller spheres of the beans, but it is not at all clear that the difference in scale between the sands of the Midi and the Atlantic sands are of the same rank as those between beans and even walnuts, let alone melons. Indeed, Azam was aware of this uncertainty. In the end, he concludes, "In view of the difference between the sands of the Gironde and those of Aigues-Mortes, the resistance is less certain, but it is still probable enough" (Azam 1882, 501). In fact, quite a few plantings were made on the Atlantic side, including on most of the Île de Ré, opposite La Rochelle.

Techniques based upon the sand barrier were many and inventive. One of the most straightforward and highly successful, but unfortunately highly impractical, was developed by the prominent Camargue viticulturalist Silvain Espitalier (Espitalier 1874). Espitalier's vineyards along the Rhône were first attacked by the phylloxera in 1868. His reaction was exactly what would be expected from a conscientious grower: he carefully sulfured and heavily manured all his vines. But, after a short revival, the vines again became observably sick. In the fall Espitalier hoed deeply, exposing many of the main roots within a fifty- to sixty-centimeter radius of the vine's main trunk. Around these roots Espitalier again piled manure, sulfured, and this time added wood ashes. In the spring, after another short revival, the treated vines languished. But the grower noted that several hectares of his vines, which were planted in almost pure sand, remained visibly healthy, indeed flourishing, a healthy island among the sick. That fall Espitalier again uncovered his vines' roots and repeated the previous year's treatment, with one exception: this year he added two different treatments with sand, on one hectare adding from six to eight liters of pure sand around each vine, and on another hectare adding between eighty and hundred liters per vine.

In the spring and throughout the rest of the year the heavily sand-treated vines prospered. Convinced that he was on the right track, Espitalier treated the rest of his susceptible eighty-two hectares. In the end, approximately fifty-four hectares were treated, with the roots of each vine covered by about eighty liters of sand. This was a huge job.[8] By the summer of 1874 Espitalier was sure enough of his method that he invited the departmental commission, with all of its usual members, to come inspect his place and verify his results, which they duly did:

> In sum, the commission has seen M. Espitalier's estate Mas du Roy, a great vignoble of eighty-two hectares, which, attacked by the phylloxera since 1868, has been and is still defended by a set of practical and efficacious procedures [involving vigorous fertilization and sulfuring] deposited on the

roots, which are then re-covered with a dense bed of sand capable of modi-
fying the milieu immediately around the vines.

Montpellier, 12 July 1874. *Le Président*, H. Marès (Espitalier 1874, 25)

It is not clear how long Espitalier continued his practice, nor whether
it was copied by many. What is clear is that his work, along with that of
the other pioneers of the sands, even though it could not save an appreciable portion of the French vignoble, provided something even more
important: hope.[9] As the 1881 commission on planting in the sands concluded, "If the hopes that we have initiated are not deceiving, France,
by reason of supreme effort, of which we have indicated the extent, can
envisage the future with confidence; she will no longer be menaced by
the sight of the spoiling of her wine production, and thus the loss of one
of the most important sources of her riches, both public and private"
(Gachassin-Lafite 1882, 141).

Buoyed by this hope, vines were planted in all the suitable sandy
places in the country, where they prospered. Yet, even though the method
was eminently successful, the grand total of replanted vines only approached a few percent of the surface planted in vines before the disaster. Another successful method—drowning the bug—offered perhaps more
hope.

SUBMERSION

Faucon wasn't the first to notice that the bug could be drowned.[10] But
he was certainly the most careful and scientific in his investigation of
the facts, and, more importantly, he was a marketing genius. From the
very first time he publicly announced his results at the Beaune conference in November 1869 (see chapter 1) Faucon was an enthusiastic
supporter of the use of submersion as the only sure way to save the classic French vines. His success, and that of the others who adopted his
methods, were the high point of the La Défense period, since they
proved that classic vineyards and classic vine culture could be defended
absolutely against the foreign invader. It was made official in 1873 that
submersion was "the only efficacious remedy known to this day against
the phylloxera" (Garrier 1989, 69). Those who were successful at the
method became known as the "submersionists"; soon a new journal,
Le Viticulteur submersionniste, appeared. As we shall see in detail below, many regions formed syndicates and *sociétés submersionnistes,* in
part to help defray the cost of the needed equipment, but mostly to

concentrate political power against competing groups or to influence the government.

Submersion could not succeed in any old place. It required satisfying some very specific desiderata. First, lots of water was needed. In a typical case, 18,869 cubic meters were required per hectare for the forty-day submersion period (de Leybardie 1882, 534).[11] Such huge amounts could be supplied only by good-sized sources, such as a major river, estuary, or other standing body of water. Moreover, water must be moved from the source to the vineyard floor, either by direct inflowing, which posed its own difficulties, or, more typically, by pumping, which was problematic in its own right. In one particular vineyard the owner had originally attempted to supply his needs with a fifteen-horsepower steam tractor, but that was insufficient, due in some part to losses between the tractor and the pump. The next year he added another eight-horsepower tractor and pump, but coverage still was not satisfactory. Finally, in the third year, he installed a fully modern fixed twenty-five-horsepower engine and pump assembly, complete with variable cutoff and steam condenser. Although the coal to power the engine during its forty days cost 1,300 francs, the owner announced himself well satisfied with its performance (de Leybardie 1882, 533).

Second, the vineyard itself had to provide a distinct environment. To begin with, it had to be either entirely flat, or very nearly so. If is was not, subdivision into essentially flat sections with sections separated by dikes was required. Obviously, water had to be transferred between diked sections, again typically using pumps. In the twenty-nine-hectare vineyard discussed above, the entire vineyard was enclosed with a 4,424-meter-long, 1.04-meter-high wall. This large area was then divided into seven basins separated by dikes. The total cost of these constructions was 8,204 francs (de Leybardie 1882, 532).

Of equal importance were soil and subsoil types. Clayey or clay-siliceous soils were the best. Sandy soils were too permeable, and overly compact soils were not porous enough for the water to move freely, thereby allowing bubbles of air to remain unexpunged. Subsoils, for their part, could not be very permeable; sandy subsoils, for example, allowed the water to flow out far too rapidly (*Compte-rendu* 1882, 185).

Other difficulties involved the water itself. It was necessary to use stagnant water, or water that was flowing with no aeration from falls or rapids (*Compte-rendu* 1882, 189). Nonstagnant water contained suspended air that the insect could breathe. Moreover, there was an extended discussion about the relative effects of "sweet" (we would call it "soft") water versus "limey" or hard water. Some submersionists claimed that softer water tended to dissolve natural nutrients and either leach them from the soil, taking them away from the plant, or, conversely, to make the previously more tightly bound nutrients available. Agreement was more general that important nutrients were provided by hard water. Also, the official recommendation was that an annual application of one hundred kilograms of available potassium and forty kilograms of phosphoric acid be applied annually per hectare "to strengthen the fruiting" (Plumeau et al. 1882b, 99).

As will be noted below, there were other problematic aspects of the method, but, in terrains where submersion was technically feasible, it was rapidly, and successfully, adopted. By 1890, more than forty thousand hectares were being submerged (Lachiver 1988, 420).

Faucon wrote the book on how to manage the method (Faucon 1874). His spirited proselytizing is evident in the introduction:

> This notice is written by a *proprietaire de vignes,* a simple practitioner. There's no need to expect herein scientific notions on the new malady of the vines, or purely entomological studies on the *Phylloxera vastatrix,* neither the technical and historical description of this insect, nor researches upon its origin and upon the exact epoch of its apparition in France. These points of the question, very interesting for science, [are] but secondary from the practical point of view of healing the vines. (Faucon 1874, v)

One major goal of the book, according to Faucon, is to put other proprietors on guard "against risky theories, which often have the result of introducing trouble in their spirit, or closing their eyes against the danger that menaces them ... or involving them in irrational operations or ruinous expenses." Moreover, he hopes "to indicate to those who are attacked by the terrible evil *an efficacious, practical, and economical means of curing their vines*" (Faucon 1874, vi; emphasis in original). Faucon was so successful in accomplishing these goals that the official submersion committee of the 1881 International Phylloxera Congress in Bordeaux declared that "in the ten years since, it appears that nothing has been added of genuine importance to the already-so-precise rules formulated originally by M. Faucon" (Plumeau et al. 1882b, 93).

According to Faucon, "a good submersion" was one "that deprived the insect of all communication with the external air" (*Compte-rendu* 1882, 181). On typical soils, this involved a layer of water twenty to

FIGURE 8. Steam-powered pump flooding a vineyard. From Barral 1883.

twenty-five centimeters deep, maintained faithfully during a minimum of forty continuous days between November and March. The duration of the submersion was dictated by the fact that the insect was less active during the winter, and hence required less oxygen to breathe. Vines were also less active in winter; otherwise, such a long submersion would have drowned them as well. Summer submersion was also possible, as Faucon had originally discovered, but the timing was crucial and dangerous. Since both insect and vine were more active during the summer, the goal was to asphyxiate the bug before drowning the vine. It was a genuine race to determine which would happen first. Few vignerons dared this technique.

The forty-day period was important. As the official Committee on Submersions warned, "It is necessary that the submersion be complete, general, and continuous" (Plumeau et al. 1882, 94). But of course there were *submersionniste* vignerons who contested this recommendation, especially in the light of success: "Presented with such remarkable vigor in the vines that had been submerged for several years, some proprietors, contrary to the opinion constantly underlined by M. Faucon, imagined that one could with no negative consequences suspend the treatment from time to time or shorten the duration of submersion to just twenty days. Experience demonstrates that suspension is bad practice" (Plumeau et al. 1882, 96). Consider, for example, the experience of M. de Meynot, who, after six treatments, interrupted the regime for a year: all his infected spots reappeared.

Although it might be said that de Meynot had it coming to him since he did not listen to the sage advice of Faucon, the same cannot be said for M. le Comte de Vassal. Twenty days into his fourth submersion, the count's pump broke. The mechanic told him that it would take eight to ten days to repair it, so de Vassal decided to halt the entire operation for the year. "Twenty days should be enough," he thought. Two or three days after the pump broke, the highest points of his property appeared above the flood. Progressively the less elevated points appeared, with the lowest point reappearing after twenty days. During the summer of that year, de Vassal observed nothing out of the ordinary; indeed, "my vines were splendid in their vegetation," he wrote (Plumeau et al. 1882, 96). But in September and especially in November, de Vassal found himself "astonished at the enormous number of *pucerons* garnishing the roots." Frightened, he made "a magnificent submersion of nearly fifty days" the next winter. Yet, in spite of that, in the summer his vines observably languished, exactly in correspondence with the level of water and the duration they had experienced the submersion before last (Plumeau et al. 1882, 96).[12] In the end, Count de Vassal retained the "well-assured conviction that we are obliged to submerge our vines every year during at least forty-five days" (Plumeau et al. 1882, 96).

One permanent problem with the method was exemplified by the experiences of M. Prades of Bédarieux, a village some forty kilometers west of Montpellier. The method, said Prades, was not absolutely certain. This was because "submersion presents at the start a considerable inconvenience: this is that one cannot know if the conditions [of terrain, soil, etc.] will be favorable in a given piece of land until one has already paid for all the costs of installation" (*Compte-rendu* 1882, 182–83). Prades tells the following sad story: He put in the best of equipment available, meticulously submerged, found live bugs, and submerged again. This routine was continued over the remainder of the submerging season, but always with the same results: "In spite of these diverse operations, the bug always survived; some vines were better, but a great number are in a piteous condition" (*Compte-rendu* 1882, 182–83). What had gone wrong, Prades asked the assembled crowd at the congress. It was not that he had not been careful, he assured everyone; indeed, if anything, he was "perhaps too meticulous" (*Compte-rendu* 1882, 182).

After a long and inconclusive discussion, it could only be suggested to Prades that his subsoil was too permeable or, perhaps, that his water was not stagnant enough. The most useful suggestion came from M. Chenu-Lafitte, who remarked that although he always observed some living bugs following a submersion, after five years of treatment, his vineyard has rebounded splendidly; perhaps Prades should keep trying for a few years more. Additionally, Chenu-Lafitte warned, "he regrets the news from Prades, whose lack of success is an exception, which could discourage some of the viticulturalists who are still indecisive" (*Compte-rendu* 1882, 187).

One interesting exchange came up during the Prades discussion. Prades' point about the present inability to predict applicable vignoble terrain was well taken, but M. de Longuerue pushed for detailed analyses to be undertaken in submerged lands where the vines had vigorously responded. He believed that there was some specific "element of defense" in these lands that could be discovered via analysis. Chenu-Lafitte retorted that, as a "simple practicing *viticulteur*," he had "more confidence in the observation of facts than in the analysis of his soil by chemistry" (*Compte-rendu* 1882, 187).

At a later point de Longuerue again insisted that a chemical analysis be made of the soils successfully under submersion. This time he was rebutted by M. Barral, well-respected editor of the *Journal d'agriculture*. "One ought not ask of science what it cannot deliver," said Barral. Moreover, "In the present state of chemistry, the analysis recommended by its honorable proponent [de Longuerue] could achieve scarcely any decent result. What that science *can* indicate would be the adjuvant elements useful to add to the practice of submersion" (*Compte-rendu* 1882, 188). It is not clear that de Longuerue's point was ever taken much further than this sort of discussion. In any case, it is highly unlikely that the failure of Prades was due to a chemical failure; much more typical would be simple (but not fixable) *physical* problems with the structure of the soil and/or subsoil.

Problems such as these, however, were not common. Submersion was reasonably successful wherever it was practiced. Once the technical problems of water supply and coverage had been dealt with, few engineering problems remained. Unfortunately, success of the method raised its own sorts of social and political problems.

Although submersion worked, it was expensive and required lots of water. Almost immediately after the technique was proved, it set up opposition between smallholders and the big estates. As M. Michel laconically noted, "The greatest successes of submersion . . . most often coincided with the grand properties" (Garrier 1989, 71). The big estates, especially in the Gironde, made expensive wine, and lots of it. For them it was easy enough to afford the expense of one hundred to three hundred francs per hectare for a proper submersion installation. But for the smaller estates of two to five hectares, especially those in the Midi whose wines sold for considerably less than that of the Girondins, affording the setup was simply impossible, even where water was in good supply. One effective response was the formation of *syndicats*. For example, at Bages, just south of Narbonne, was born in 1891 the Free Syndicate of the Maritime Agricultural Colony of the Bay of Capitoul (Garrier 1989, 71). As a typical syndicate this group excavated the bay, built dikes, and, most importantly, bought a steam tractor–powered pump, which could be moved from one place to another.

Fights between *syndicats* and large holders were not uncommon. At Sigean conflict arose between the Duke of Sabran, who pumped directly from the bay, and the local syndicate, which had built a network of canals fed by the bay. If the duke pumped too much water, the canals started going dry.

And, of course, as always, it was exceedingly difficult to convince the independent French rustic that it was in his own best interest to form a collective. In the lower valley of the Aude, Maurice Bouffet, the engineer-general of the famed Ponts et Chaussées,[13] had to personally intervene to bring about formation of several *syndicats,* necessary preliminaries to the construction—at the cost of the state—of a network of canals. Even when the main startup costs were to be borne by the public powers, the French vigneron resisted signing up for the benefit (Garrier 1989, 71).

Internecine warfare wasn't uncommon. M. Miquel-Paris offered the following motion to the 1881 Bordeaux Congress:

> Whereas submersion is a certain means of combating phylloxera; and thereby there is an interest in the widest possible generalizing of this mode of defense; but, in fact, certain properties susceptible by their topography of using submersion have not however been able to receive it, because, since they do not immediately border on the river or other water source, they have the necessary preliminary of obtaining permission from their neighboring proprietors, who immediately border on the source, for the water to pass over their property, but these latter have refused their permission in many cases; since this is a grave attack on the public and private wealth, this congress expresses the vow that this lacuna in legislation be remedied promptly. (*Compte-rendu* 1882, 433)

The motion was adopted. As always with submersion, "the technique, so simple in appearance, in reality relied upon the state" (Lachiver 1988, 421).

Unfortunately, the state, when it really counted, could not seem to act with any due speed. Beginning in 1874, the National Phylloxera Commission began passing resolutions that the government provide the funds necessary for a major network of canals to provide water sufficient to inundate a major part of the southern vignoble. In 1877, Ferdinand de Lesseps, the famed canal builder, signed on to the largest project of all—the canal linking the Rhône and the Gironde (Morrow 1961, 10). Had this canal been built in a timely fashion, it would have served to save hundreds of thousands of hectares of the vast southern vignoble. But the government dithered, local political forces warred with one another, and, in the end, the canal was not begun until 1889, and even then work went too slowly to achieve much toward defense against phylloxera (Morrow 1961, 11).

By 1900, forty thousand hectares of vines were under submersion. But this is only a small part of the estimated two hundred thousand hectares eligible between lower Provence and the lower Loire. Why did the number end up being so small? Michel cites three reasons: first, the high cost; second, "the nonparticipation of the state in building the huge infrastructure that was a necessary preliminary to getting water from the big rivers"; and, finally, "the efficacy of the reconstitution of the vignoble by hybrids and grafting, which, after 1890, rendered submersion projects unuseful" (Garrier 1989, 70). In the end La Défense by submersion failed to avert the ever-accelerating invasion of the American vines. But La Défense had one final weapon against the bug: gas warfare.

SULFIDING

From the start of La Défense, official Paris argued vehemently in favor of insecticides. Once carbon disulfide (CS_2) and its salts—the so-called "sulfocarbonates"—had been shown to be effective against the bug, national officials were unstinting in their support. It is a great irony that, although their support put the brakes on the initial experimentation and use of the hated American vines, in the medium to long run, it was self-defeating; in the end, the American vines—a major target of La Défense's officialdom—won a total victory.

La Défense counted some strong allies. Dumas, Cornu, and Mouillefert, the "most ardent and zealous" among *les sulfuristes* (or, alternatively, *"les chemistes"*), "figured among the most influential members of the Superior Phylloxera Commission." From their position they were able to do some real harm: "Persuaded that only insecticidal treatment, to the exclusion of all other procedures, was capable of battling effectively against the phylloxera and making it definitively disappear, they adopted an intransigent attitude and proclaimed obstinately their hostility toward the American vines and their partisans, the 'Americanists'" (Pouget 1990, 41). Most certainly, the blocking activities of "the partisans of insecticides ... contributed in large part to retarding the replanting of the vignobles after 1880" (Pouget 1990, 41).

Many of the technical problems associated with submersion accompanied sulfuring as well. First, not every terrain or vignoble was suited to it. Second, it required specialized equipment and setups. The social and economic factors also resembled those of submersion: it was costly; and for several reasons, which we shall see soon enough, big growers were again frequently set against the smallholder. Unlike submersion, however, it required an expert workforce for application.

CS_2 is an oily liquid with a nauseating odor—anyone familiar with a high school chem lab is familiar with the substance. Because it is heavier than air it settles into the soil, setting up an asphyxiating layer that soon enough kills many of the bugs, especially phylloxera. But it does not kill all the bugs.[14] Thus it is only a palliative in the long run; applications must be renewed at least annually, and, in some locations, two annual applications are required. Finally, and this is the saddest part of all, unless exquisitely careful steps are taken to strengthen the vine after treatment, continued use of the insecticide "has the effect of weakening the vine in a certain measure, and sometimes even opposing its rapid restoration" (Plumeau et al. 1882a, 78).

Problems associated with the substance included its flammability and its toxicity. According to some reports, explosions, although infrequent, were not unknown (Ordish 1972, 83). Wealthy proprietor M. Prosper de Lafitte, along with inventor and manufacturer M.F. Rohart—who, incidentally, had developed an alternate, nonliquid means of application—in 1877 and 1878 set up a traveling show focusing on the bad features of liquid application of the chemical. The two would announce that the chemical "is deadly for both cold-blooded and warm-blooded animals." Demonstration of this fact involved putting a few drops of CS_2 on a sponge and placing it in a glass cage containing a bird; the bird soon died. At one point during the show Rohart produced an explosion. After everyone settled down the two immediately set to business. "You can see it is dangerous stuff," said de Lafitte. He continued, "On the other hand, combined with gelatin or combined in sulfocarbonates, it is safe and effective." At this point the gelatin cubes manufactured by M. Rohart would be produced (Ordish 1972, 90).

In addition to its explosive properties, the toxicity of CS_2 required its application to be carefully managed. Even then, accidents happened: "One is reminded of the emotion produced in the spring of 1879 by the news of a disaster occurring, it was said, in one of the first vineyards treated by CS_2 in the environs of Libourne. Verification made, the accident reduced itself to two thousand vines destroyed out of a total of one hundred thousand treated. True, that was still much too much; but the word 'disaster' was genuinely excessive" (Plumeau et al. 1882a, 82). This accident was just another in a long string of problems facing use of insecticides.

At first insecticides, even CS_2, universally failed. Baron Thénard's original experiments in Bordeaux managed to kill all the vines along with the bugs. Other substances, including carbolic acid, arsenic sulfide, arsenious acid, potassium arsenate, tobacco, naphtha, mercury sulfide, and even quinine were tried, but each was found as useless as the other. Émile Duclaux, professor of chemistry at Clermont, wrote in 1872 that "all the insecticides employed till now have shown themselves completely ineffective or highly contestable" (Duclaux 1872, 43). In a November 1873 review article of insecticidal effects, Felix Sahut came to the same conclusion. So far as Sahut was concerned, the substances for the most part failed simply because they were not getting down to the deeper roots where the insect lived (Lecouteux 1870–73, 30 November 1873, 722–29).

Reliable success with CS_2 did not happen until the series of experiments and workshops conducted by the Compagnie des chemins de fer de Paris à Lyon et à la Méditerranée, the famous PLM railway. PLM was presided over by the redoubtable seventy-five-year-old M.P. Talabot. Talabot had two motivating forces in this enterprise. First, his line was beginning to lose considerable revenue because of the ever-declining amount of wine needing delivery from the south to Paris. Second, he had a genuine personal interest in eliminating the pest: he was a wine aficionado who hated seeing the ravages resulting from the bug (Morrow 1960, 34).[15] Once Talabot heard about the positive results of the experiments of Cornu and Mouillefert in Cognac, he got two other railroads to join him in lowering the freight costs for transport of CS_2. Then, through his own initiative, a regional committee was set up in Marseilles

FIGURE 9. Sulfide treatment team using Pal injectors, which pressurized and pumped the sulfide into the ground. The Pal was the standard equipment for sulfide treatment, requiring a large tank of the insecticide and a team of workers. In this picture it appears that the team leader is smoking a cigar! From Barral 1883.

in 1877, with the goal of focusing upon insecticidal responses to the disease (Ministère de l'Agriculture 1877, 7–8). Their first meeting hosted Baron Thénard himself, who plumped for CS_2 as the only insecticide with a chance for success.

In spring 1876 PLM instituted a series of experiments on uses of CS_2 and its salts. The experiments were headed by Professor Antoine Marion, professor of chemistry at Marseille. Marion's work was well financed and careful. Very soon his team rejected potassium sulfocarbonate ($KSCS_2$, the leading nongaseous contender) because of its huge requirements for water (Marion 1879, 135). They then focused upon CS_2 for the remainder of their work. By two years later, not only had Marion and his colleagues worked out an effective protocol for application of the gas, they had also instituted a series of workshops in a traveling "extension" course, worked with PLM in setting up an efficient distribution system, and written a detailed set of instructions for applying CS_2 (Compagnie 1878). In addition, the railroad stabilized the cost of the insecticide at 50 francs per hundred-kilo barrel, provided Pal injectors at stations everywhere on their route, and insured that all of the principal stations had agents trained in the application procedure (Morrow 1961, 35).

FIGURE 10. Pal injector, detail. From Barral 1883.

Slightly earlier Cornu, Mouillefert and Dumas had carried out experiments with $KSCS_2$ in Charentais-Inférieure, research that was sponsored by the big local Cognac firms. For their part, the three settled upon that compound as being far superior to the gaseous form (Dumas 1876). Not only was the salt far less toxic—and therefore far less dangerous in practice—after the CS_2 was liberated from the salt, the residue was immediately available as a strong fertilizer for the vine. But the salt had two significant liabilities: it was quite expensive, and it required excessive dilution in water in order to be applied. Approximately 450 kilograms per hectare of $KSCS_2$ were required to be dissolved in about 125,000 liters of water—equal to roughly 1.25 centimeters of rain! (Ordish 1972, 93). Irrespective of the water required, a "cost of 600 francs [per hectare] is not rare, and that of 450–500 francs is much more common"; this latter number "can be counted upon as being much more realistic." But that was not the end of the expenses: although the potassium salt provided much of the fertilizer needed by the treated but weakened vines, additional fertilization was needed. In the end, "700 francs of expenses per hectare during the first year of treatment by sulfocarbonates" was expected (Plumeau et al. 1882a, 77).

But it was the water requirement that most severely limited the use of the better $KSCS_2$ treatment. Only areas producing valuable wines and with proximity to a large water source could afford the sulfocarbonate treatments. In practice this meant that only the wealthiest properties in Bordeaux and Burgundy could utilize the treatment. Unfortunately, however, even there the procedure was not certain. Consider the case of poor M. Piola from the premier Saint-Émilion region:

On Saint-Émilion's plateau of pulverized chalk M. Piola had established a vineyard, which thrived perfectly in earlier times; this was the first region of his domains attacked, the first defended by carbon disulfide, then cultivated with love accompanied by perseverance and generosity; but there, while it cannot be said absolutely that the battle has been lost, the plague is far from being vanquished. Will the French varieties there ever again give productive harvests? It is permitted to doubt whether they will. (Plumeau et al. 1882a, 69)

Just as with submersion, successful treatment with $KSCS_2$ and CS_2 depended upon the soil structure and type. By a painful irony, it is again a case of "them that has, gets." The 1881 congress committee's recommendations for sulfide treatment noted the following fact: "Everywhere that we have seen weakened vines regenerated completely in fruit and wood," or where attacked but still vigorous vines "maintain their normal state," the soil there "is deep, and of a definite fertility" (Plumeau et al. 1882a, 64). Because of this fact, the committee's final recommendation is a warning: "To our minds, it is a hazardous enterprise that a viticulture in distress cannot attempt, to pour insecticides, fertilizers, labors, and all their surrounding resources into poor and shallow soils" (Plumeau et al. 1882a, 65). Obviously the smallholder, the peasant on poor land up on the ridge or nearly anywhere in the Midi, is excluded by this solemn warning. For this sort of person there was only one way out: syndication. But, as already noted more than once, for the individualistic French rustic, joining a group effort is one of the hardest things to do.

The laws of 15 July 1878 and 2 August 1879, in addition to declaring zones where American vines were forbidden, allowed the formation of *syndicats de défense* against the phylloxera. This was something new: prior to this law, syndicates had not been allowed in France (Lachiver 1988, 427). But the laws were explicit and detailed in their prescriptions; as Garrier notes, "in fact, it was the state itself that fixed the administrative rules, and hence the modalities for using carbon disulfide." Although not obligatory de jure, de facto "the treatment must be a collective work" (Garrier 1989, 91). The law specified subsidies for the purchase and application of the chemical, but only syndicates could administer the forms and terms of the subsidies and dispense the money itself. The process was long and involved. First the mayor had to claim that a certain vineyard or region was invaded. This proclamation brought about inspection of the afflicted area by the local antiphylloxera committee. If the committee agreed that the bug was present, they so reported to the mayor, who then asked the affected winegrowers to consent to the CS_2 treatment and to form a syndicate to apply for the subsidy, receive the chemical and equipment, and oversee its application.

Resistance to this process occurred, at least initially, nearly everywhere, but it was worst was in Burgundy, a region whose vineyards were extremely small and scattered among a vast and wildly diverse number of owners, even by French standards.[16] As Morrow notes, "In Burgundy the peasants simply hated the features of the act which required chemical treatments if the majority of winegrowers in a *commune* demanded that they be tried" (Morrow 1961, 33). In some places there were riots. In July 1879, at Bouze, demonstrators carried on even in the presence of the prefect and the gendarmes. It was worse at Chenôve, where 160 growers chased the treatment team out of the area, "saying that they were scoundrels, more to be feared than the phylloxera" (Laurent 1958, 333). When the winegrower involved was later hauled before the court at Dijon, he won on appeal.

Yet, even in face of this resistance, desperation won out. From only two syndicates in 1881, the number in Burgundy's premier Côte d'Or region grew to a peak of 198 in 1893 (Morrow 1960, 33). Following this peak, however, the number of syndicates using CS_2 treatment dropped rapidly. Two reasons underlay the drop. First, the government subsidy declined from eighty francs per hectare in 1882 to twenty francs per hectare in 1891, which made it difficult for all but the largest owners to afford the treatment. Second, the syndicates increasingly turned their activities toward the procurement and use of American vines, even though the practice was at first outlawed.[17] Syndicates justified their behavior in class terms, "and the conflict took on the look of a 'species of class warfare'" (Bouhey-Allex, quoted in Laurent 1958, 347).

Early in the turmoil over the use of CS_2 the Bonapartist paper *Bien Public* argued that the committees forced treatments that "were a plot to destroy the common wines for the benefit of the luxury wine growers" (Ordish 1972, 125).[18] Indeed, because of the costs, only the largest estates could afford the treatment, even with a subsidy. Hence, as the "bulk of the winegrowers lost their old French staples, they demanded, against the protest of the big proprietors . . . that American vines be admitted" (Morrow 1961, 38). Only in the grand crus of Burgundy and Bordeaux, "where the higher price of the wine allowed the costs of repeated applications to be recovered," was CS_2 use an economic possibility (Pouget 1990, 35). A grower with "only a modest production of table wines" could not support the cost of the insecticide because, "when its cost was added to all the other costs of production, the treatment cost more than the price of the wine" (Lachiver 1988, 422). Laurent masterfully summarizes this aspect of the "class warfare":

> Public opinion was profoundly divided about action in the face of the plague. The *"sulfureurs,"* among whom numbered the proprietors of the grand crus, are partisans of forced defense and even to excess manifest the most vigorous hostility to the very idea of foreign vines. The *"américanistes,"* who comprise all the small proprietors of ordinary vines, having abandoned all defenses, or practicing them only half-heartedly in their best regions, desire on the contrary the reconstitution of the vignoble using the American vines. (Laurent 1958, 344)

Small winegrowers also had other, related problems. In most of France viticulture was just one aspect of a flourishing polyculture. Land use among smallholders was inevitably mixed in order to meet the needs of the grower families. In Provence, for example, vines were traditionally interplanted with fruit trees and bush fruits, neither of which could withstand the sulfur treatments (Garrier 1989, 94).

In the end, the moves made in 1885 by Burgundy's newly formed Société vigneronne de Beaune—a rapidly growing syndicate of smallholders—were the pragmatic model for all subsequent attempts to modify the laws in order to allow the free transport of American vines. The committee's secretary, M.L. Latour (even today a family name of high regard in Burgundy), toured Beaujolais in order to view some of the vineyards reconstituted with both American direct producers and graft stock. On his return he issued a report recommending that insecticides be used for defense of the fine Burgundy estates but that devastated areas be allowed to be reconstituted by direct producers and grafted vines, following the example of Beaujolais. He printed eight hundred copies of the report, which he then distributed to all the usual suspects. Journalists got hold of the story and gave it the expected twist: "If the bourgeoisie would not let the small proprietors replant with American vines, it was only to get cheap labor for their own vines" (Laurent 1958, 347).

Ultimately, insecticidal treatment was a general failure. It was costly, it required skilled labor, and, worst of all, it was no permanent solution: "sulfiding can prolong the life of vines for a few years, but it can't cure them" (Garrier 1989, 94). Moreover, because of its dangers, sulfur treatment was, as Bazille remarked, "a razor in the hands of an infant" (Pouget 1990, 5). By the mid-1890s insecticidal use disappeared from all but the finest properties; "this means of defense, having caused many disappointments in most types of soils, ceased to be applied" (Zacharewicz 1932, 279).

In the end, La Défense failed: there wasn't enough sand for sand planting, there wasn't enough water for submersion, and sulfiding was too tricky and too expensive. There was no other way than the American way.

CHAPTER 3

La Reconstitution

By 1882 it was clear that La Défense had failed. As much as official Paris wanted to keep American vines out and traditional French practices in, it wasn't going to happen. Defending traditional French practices against the American insect scourge was simply too expensive and ineffective in terms of time, environment, labor, and finance. Luckily, even as La Défense was being proclaimed from Paris, alternatives to it were being tried and tested in the provinces. These alternatives all involved vines from America in one way or another. American vines functioned in three roles: as direct producers, grown on their own roots to make wine;[1] as graft stock for traditional grapevines; and as parents of genetic hybrids. Each role had its successes and its failures; in the end, decisions about which role to feature in a given vignoble rested upon an evaluation of the trade offs between aesthetics, economics, and biological viability. What made the decision so difficult in each case was the fact that "success" was a moving target. At first, when nearly every vigneron in a region was devastated, success simply meant surviving, making some wine for your family to drink, with perhaps a little left over to sell. Then, once that goal had been accomplished, and progress had been made in the university labs, at the research station, or in a private concern in the next arrondissement, success meant better wine, better production, or even production on pieces of your property that couldn't grow the first wave of vines.

By the 1890s the original tide of American vines—the direct producers and wild species rootstocks—had ebbed and the second wave was starting: genetic hybrids among American species and European grapes came into use in two roles. First, second-generation graft stocks, hybrid rootstocks,[2] were being rapidly disseminated from the École in Montpellier and research stations in Bordeaux, the Charentais, and Provence. Among these graft stocks were candidates suitably adapted to almost every terrain in France and compatible with almost all the traditional varieties. But competing with the grafting campaign run by Montpellier was an alternative program emphasizing hybrid direct producers, an entirely new class of vines whose genetic foundation combined elements from both American and French varieties. This campaign's theoretical base was the University of Bordeaux. In practice, however, many of the most valuable additions to the armamentarium of available hybrid direct producer varieties were developed by private individuals, most of whom were located in the southeast, particularly around Ardèche.

Noisy conflict between the two hybridizing programs lasted for at least twenty years. But even then it didn't go away; rather, it just got less noisy, even as it spread far beyond France. Indeed, it is quite fair to say that principals of the two sides are still vigorously competing after the turn of the millennium, more than a century later.[3] The issues that divide the two camps are complex, subtle, and have deep roots, roots that are firmly embedded in the ever-opposed terrains of tradition and modernity.

But let us begin in the Midi in the early 1870s, at the very start of the "period of direct production" (Sahut, quoted in Pouget 1990, 58).

PHASE ONE: THE AMERICAN DIRECT PRODUCERS

After two centuries and innumerable failed attempts to transplant European viticulture to their soil, Americans finally gave up and in the early 1800s turned to the wild vines growing everywhere in their forests. Beyond being quite evidently different from traditional European wine grapes, the wild vines exhibited a wild and confusing diversity all their own. In the Northeast, the wild grapes called "fox" grapes—for reasons opaque even today—produced large berries with tough skins and a unique taste. In the Midwest flourished the "riverbank" grape, the "bush" grape, and, in more southerly regions, the "sweet summer" grape.[4] As the botanists began to sort out this plethora of vines, it wasn't long before the realization sank in that America's suite of grapes represented the richest treasure trove of species and varieties in the world (discussed in Appendix B).[5] The question was how to take advantage of them to make decent wine?

Practice overtook science from the very start. American grape enthusiasts of all kinds took to the woods in a search for vines to suit their interests and needs. Once in hand, these selections formed the basis of a nascent indigenous viticulture, whose further selection and deliberate breeding spawned a successful national industry.

One happy circumstance favored these pioneering viticulturalists. Unbeknownst to them, at least at first, was the fact that the long, failing effort to transplant European vines into America had produced one quiet success. Even while the European parents were dying from ravages of climate and disease, growing in the woods were their mixed-blood offspring, offspring resulting from the accidental crossbreeding of European and wild American vines, the results, perhaps, of pollen blown from one parent to the next, seeds finding their way into the forest via the intestinal tract of a bird, a strong fall rain, or some such scenario. Whatever the reason for their existence, accidental crossbreed offspring of miscegenous parents survived down through the years, awaiting discovery by the grape seekers.[6]

Euro-American hybrids typically showed some features of each parent. That they survived at all while their European parents didn't revealed their expression of American genes. But European characteristics were expressed as well, particularly in berry size, color, sweetness, or, especially in the *labrusca* regions of the Northeast, a diminution of the typical "foxy" taste. Even though the vines' original finders did not recognize it themselves, the presence of European characteristics was frequently responsible for the attractiveness of these "wild" grapes.[7]

By the 1850s the burgeoning American industry in New Jersey, New York, Ohio, Missouri, and other far-flung places had attracted interest from numerous European private collectors. Since several of the varieties had very desirable features—Isabella, for example, was simply beautiful on a decorative arbor—collectors lusted to have them in their gardens. Orders for French gardens and collections were sent, frequently via the French commercial consulates in New York, Chicago, and St. Louis, and sometimes via commercial or scientific institutions; the grapes were shipped, received, planted, and tended. All available evidence suggests that these collections infected Europe with the phylloxera plague. But it was from these same collections that Europe also found the way to cure that plague.

One of the most avid of the collectors was Bordeaux's Leo Laliman. Laliman began adding American vines to his collection sometime in the early 1860s.[8] By the time the plague started, his vines were well established, indeed flourishing. And, as his nearby European vines sickened, failed, and finally died, Laliman closely watched the Americans flourish, flourish, and flourish. When the first big national viticultural conference focused on the plague took place 8–10 November 1869, Laliman was there, showing his vines and lush crops to everyone (Barral 1869, 666–67). Obviously, the American vines were somehow resistant to the attacks of the insects; perhaps varieties for winemaking should be sought. Laliman also suggested that the American vines might very well be suitable for grafting as rootstocks for the European vines. As he explained later, even though "it is evident that most of them [the American vines] only produce an undrinkable beverage," their roots "brave the venom of this insect well enough to continue to give fruit" and "there is in this providential fact a means of saving our most precious species by grafting them thereupon" (Laliman 1872, 23). Laliman then quotes with approval Riley's recent view that the American bug and the French bug are one and the same, and it is thus probable that the American vine's resistance would be the same in both cases (Laliman 1872, 21).

Laliman wasn't the only viticulturalist with his eyes on the American vines. From the moment that he, Planchon, and Sahut had first spied the yellow louse on those vine roots in Provence, Gaston Bazille had been convinced that the insect was the unique cause of the malady. Bazille—an intelligent, resourceful, and energetic man and longtime president of the SCAH—had thenceforth focused upon finding some way to defeat, or at least to deflect, the enormous appetite the beast had for the roots of French grapevines. In 1869 he had published a suggestion that the grape should be grafted upon *la vigne vierge,* the "vine of the woods," *Parthenocissus tricuspidata* (Convert 1900, 512).[9] A year later, having been convinced by Laliman's demonstration that some American varieties survived the depredations of the insect, he obtained some of the foreign vines from the Bordelaise and planted them. In May of 1871, Bazille put his two ideas together and grafted some European tops onto American bottoms. His hypothesis was that the ability of the American roots to withstand the louse would continue even while they supported European scions. Bazille's success that season was good enough that he placed a large order for American vines with Bush and Sons and Meissner in St. Louis, to arrive in the spring of 1872 (Morrow 1961, 22–23). His order consisted of one hundred vines divided among thirteen varieties. At the same time he ordered a shipment of new specimens from Laliman (Convert 1900, 512–13). All the vines were immediately planted upon arrival and, where possible, grafted. From these experiments would come the knowledge that would eventually lead to the salvation of wine growing in France. But incredible difficulties would be faced along the way.

The question of phylloxera resistance was central to the initial work with American vines. At issue was the nature of the resistance. But how the nature of the resistance was conceived depended in part upon the disease theory one held—phylloxera cause or phylloxera effect. Additionally, the hypothesized nature of the resistance was linked to two general biological theories, Darwin's theory of natural selection—that only the most adapted survive in a given environment—and some general theory of inheritance.[10]

Links to Darwinian theory are the easiest to explain. Riley argued in 1871 and 1872 along straightforward Darwinian lines that the American vines would be permanently adapted to resist the depredations of the bug because the two had coevolved (Riley 1871, 91; 1872c, 624). Although the French workers, including Planchon and Laliman, were far from being staunch Darwinians, they were apparently familiar enough with the theory to find Riley's views agreeable: "Among those varieties accepted for culture, it could only be that natural selection has little by little eliminated those that could not fight off the insect enemy. This hypothesis, cautiously uttered by Riley, accounts for the persistence of certain American vines, for the semi-resistance of others, and for the relative decline of others" (Aîné 1877, 939). Yet, as we shall see, Planchon, for reasons of normal scientific caution, was not yet prepared to claim that the resistance was permanent. In part, his hesitation, as well as that of other contemporary scientists, was only to be expected given their lack of a suitable theory of inheritance. Mendel, unfortunately, was still unavailable to the great mass of biological scientists.[11]

Without Mendel, Darwinian theory remained somewhat mysterious. Variability, according to Darwin, is natural; selection picks out certain variations as successful, and those successes are passed on to succeeding generations. If this idea is applied to American vines, it might be expected that succeeding generations would continue to resist phylloxera, as had their successful parents. But how? That was the question. Absent some suitable theory of inheritance—and, in particular, some principle of stability or notion of stable inheritable character—Darwinian theory alone is deficient. Insofar as he accepted Riley's argument, Planchon was committed to the existence of some heritable mechanism of resistance being present in the lines of the successful American vines. But beyond that he could not go, hence his caution about the permanence of phylloxera resistance.

But more important in this regard than either Darwin or inheritance is the issue of disease theory. According to the physiological theory, environmental conditions are strongly causal with respect to disease etiology. It is quite plausible for proponents of this view to argue that the resistance of American vines holds only in the American environment; once the vines are transported to France, their resistance will dissipate in the significantly different environment. Indeed, this view was brought against the Montpellierians more than once. Unfortunately, Planchon and his colleagues had no good answers to these questions—and until neo-Mendelian genetic theory, there quite simply were no good answers. But the problem was even more general than evolution, inheritance, and disease theory: at bottom, there was an undeniable lack of knowledge about the American vines, their aptitudes, and their responses in various environments.

Paris realized that the problem was a serious one. Two courses of action were taken in late 1872. First, officials from the Ministry of Agriculture got together with members of the SCAH and assigned Planchon to travel throughout the viticultural regions of the United States to gather whatever information he could about the American vines. Secondly, Bordeaux botanist Alexis Millardet was directed to make a thorough study of the botany of the American grapes, classifying them taxonomically and, so far as possible, discovering their aptitudes.

Planchon left for America in late August of 1873. As expected, Riley took charge of his visiting French colleague, making sure that the three-month visit was as rich as possible. Riley met Planchon at the dock in New York and the whirlwind tour began immediately. Between late August and early October they visited in succession vineyards, nurseries, and botanical gardens in New Jersey, Pennsylvania, Maryland, Washington, D.C., North Carolina, Ohio, Missouri, western New York, and eastern Massachusetts. Everywhere he went Planchon took copious notes about the vines, wines, and people he met. He particularly enjoyed a visit to a colony of French viticulturalists in Ridgeway, North Carolina, where he was treated as something of a celebrity, and he was even invited by the North Carolina Agricultural Society to visit the state fair (Morrow 1960, 73).[12] Some of his most valuable time was spent with

Riley in Missouri, especially when the two met for an extended time in St. Louis with Dr. George Engelmann, probably the foremost grape man in the country.[13]

At that time the Missouri vignoble was far ahead of almost everyplace else—even California—in making progress toward a high-technology wine industry.[14] Engelmann and Riley shepherded Planchon around many of the vineyards in eastern Missouri, where Planchon observed a wide variety of cultivated vines representing several important species and their many hybrids. Unfortunately, something went wrong for Planchon and Riley in their Missouri fieldwork, which, as we shall see shortly, was to cause awful pain back home in France over the next several years.

Planchon left St. Louis on 26 September for Sandusky, Ohio, and the wines and vines of Lake Erie's islands. At that time there was a large and flourishing wine industry on the archipelago of islands curving out from Sandusky, especially on Kelleys Island.

Planchon was interested to see there the last vestiges of the Ohio Catawba wine era. Catawba, an accidental *labrusca-vinifera* hybrid most likely discovered in the 1820s in the Carolinas, had formed the basis of an enormous industry founded by Nicholas Longworth around Cincinnati in the 1830s and 1840s. But over the next two decades, the Catawba had foundered, diminished, and died; the Cincinnati wine industry died as well. Planchon and Riley both attributed the death of Catawba to phylloxera attacks: "What, then, is the author of the sickness that affects these final plants? We firmly believe, with Riley, that it is the phylloxera" (Planchon 1875, 82). But on the Lake Erie isles, Catawba had been somewhat isolated, and thus safer from the insect, at least until the bug had become well established there in the late 1860s. What Planchon observed were the last remaining major plantations of the grape.

Yet, right alongside the failing Catawbas, the Concords were flourishing. They were, in a word, *"superbe!"* (Planchon 1875, 83n). This observation of Planchon's turned out to be extremely unfortunate.

Following his visit to the Erie isles, Planchon traveled to Niagara Falls, which he much enjoyed, especially since there he was able, he claimed, "to escape the obsession with phylloxera and vineyards." This claim, however, is immediately followed by more than a page of description of the wild *riparia* vines he saw growing everywhere, covering the trees with their deeply lobed leaves (Planchon 1875, 92). Naturally enough, even though he had "escaped" his phylloxera obsession, Planchon dug up a number of the wild vines only to find them covered with the bugs. Yet, in spite of the infestation, the vines were totally healthy. These informal, unplanned observations on wild *riparia* would serve Planchon and France both well and badly in the next decade.

From Niagara Falls Planchon traveled to Boston—which he found most civilized—and then on to New York, where his cares about departure were all amiably and helpfully addressed by Riley's botanist friend George Thurber. After a twelve-day journey Planchon returned to his beloved Midi on 16 October; there his "wife Delia was able to make up for these deficiencies in the cuisine" of the New World, a place where dishes were rarely seasoned at all, there were no ragouts, sometimes good steaks, and lots of very inferior ice cream (Morrow 1960, 75). Upon his return, Planchon immediately wrote and published a twenty-four-page report on his trip (Planchon 1873). Over the next two years two more reports on the trip were published, the first a somewhat popular but very lengthy treatment in the *Revue des deux mondes* (Planchon 1874), the second (in early 1875) a two-hundred-page technical volume, *Les vignes américaines, leur culture, leur résistance au phylloxéra, et leur avenir en Europe* (Planchon 1875).[15] This latter volume—whose first half consists of a detailed day-by-day journal of his travels, and whose second half carefully reviews all the information known about the American vines— instantly became the bible of the expanding group of *américanistes,* those focused on reconstituting the French vignoble via American vines.

From everything he had seen, heard, and tasted, Planchon concluded that most of the American vines were indeed more or less resistant to the bug. He suggested that three categories of resistance be established (Planchon 1874, 940–42). First were the immune —"those that were not even attacked"—a class consisting solely of the *Vitis rotundifolia,* or so-called muscadines, from the deep South. Second were the resistant, those varieties that were attacked by the pest but seemed to live a normal life nonetheless.[16] Finally there were the nonresistant. Obviously, the French would focus upon the first two groups, but since the muscadines essentially couldn't survive the climate anywhere in France, they were of research interest alone. Planchon's discussion of the resistant varieties thus bears the mark of his efforts in America; unfortunately, it was here that he and Riley make their mistake.

Laliman had warned several times in print that the *V. labrusca*–based varieties, including Catawba and Isabella, had perished in his vineyards. According to him, only varieties based upon *V. aestivalis* could be trusted to resist the insect. In his reports, Planchon rejected Laliman's advice because Riley had: "Better informed on the true varietal names, better instructed by field observations in Missouri, Riley corrected these data on several points" (Planchon 1874, 941). Based on "his observations and on mine . . . we can produce a scale of resistance of various varietals, in which we pass from the most resistant to those that are less so," wrote Planchon. At the top of the scale he and Riley agreed with Laliman: the *aestivales,* including Herbemont, Cunningham, and later Jacquez, were most resistant. Following these, however, Planchon placed, among other "members of the *labrusca* group," Concord, "the rustic and vigorous plant par excellence." Interestingly enough, he agreed with Laliman that some of the *labruscas,* in particular the Catawba and Isabella mentioned specifically by the Bordelais, were obviously susceptible. Indeed, as Planchon remarked, the much-vaunted and highly desirable *labrusca*-based Delaware was nearly as susceptible as the European vines: "The only vineyard in America to offer me the same desolated aspect of dying vines as our Midi was a plot of Delaware on Kelleys Island" (Planchon 1874, 941). But, he added, they were planted "side by side with a plot of luxuriant Concords." Planchon's addition here is important. It clearly is one of the several points of evidence that Planchon personally added to those of Riley in his argument that Concords are resistant to phylloxera. Whether or not they genuinely are resistant, given the recommendation of Planchon and Riley, they would be treated as such from this moment on, with disastrous results.

During the winter of 1872–73 Bush and Sons and Meissner in St. Louis received orders for more than three hundred thousand cuttings and rooted vines. Concords predominated the orders. The following year the number of orders nearly doubled, and this time Clinton—also based on *labrusca*—was the preferred variety.[17] In and around Montpellier much of the vignoble was replanted with these and other American vines. At first they flourished, but within two years many had perished, some, apparently, from the insect, no matter what Planchon and Riley had said about resistance.

But in addition to their questionable resistance to phylloxera, there was a second and confounding problem with the *labrusca*-based imports. As the later French terminology would have it, these vines were not "adapted."[18] Put quite simply, these vines, whose ancestors evolved in the cool Northeast of the United States, fried under the Mediterranean sun of Provence and the Hérault. As Sahut later put it, "Wherever olives flourish, *labrusca*-based vines won't." The result was a disaster.

I take pity upon the poor grower who, having been half-ruined the first time by the phylloxera, used all his remaining savings to reconstitute his own small vineyard, only to have to watch the annihilation of all his hopes. It was said without doubt that, as it was in America, so it will be in France[19] . . . and in the era of which I speak, it is said that budsticks and rooted plants of Concord and Clinton were imported by the millions. These two varieties . . . have thus been the cause of ruin for most of those who imprudently adopted them. (Sahut 1888, 17)

Imagine the situation: Your vines are killed by the phylloxera. Nothing you hear or read indicates that there is any possibility for resurrection. Then, one hears from Laliman and others that certain American vines resist the plague; moreover, it is possible to order these vines from American suppliers. You duly order the vines recommended by Planchon and Riley, Concord and Clinton, which are at that time the most popular vines in America. The vines arrive from St. Louis. You plant them. Initially, they flourish. But within two or three seasons they decline, and they eventually die. You are for the second time vanquished by a scourge from America.

Whether the vines died from the insect, the scorching sun, or a combination of both meant little to the vignerons. Dead vines are dead vines. Yet to the scientists involved, the reasons were crucial: if the vignerons and the general viticultural public were to be convinced that American vines were their salvation against the phylloxera, they must not be allowed to conclude that phylloxera killed their Concords. Planchon, smarting from the failure of his recommended varieties, argued vigorously in an 1877 follow-on article in *Revue des deux mondes* that the failure of the Concords should not be generalized to include all American vines.

He began by noting that M. Fabre, of Saint-Clément, near Montpellier, had repeated the view of Laliman that "all the *labrusca,* most notably the Concord, succumbs and *must* succumb to the attacks of the phylloxera." But the truth, he said, "is that the Concord, so vigorous, so healthy, so resistant in the United States, accommodates quite badly to the hot sun of our climate of the olive." Indeed, although the vine suffered many ills here, "These ills are absolutely independent of the phylloxera, and if I announce this here, I do so in order to refute a false notion that attributes to the insect that which belongs to the soil, to the climate, and to other conditions still unknown" (Planchon 1877, 273). His ultimate conclusion was fairly stated: "The partial check of the Concord in certain places in France must not shake the confidence of the agriculturalists in the overall resistance of the American vines, and more particularly in that of the *aestivales* and the *cordifolia.*" Clearly Planchon had given up on the *labruscas,* especially Concord and Clinton. In the other species, however, he still had a strong faith, which later would be justified.

Planchon musters three arguments in favor of the resistance of the American vines, excluding the *labruscas.* First, he says, consider the existence in America of indigenous vines cultivated over a long time, "even though the phylloxera is everywhere, and the European vine, introduced a hundred times, has always perished under the attacks of this invisible enemy." Although Planchon is somewhat justified here, he is glossing over things he knows to be true because he witnessed them himself during his travels in America. For example, as much as he admired the wine industry around Cincinnati, he was well aware of the fact that it had declined because its star, the *labrusca*-based Catawba grape, was succumbing to the dread insect. Moreover, he remembered the Delawares, dying on Kelleys Island. Thus, in point of fact, Planchon was quite aware of the reality that American viticulture exhibited much more phylloxera-induced failure than successful phylloxera resistance in its brief history.

This is not to say, however, that Planchon had not been exposed to successful long-term viticulture in America. Around Hermann, Missouri, for example, the Norton and Cynthiana, both *aestivalis*-based grapes,[20] had provided more than thirty years of burgeoning success. And further south, as he was also well aware, Jacquez and Herbemont, again *aestivales,* secured a very successful vignoble. So it is not that there were no successes that Planchon had witnessed; rather, his first argument is flawed more by what he does not say than by what he does.

Planchon's second argument—or "second proof," as he calls it—is obviously more germane. Consider, he says, "the vigor of those diverse American varieties (Jacquez, Clinton, Taylor, and others) taken with phylloxera fifteen, thirteen, or twelve years ago, in gardens or nurseries where the French vines are dead or heavily enfeebled." Planchon's point is that these vines have been living with phylloxera infestation for quite some time in France and yet have retained their health. Moreover, he points out, this doesn't even "speak about the plantings of three to five years duration, abundant examples of which are to be found in the departments of Var and the Hérault." Here Planchon knows precisely whereof he speaks. Two years earlier he and Vialla had been commissioned by the SCAH to set up a committee to survey the American vines planted in the Hérault. Their objective was to answer the question whether there was genuine phylloxera resistance in the American vines; their methods involved a census to determine the success of the vines plotted against geography and terrain. In general, the two concluded, "in certain vines, this resistance is very probable, and even our reservations are accompanied by very great hopes" (Vialla and Planchon 1877, 41).[21] From this work, Planchon had collected the evidence to back up his second argument against generalizing the Concord failure.

Planchon's third "proof" at first seems to offer nothing new. Consider, he says, "the comparative experiments planting American varieties and French varieties side by side in the same phylloxerated soil, in which the former have generally prospered and the latter perished" (Planchon 1877, 273). Although this argument seems to resemble the second, it is in fact different. In the second argument, the vines grew on properties together purely by happenstance and they were not necessarily planted side by side. Here Planchon is referring to deliberate experiments involving the replanting of vines in soil that has already killed an initial cohort of vines. In these soils, European vines' buds push the first spring, but the vine is dead by fall. Their adjacent American neighbors, equally newly planted yet apparently impervious to the depredations of the bug, continue to flourish into the next year and the next. This argument is especially important because most of the reconstitution will consist of replanting in already ravaged fields.

Taken together, Planchon's arguments were certainly sufficient to block any anti-American generalizations based on the Concord failure. Of course, the success of his arguments did nothing to recompense the poor vigneron who had plowed his entire nest egg into Concords, but at least they would encourage him, when (or if) he recovered sufficiently, to replant a second time, and finally successfully.

Planchon concluded his report with some positive remarks about the destiny of the American vines. He noted that "the first distinction to make in this regard is whether the varieties in question should be cultivated as subjects upon which to graft the European vines, or cultivated for themselves, for their wine" (Planchon 1877, 274). Grafting, so far as he could see, was only a question of the vigor of the vine under consideration. As we shall see later, the issue is far more complicated than that.[22] But in the meantime, Planchon predicted that *cordifolia*—a species centered upon the Tennessee Valley—would furnish a very resistant graft stock.[23] Additionally, he could predict that the *labrusca*-based Clinton and Taylor, although they might suffer a bit from sunburn, and Taylor's fruitfulness was lower than expected, would function as well as graft stock as they did producing directly. Planchon at this point made with certainty a remark of great importance: "The foxy taste of the *labrusca* [in Clinton and Taylor] never passes in the slightest degree into the fruit of a variety inserted into this foreign nourishment." This remark addressed a widespread fear that the American rootstock would impart a bad taste to the European scion.

But this question of taste was *the* major issue in the use of American vines as direct producers, and it is one that Planchon had to confront directly. Laliman had already claimed, in 1869 at Beaune, that the wine from most of the American vines was undrinkable (Laliman 1872, 23). As far as the *labruscas* were concerned, there weren't many in France to dispute him, even as they imported Clinton cuttings by the millions and planted the foxy Noahs and Othellos. But Planchon here moves right on past the problematic *labruscas:* "The *aestivales,* at least the majority of them, can be adapted to direct culture and give to Europe remarkable wines of a diversity of styles." Herbemont, for example, results in a wine of finesse and brilliant color; Jacquez, on the other hand, provides a wine with alcoholic force and, above all, an intensity of color that should make it a blending wine of the first order. Finally, he says, there are "Norton's Virginia and Cynthiana, which, in addition to a coloration three or four times more intense than the Roussillon, add a particular bouquet recalling, with a bit less finesse, the wines of Burgundy" (Planchon 1877, 275). Clearly, Planchon seems to say, with this sort of foundation available, the reconstitution of the French vignoble using American vines can proceed apace.

At just about the same time that the ministry was sending Planchon to explore American viticulture, Millardet was beginning his work of bringing some taxonomic sense to the American grape species. Millardet's task, as he put it, was "to understand: *the resources that the American vines can offer to French viticulture*" (Millardet 1877, 5; emphasis in original). He published two memoirs relating his findings, one in 1874 and the other in 1876. Millardet almost immediately collected his research together under one title and published it as *La question des vignes américaines au point de vue théorique et pratique* (Millardet 1877). The first chapter of the book and the first section of chapter 2 were strictly theoretical and focused upon the central issue, resistance, which Millardet tied to his principle of "heredity": "I know that in effect it [the principle of heredity] will be for me Ariadne's thread to use in a milieu of obscurities and apparent contradictions.... The property of resistance to phylloxera is strictly hereditary" (Millardet 1877, 23).

FIGURE 11. Statue of Alexis Millardet
in the Botanical Gardens, Bordeaux.
Photo courtesy of Jeffrey M. Davies.

The "milieu" Millardet here refers to is, of course, the controverted question of the resistance of the *labruscas*. He did not hesitate for a second: section 2 of chapter 2 is devoted to the classification and characteristics of the American vines, beginning with the *labruscas*. What Millardet found is quite interesting. Employing his principle that resistance is a hereditary trait, he focused upon analysis of pure American species rather than cultivated varieties. For this, he needed seeds from plants collected from the wild, which he got either from America or from the specimens growing in the botanical gardens, particularly those in the collection at Montpellier. In cooperation with Fabre—whose phylloxera-devastated vineyard near Montpellier was used as the test plot—Millardet planted one thousand seeds from wild vines of the species *labrusca, riparia,* and *aestivalis*. From these seeds, two hundred vines sprouted. The sprouts from *riparia* and *aestivalis* did exceedingly well, many of them in one year growing eighty-centimeter shoots above ground and root systems about half that size. The wild *labrusca* seeds failed.

Millardet reached two extremely important conclusions based on this experiment. First, the pure species *riparia* are quite resistant to phylloxera, and the *aestivales* are not far behind. Second, because he now had some idea about the range of morphological variation in the pure species, he saw clearly that the varieties then being imported—Clinton, Concord, Taylor, and Othello, for the most part—are in fact *labrusca*-based hybrids. Clinton, for example, is a hybrid between *labrusca* and *riparia*. Concord apparently has some European *vinifera* in it mixed up with the American *labrusca*. In Millardet's view, whenever a resistant species is crossed with a nonresistant one, the likelihood is that resistance will be weakened.[24] From this he reached a new practical conclusion: rather than using the imported vines, the direct producers, as grafts—a practice heretofore much followed, especially since it had been endorsed by Bazille, Laliman, and Planchon—it would be a much sounder policy to rely upon the pure wild species types, especially the *riparia,* as rootstocks. In other words, nurserymen, botanists, owners, and even the smallholder vigneron should plant seeds of the pure wild types of *riparia* in their vineyards or seedling areas, grow the seeds, and then transplant and graft in place during the second year of growth.[25]

Millardet's ideas did not go unnoticed. By the end of the 1870s, use of the direct producers as graft stocks was essentially over: "One can therefore say that the reign of the American direct producers employed as graft stocks is passed. In reality, the issue has entered a new phase, that of the graft stocks from wild species, and, since this evolution, things progress in a more rapid and assured manner" (Millardet, quoted in Pouget 1990, 65). Obtaining the seeds from wild vines was sometimes a bit problematic. Millardet found one reliable source, however, in Missouri's Dr. George Engelmann, who sent Millardet somewhere between five hundred and a thousand kilograms of seeds in late 1881.[26] Of course, once the French nurseries got into the business, it was easy enough to propagate the wild vines from homegrown seeds. When the resistance of the species *rupestris* was discovered in 1879—again by Millardet—the machinery of nursery propagation was already up and running. It is clear that the work of Millardet between 1874 and 1880 markedly changed the world's understanding of the biology of the American vines, which was exactly why the academy had chosen him for the job in 1874. As Pouget remarked, "His precise and exact observations are remarkable for the era and testify to the observational sense, the just reasoning, and the intuitions of Millardet" (Pouget 1990, 64).

Millardet's information was bolstered by that from other sources. One chief source appeared in late 1876: *Les Vignes américaines: Catalogue illustré et descriptif* (The American vines: An illustrated descriptive catalogue), by Bush and Sons and Meissner of St. Louis, was translated by Louis Bazille and reviewed and annotated by Planchon. The catalogue was an encyclopedic source—a compendium, really—containing all the current and attested information for the vines available from the Bush and Sons and Meissner nurseries. At that time, these were certainly the dominant grape nurseries in the United States, and their catalogue provided the best available guide to the American vines. Needless to say, the French edition was gobbled up immediately.[27]

Planchon's short introduction leads into a fifty-six-page "Manual of Viticulture," a handbook for grape growing that covers everything from taxonomy to grafting, propagating, and training, with pathology and insect control thrown in as well. The hand of Engelmann is clear, and Riley's entomological contributions stand out in his always-recognizable drawings of the beasts and their depredations. What is somewhat surprising to find in the manual is the international foundation of the data. All the significant discoveries made in America and France in the years just prior to the volume's publication are recounted. On the cutting-edge topic of grafting, for example, we find on page 32 a discussion and diagram of the system of *Gardener's Monthly*'s M. Cornelius, which precedes a discussion and diagram of the "quite recently proposed system of M. Henry Bouschet, of Montpellier (France)." Page after page of technical discussion reveals that communication of research results and anecdotal experience took place at incredible speed, and apparently freely and easily, between St. Louis and Montpellier. All the usual suspects are found in these pages.

In his introduction, Planchon makes even more explicit the cooperative basis of the volume. After noting that there are a great number of works on the American vines, certainly enough to satisfy the pressing needs of the French viticulturalists, Planchon remarks that "the work that provides the best résumé of this vast subject, and, above all, by reason of its numerous picturesque and descriptive illustrations, most satisfies the eye, is this one, which Louis Bazille, wishing to profit the French public . . . has taken on the thankless task of translating" (Bush and Sons 1876, v). Planchon notes that, although his own recent book (Planchon 1875) contained many contributions from the Bush and Sons and Meissner catalogue, "in order for this important work to be fully useful for our compatriots, it is necessary to reproduce it totally complete; and for this, to obtain from the authors and Riley the plates for the lithographs." While the scientists strongly desired the catalogue to be widely available, copying the plates would have made the cost prohibitive. But "the generosity of MM. Bush and Riley" in donating the plates "had made possible the realization of this desire." It is quite amazing today to look back and find such close cooperation between the Montpellier workers and their Missouri colleagues; this is truly a model of scientific collaboration.

Planchon concludes his introduction with an interesting remark. "It is hardly necessary to say," says Planchon, "that the commercial nature of this catalogue, in our eyes, detracts nothing from its scientific value." Indeed, the "collaboration of men so distinguished as MM. Engelmann and Riley proves all by itself the intrinsic value of the work, even if this value weren't already fully asserted by the tone of sincerity that is the dominant note of the work" (Bush and Sons 1876, vi). Planchon's insight was correct: very soon after its translation and publication, the Bushberg catalogue (named after the Missouri town where the nurseries were located) had become the standard manual of the new viticulture based upon the American vines.[28]

THE *AMÉRICANISTES* IN BATTLE

But the new viticulture was by no means unopposed. Controversy and turmoil were the hallmarks of the times. From 1875 to 1881 two major and independent campaigns were in action against the hated insect. First and foremost, if for no other reason than its support from official Paris, was La Défense, the effort by means of submersion, sand, and sulfur-based insecticides to stop the pest dead in its tracks. Especially active, not to mention vociferous, were the proponents of chemical treatments. Backed by the great railroads and chemical manufacturers, those associated with this arm of La Défense came to be called *"les chemistes"* or, sometimes, *"les sulfuristes."* But as we have seen, when it came to practical applications—especially by the small to medium-sized vignoble—each of La Défense's three weapons had serious limitations: constraints posed by the lack of suitable terrain or water, shortcomings in technique or equipment, an ill-trained or otherwise inadequate workforce, and, in all cases, high expenses. Nevertheless, the publicity and push for La Défense during these years was unceasing.

But even while La Défense was gathering steam, folks in the Midi, especially around Montpellier, were acquiring the knowledge and experience to back up their hunch that American vines offered a practical alternative to the expensive high-tech solutions proffered by the *chemistes*. Planchon was everywhere, speaking, demonstrating, arguing, always pushing the *américaniste* solutions. Since some of the American vines have practical resistance to the bug, he argued, let us use them in either of the two effective ways, either as direct producers of table wine, or as rootstock for the classic French varieties. By the late 1870s, both Montpellier's SCAH and the École itself had instituted workshops in grafting, planting, and pruning—all the essential skills required to grow American vines. Moreover, and perhaps more important, "all the observations and results" of work on the American vines "were announced and discussed in conferences organized" by the two institutions (Pouget 1990, 57). Publication followed immediately: "Reports and accounts of experiments were published in the local newspapers and in technical reviews, such as *La Vigne américaine*,[29] by the most diverse collection of experimenters: savants, professors, lawyers, doctors, deputies, senators, aristocrats, proprietor-growers, amateurs, etc." (Pouget 1990, 57). Visits to experimental vineyards were organized regularly; overall, "there reigned everywhere a very frank atmosphere of collaboration and solidarity, without distinction of social rank." This was, of course, because "everyone was equally struck by the common enemy, the phylloxera" (Pouget 1990, 57).

One of the biggest *américaniste* "festivals" was sponsored by Pulliat at Lyons, in the heart of the most dedicated *chemiste* stronghold. Described as "the great international court of the reconstitution," famous Beaujolais grower Victor Pulliat's September 1880 affair had more than one thousand participants, of whom more than a hundred were foreigners—including several of the Missouri nurserymen (Garrier 1989, 110). Three years later Pulliat put on the train 120 proprietors and vignerons from Beaujolais and the Mâconnais for "a trip to Languedoc, more particularly to the nurseries and the grafting workshops of the École of Montpellier" (Garrier 1989, 110).

Pulliat's actions here were not unusual: he was a dedicated *américaniste,* willing to use just about any means to convert his stubborn Beaujolais neighbors from the chemical campaign to the side of the botanists. One lovely story involves Pulliat and his good neighbor and friend Émile Cheysson, who, like Pulliat himself, was a proprietor and viticulturalist at Chiroubles. Cheysson was an inspector for Bridges and Roads, a professor at the National School of Mines, a great scientist, and a friend of Baron Thénard, the apostle of CS_2. But Cheysson was also an apostle of Le Play, the Catholic militant theorist, and thereby saw in the *chemiste* position "an occasion to cement the union of the capital of the proprietor and the labor of the vigneron" (Garrier 1989, 111). Not entirely innocently, Pulliat succeeded in convincing his friend that the reconstitution, given its high cost, was "a better laboratory of 'social peace' " than was La Défense. Cheysson's "spectacular rallying to the *américaniste* position completely rocked the departmental commission." Cheysson would always remain loyal to his persuasive friend. In his dedication of a bust of Pulliat in Chiroubles in 1896, soon after the latter's death, Cheysson remarked that "if all those whom he had persuaded, all those whom he had rescued, had been there, there would have been an immense throng in among the green slopes of vines, constituting the most beautiful of crowns" (Garrier 1989, 111).

Resistance to the *américanistes* and their message, perhaps in part due to jealousy or even regional ill will, was widespread. In late 1879, the presidents of two agricultural societies in Isère were sent by their departmental prefect to visit the Midi. On their return they issued a report that was not terribly favorable to the American vines; in particular, they noted that there were a lot of yellow American vines around Montpellier,[30] and that the grafts of Aramon on Taylor didn't succeed. According to the story as reported in the journal *La Vigne française,* the two presidents were immediately attacked in the pages of a local Midi newspaper, *Le Sud Est:* "The conclusions of their report have been combated with extreme violence in *Le Sud Est* by the *américanistes.*" Moreover, the journal notes, the workshop at Nîmes "was, in fact, a veritable congress of *américanistes.* There, it's always the same speakers come to pronounce their small speeches or their reports prepared for the circumstance; it's always the same denigrations addressed to the insecticides, and the affirmations in favor of the American plants" ("La crise phylloxerique" 1879, 57–58).

The controversy did not end there. In the next issue the editor of *La Vigne française* reported that "one of our subscribers in Vaucluse has written us a propos of our column [about the attacks upon the two from Isère] in order to ask us 'if we are hostile to the culture of the American vines, and if our intention is to make a campaign against the propagation in France of the American vines' " ("Réponse du editeur" 1880, 135). The editor responded that the journal had no hostility at all, and that it would campaign only for continual and better research. Moreover, he argued, "We must take note of the success of immersion and the CS_2 treatments." In the end, the journal's position was that "free access to American vines" should be allowed only in the cases where it was clear that the European vine was destined to perish, and after it had been demonstrated that "these exotic varieties are immune to the pricks of the insect" ("Réponse du editeur" 1880, 135).[31]

Later in the same issue, the editor remarked that his story about the visit of the two presidents had come under fire from the editor at *Le Sud Est,* who complained that he was in some places misquoted in *La Vigne française*'s earlier article and invited the offending editor to visit the Hérault. He concluded with the remark that most likely a combination of grafting and CS$_2$ would solve the problem ("Réponse du editeur" 1880, 135).

At times opposition between the chemists and the botanists took a more lighthearted form. For example, the *Journal d'agriculture pratique* carried a debate between the two sides in the form of a dialogue between personifications of the two opponents. The first, written in late 1879 by Aimé Champin, was entitled "Letter from an American vine to the boss of the insecticides"; the response, by Vincendon-Dumoulin, appeared in 1880 under the title "Response from carbon disulfide to the American vine." Obviously, given the venue, the dialogue was not only of regional interest; rather, it was being carried on "at the level of the whole country, if not all of viticultural Europe" (Garrier 1989, 107).

Sometimes the resistance came from unexpected places. The Var, just a few short kilometers east of some of the very first infection foci, held out for years before introducing the American vines. In 1875, for example, the consul general refused to vote for the finances necessary to buy American vines for experimentation. He argued that he didn't want to encourage "the use of a special remedy whose effectiveness hasn't been demonstrated," and, moreover, something that could cause "the malady to break out in a region heretofore spared." Two years later the Phylloxera Commission of Var's Agricultural, Commerce, and Industry Society underlined this position strongly when they argued that there were still "many studies to make, many experiments to try, in order for us to be able to say that these vines from the New World offer salvation to French viticulture (Garrier 1989, 108). It is hard to explain how this attitude in 1877 Var could possibly square with the obvious successes of the American vines in 1877 Montpellier, no more than 170 kilometers to the west. Yet there is no question that feeling against the American vines ran high at all levels of viticultural society. At the very top, the Phylloxera Commission of the Academy of Sciences declared, in an official publication, that the American vines remained permanent foci of the disease (Commission 1877). At the very bottom, at the level of vine workers and smallholders, there existed "a durable distrust of the American vines as vectors of phylloxera" (Garrier 1989, 112).

Américaniste counterattacks against this resistance were continuous. One of the most important forays against the opposition was made by Gaston Bazille in Paris on the occasion of the national meeting of the viticultural section of the Société des agriculteurs de France on 14 June 1878. Before an audience of crucial significance to his cause, Bazille countered the opposition's sallies—some of which were quite nasty— and laid down markers of his own.

Bazille first noted some very grim statistics: according to the latest official statistics, twenty-eight departments were more or less invaded, with 288,000 hectares of vines completely destroyed. Against this in-creasingly devastating evil, the viticulturalists hadn't been standing around with their hands in their pockets; rather, they had been trying everything and anything, and, after seven to eight years, they had won some small successes, which he would now report.[32] Sand and submer-sion are discussed, and their success noted; the long-term failure of all types of fertilization and special nutrition is regretfully remarked. Ba-zille then vigorously condemned all the various "bizarre procedures," such as treating the vines with camphor, liver of sulfur, chalk mixed with salt, and goat urine. There are only two insecticides, he said, that merited further testing: carbon disulfide and potassium sulfocarbonate. He acknowledged as well the PLM Railway for setting up the distribu-tion, keeping the costs down, and generally helping with the use of in-secticides. Nevertheless, he noted, perhaps because of climate, terrain, or season, the insecticides are not everywhere successful, especially in the Midi: "In the Midi, the dryness and the amount of sun often make the ap-plication absolutely impossible." Bazille then went on the attack against the *chemistes*:

> In spite of all these difficulties, all these uncertainties, experts, men in high places, have wished for diverse reasons to arm the government with legisla-tive decrees that render obligatory the employment of carbon disulfide or the sulfocarbonates. The recently passed law carries traces of this preoccupa-tion, and the minister can, if need be, force a proprietor to apply a treatment called for by the Superior Phylloxera Commission. . . . Is it necessary to send the gendarmes after the *viticulteurs* in order to oblige them to sulfur their vines? (Bazille 1878, 9)

Rather, Bazille thought, "it would be much more reasonable to leave each to his own free decision and not to appeal to law." When, in the end, an "incontestably efficacious and practical treatment was recognized, it will be applied by all interested parties, with no other type of pressure."

Bazille followed his attack upon the *sulfuristes* with an immediate turn to a "vividly controversial question: I wish to speak of the American vines." His opening words show precisely how nasty the controversy over the American vines had become: "Right from the start, let me reject a reproach, an insinuation, that never fails to show up whenever talk turns to these vines.—'Plant salesman!' it is cried whenever a *viticulteur* proposes use of the American vines. 'Plant salesman!' it is easily said; but these two words do not have the power to prove that the proposal is a bad one." (Bazille 1878, 9).

Although it might seem strange to us today, it is clear that Bazille and the other *américanistes* felt stung by the accusation that they were recommending the American vines just because they wanted to profit from their sale.[33] While admitting that, yes, many of the *américanistes* do in fact sell cuttings and vines, Bazille noted that they had taken serious risks in experimenting with these new vines, and that, in his estimation, "these plant salesmen do useful work, and they ought not have stones thrown at them." Moreover, he concluded, they are elite people: "One sees among the propagators of American vines members of our legislative assemblies, eminent professors, dignitaries from the chambers and tribunals of commerce, and agricultural societies" (Bazille 1878, 10). This should suffice to settle the "plant salesman" criticism, Bazille said, before he turned to substantive matters concerning the American vines. It is no wonder that "the American vines have their most convinced partisans in the Midi," Bazille claimed. "Our vineyards, the first attacked, are in great part destroyed; for eight years it has been impossible for us to save them using the means science has put in our hands." But, as the resistance of the American vines became more and more certain, "how could we not grab onto this last possible saving branch?" (Bazille 1878, 9).

Bazille's argument was detailed and rhetorically solid. He noted how a colleague had put his faith in insecticides to save his traditional Aramons and Carignans; but, "unhappily, my honorable colleague now has neither Aramons nor Carignans." And, he was pleased to say, "several adversaries of the American vines have become their partisans": two men of high stature—M. Loubet, president of the civil tribunal and *comice* in Carpentras, and M. de Mortillet, distinguished savant and director of the well-known agricultural journal *Le Sud Est*—were good examples. Additional strong support took the form of nearly unanimous recommendations favoring American vines from the agricultural societies of the Hérault, Gard, Vaucluse, and Var. The clincher came from the Gironde: the previous December, M. le docteur Micé, president of the agri-

FIGURE 12. Gaston Bazille (third from right), painted by his son Jean Frédéric Bazille, in the painting *Réunion de famille*, 1850.

cultural society of the Gironde, had forwarded the following recommendation to the prefect of the department: "In summary, M. le préfet, the conclusions of the Society of Agriculture of the Gironde, after the year that has just passed, are the following: less confidence in the insecticides, more confidence in the American vines" (Bazille 1878, 11).

Bazille's argument was a strong one. As he noted, it was the viticulturalists in the south who bore the brunt of the destruction; it was they who paid the costs, expended the energy, and shed the tears. Who were the others, whether unaffected growers or officials in Paris, to tell those suffering what they should or should not do? If, as the rapidly accumulating evidence attested, the American vines offered some last hope to otherwise hopeless viticulturalists, then there would be no justice in trying to prevent them from seizing upon this final course of action.

A further strength of his argument came from the diversity of those who shared his recommendations. Northerners might have tried to argue, "Wait, we're only talking here about the mass-produced table wines of the Midi and not about high-quality wines," but this riposte was deflected by the Gironde's presence among the allies of American vines. If the official agricultural society in Bordeaux—which, along with Burgundy, is the region producing France's highest-quality wine—added its recommendation of American vines to that of the departments of the Midi, then the case was essentially made.

But even if the Americans vines seemed to be succeeding, the original doubts still remained: What was the nature of their resistance? Was it permanent and biological, that is, essential? Or was it only accidental, induced by their cultural environment? Bazille dealt with this worry in witty—and devastating—fashion, playing the true naturalist philosopher. "Let us discuss this question from the purely theoretical point of view," Bazille said. Given the observed facts, it could be said that "a priori, it would seem . . . there must be vines resistant to the phylloxera." In the first place, the bug survives only on the grape vine and on no other plant. Therefore, "if there weren't, somewhere on the globe, resistant vines providing the bug the appropriate nourishment and allowing it to breed, it wouldn't be able to live and perpetuate itself." But the bug isn't dead. This allowed the elegant and direct conclusion, "We are able to say, a priori, following the example of a celebrated philosopher of the seventeenth century, who remarked in a well-known syllogism, the entailment of thought and human existence, *The phylloxera exists, therefore there are resistant varieties*" (Bazille 1878, 12). It has been well said that "Scratch a Frenchman, find a Cartesian," hence the reference to Descartes' famous "cogito." Indeed, using it in an argument stood Bazille in extremely good stead rhetorically, philosophically, and sociologically! Bazille's argument obviously depends entirely upon acceptance of the Darwinian theory of evolution. Only according to this theory can the coevolution of parasite and host be derived from the principles. But Bazille's appeal here is more than scientific.

The rest of Bazille's presentation to his hostile audience remained on the same high level. He rehearsed empirical facts, such as results from the École's Professor Foëx's experiments comparing American and French root behavior when piqued by the bug; he cited authorities, such as Dumas' view that since destroying the insect is impossible we must learn to live with it;[34] he attacked anew the alternatives to American vines, for example, by emphasizing the extremely high cost of insecticides in relation to their effect; and, finally, he held out the promise of the American vines, praising the quality of wine that resulted when they were used as direct producers, claiming, "I will not be surprised if Jacquez remains numbered always among the vines cultivated in the Midi of France."[35] Bazille's conclusion is a rousing one that does not just defend the *américanistes* and their vines, but also vigorously puts down their opposition, especially the officials in Paris:

The propaganda that we generate in the Midi, the truths that we seek to distribute, finds, I know, especially outside our region, incredulous and refractory minds. One finds still discussed in Paris, and certain agricultural journals, the merits of the American vines; one proscribes them in the name of principles, whereas, in the Midi, we have already reconstituted a great part of our vignoble with the aid of these precious vines. (Bazille 1878, 16)

Clearly, Bazille and the *américanistes* had become a force to be reckoned with, no matter who the opposition might have been. Sometimes, however, the opposition was resolute, indeed, even downright heroic.

THE PERSONAL CRUSADE OF PROSPER DE LAFITTE

Prosper de Lafitte waged a vigorous four-year personal battle against the American vines beginning in 1879. No small-timer, de Lafitte was an influential owner, graduate of l'École Polytechnique, and president of the Committee of Study and Vigilance of Lot-et-Garonne. Although de Lafitte frequently claimed that he "was not an adversary of the American vines" but rather thought that "an elegy had been made to these vines" that seemed to him to surpass their just deserts, and he thus believed it necessary to say: "You have gone too far!" (*Compte-rendu* 1882, 274), his many writings proved the contrary. When a compilation of his works appeared in 1883, his crusade against the American vines was announced explicitly for the first time. The book's title is revealing: *Quatre ans de luttes pour nos vignes et nos vins de France* (Four years of struggle for our vines and wines of France) (de Lafitte 1883b). In conjunction with his attack upon the American vines, de Lafitte argued in favor of two important measures: CS_2 and Professor E.G. Balbiani's attempts to interrupt the phylloxera's life cycle by obliterating the "winter egg."[36]

In the foreword to his book, de Lafitte makes clear what his strategy has been. Balbiani's theory had been proposed in the late 1870s, but it had not received any support, especially from the administration. Indeed, de Lafitte claims, "Balbiani's eggs had against them an ensemble of bad will that we have had no success in disarming. . . . During four entire years, I have been alone in defending him" (de Lafitte 1883b, xii).[37] Throughout the four years in question, de Lafitte had forwarded the Balbiani theory in two ways: "Above all, I have crusaded for the theory itself; then, by studying closely the remedies that our adversaries have put on offer; and then, taking the offensive, I have shown that our vignobles—at least three quarters of them—have nothing to expect from the known treatments alone, and still less from the American vines" (de Lafitte 1883b, xii). De Lafitte did not believe that CS_2 has been tried rigorously or correctly; to change this, he developed his own process for applying CS_2 effectively, as we shall see shortly. But what is of interest to us here is his attack upon the American vines.

When we follow de Lafitte's arguments closely, it becomes evident that he opposed any entry of American vines into the French vignoble. He argued strongly against their use as direct producers, and just as strongly against their use as graft stock. His attacks hit every possible target: individual American varieties (e.g., Clinton), American species (e.g., *labrusca*), grafting on cultivated varieties (e.g., on Solonis), and grafting on wild varieties taken from the Missouri woods, either as cuttings or as seeds (e.g., on *riparia*). If there were a way possible to use American vines in French vineyards, de Lafitte was against it, and volubly so.

A superb example of de Lafitte's rhetorical skill appeared in early 1881 in the *Revue des deux mondes*. Given the status of the *Revue*—then France's leading periodical of political, literary, and general cultural commentary—it is quite clear that de Lafitte's purpose was not just to carry the fight against the American vines to the growers, owners, and smallholders; rather, he hoped to take the fight right into the homes and offices of the officials in the administration.[38] The twelve-page article is entitled "La question du phylloxéra et le rôle des vignes américaines." In it, de Lafitte's theme is that even though the remedies then known—CS_2, inundation, and sand—were effective, they were too costly to apply to the majority of vignobles, where the value of the wine crop per hectare was frequently less than the cost of saving it. Hence, or so the argument went, American vines were the only cost-effective solution.

De Lafitte calls this argument into serious question. Before he spends what remains of his wealth, de Lafitte argues, the prudent vigneron or owner should perform a very careful analysis of the pros and cons of moving to American vines. Doing so would reveal that estimates of the *"chances favorables"*—the pros—were based solely upon what had been written or said on the matter, but which, de Lafitte claimed, "contain some theories that appear to me dangerous, because those imprudent persons who confidently accept them without examination could find ruin through an enterprise that is premature" (de Lafitte 1881, 197). As for estimates of the *"chances contraires*—the cons—in adopting the American vines solution, well, on this issue "many diverse circumstances have been left far too much in the dark." His article would analyze and criticize the pros as well as shine upon the cons the light that they deserve. In the end, of course, we will see that de Lafitte's conclusions contain no pros and many cons with respect to the American vines solution.

De Lafitte's arguments are inevitably theoretical. He states a claim about an advantage of using the American vines, and then counters that the claim was exaggerated, or failed to consider all aspects of the situation, or, frequently, that experience with the American vines was far too limited to be relied upon in this instance. The overall argument focuses first on the notion of "resistance."

According to de Lafitte, proponents argued that there were two categories of vines, those *"cépages résistans"* that resist the phylloxera's assaults, and those *"cépages non rèsistans"* that do not; American vines belong to the first category, while European vines do not. But this claim, de Lafitte argued, is false. Resistance, assuming that there is such an inherent property, varies in infinite degree: "Between the wild species most refractory to the insect and the most vulnerable cultivated variety, one could arrange all the known vines along a series, between any two members of which there is very little difference from the point of view of resistance" (de Lafitte 1881, 197). The conclusion, then, would seem to be that there are not simply two classes; indeed, there are no classes at all, only endless variation among vines.[39] Obviously, de Lafitte hoped that the *américanistes* would slide quickly down this slippery slope.

Moreover, de Lafitte asked, "What will be the average lifetime of diverse *cépages* in the typical cultural conditions of various regions given the presence of phylloxera?" The answer to this most important question simply is not known; worse, the most interesting and plausible American candidate vines are precisely those with which our experience is the least.

The central argument in favor of the reality and permanence of the American vines' resistance may be called the "survival argument." We have seen it before: beginning with Riley and Planchon—and, more recently, in Bazille's 1878 presentation—*américanistes* argued in quasi-Darwinian fashion that, were the American vines not well and truly resistant to phylloxera, they would have long since vanished from the earth since their homeland was rampant with the bug.

"Nonsense," argued de Lafitte. It is not necessary that any given species, nor even any given individual, be resistant. "It suffices," he asserted, "for any given individual to live long enough to reproduce" (de Lafitte 1881, 198). In France, of course, there would be no sanctuaries where this could happen—every patch of land belonged to someone. But in America there were vast forest reserves where all it took was a bird dropping a grape seed in some area not yet infested with the bug, thereby assuring survival for yet another generation.

In de Lafitte's eyes, his counterargument destroyed the foundation for any general claim that American vines' existence implies their resistance. So what is left of the claim for resistance? Scientific studies, he claimed, and particularly anatomical and chemical analyses, are frequently used by the *américanistes* to explain why phylloxera resistance is inherent among certain of the New World vines.[40] But de Lafitte would not buy this kind of argument either. Although the scientific theories were presented with all due reserve by their authors, "these theories have been accepted far too eagerly and without a severe enough critique" by the *américaniste* world beyond the laboratories. In actual fact, he proclaimed, we do not know yet what the vine offers to the bug, nor what the bug actually does to the vine to cause injury. So to refer to anatomical or chemical features as the basis of inherent resistance is simply premature. Concludes de Lafitte, "The theoreticians seem to try to establish between the hardness, the density of the tissues, and the resistance relations of cause and effect, without remarking that the coexistence of two phenomena is not proof in itself that the one is the consequence of the other" (de Lafitte 1881, 200). On this point de Lafitte was surely correct. Correlation is not causation, and thus to claim that thick cell walls account for the American vines' resistance is to mistake the former for the latter.[41]

In the end, so far as de Lafitte could see, there was no good theoretical reason to believe in the inherent nature of phylloxera resistance. One was thereby limited to the very local claim that "such and such a *cépage* has lived this many years with the phylloxera in these conditions of climate, of soil, of culture; but nothing permits us to speculate how much time it has left to live" (de Lafitte 1881, 200). This brings us to the concept of adaptation.

As early as 1874, Planchon noted that Concord and other *labruscas* did not do well in "the land of the olive." Much more formally, Millardet had observed in 1877 that "resistance is the resultant of two actions combined together: the action of the conditions intrinsic to the plant (its nature, temperament, etc.) and the action of the extrinsic conditions (phylloxera, soil, climate, etc.)" (Pouget 1990, 67). Yet, sensible as the notion might seem, de Lafitte was having none of it. So far as he was concerned, adaptation was being used simply to excuse the absence of the resistance expected in certain varieties. To prove his point, de Lafitte rather wickedly quoted a story told by "a very distinguished botanist, no enemy of the American vines, no more than I am myself." (It is Millardet, of course, but speaking in another context.)

> Let's suppose that we have arrived with the commission on a parcel of several hundred Taylors planted three to five years ago, where there's a weakness, and a "spot" has been declared.[42] . . . Three or four vines are pulled up, the roots devoured by phylloxera. . . . The proprietor watches with an anxious eye the face of the president of the commission, fearing to read there his death decree. Life is no bed of roses. The face brightens: there is happiness! . . . He declares that it is necessary, before making the pronouncement, to make an analysis of the soil. . . . In his joy at this unexpected response, the proprietor forgets to curse the vine, the earth, and the phylloxera. However, six months later the report appears, with the analysis of the soil and subsoil correct to the fifth decimal place. The conclusion of this eminently scientific piece of work is that it is infinitely probable that the Taylors, if they succumb, which is not certain, will succumb not to the phylloxera, but to a failure to adapt to the soil and the climate. (de Lafitte 1881, 201–2)

"Beneath this humorous form," de Lafitte remarks, "which leaves intact the sincerity of the fictional dramatis personae, there is most assuredly a basis in truth." And most assuredly there is.

American vines grow across an incredibly diverse range of climates and soils. Yet, even given their diversity, few American environments come very close to those of France. Consequently, when the American vines were put into the French environment, they encountered conditions quite different from those they had evolved to fit. Matching vine to environment is a variety-by-variety task. To his credit, de Lafitte agreed that "the most useful employment of the scientists" at this point would be to "institute and pursue with perseverance the study of this very difficult question" of "which conditions of climate, terrain, and culture are necessary to be put in place in order to obtain the most that these vines can offer." But it wouldn't be easy. Objectivity and disinterest had to be foremost, such that the scientists' "interpretations of their observations and those of others take place without allowing even the slightest consideration foreign to agricultural science" (de Lafitte 1881, 202).

Until this point de Lafitte had for the most part contented himself with a broad attack upon the direct producer American vines, those cultivated for their wines. Lest anyone think, however, that these vines were his sole target, de Lafitte now attacked on another front, combating the American vines being tried as graft stock for the traditional French varieties. As he immediately made quite evident, de Lafitte rejected these vines as well.

De Lafitte's basic argument is captured in one apparently innocent principle: "The grafting operation puts the American vines in conditions so different from those they have ever grown in before that it is necessary to pause a moment to think about the possible consequences of this transformation" (de Lafitte 1881, 204). Basic to the whole set of issues that follow is the term "possible." De Lafitte focused upon suggesting hypothetical mechanisms that might limit the life of the grafted vine. For example, he pointed out that clearly there must be some important interactions between roots and foliage: "Between the roots and foliage there is an incessant exchange in which, depending on the circumstances, each system gives its associate more or less what it itself receives." These systems are obviously in harmony with each other. Suppose, however, that a root system of a variety accommodated to a luxurious foliage system is grafted to a variety that has only a moderate amount of foliage, or vice versa (which, he implies, is typically the case). How can there possibly be a harmony in these situations? Only disaster can follow. As a consequence, since "the physiology of the graft is, in effect, one of the questions least understood in botanical physiology," our ignorance of the conditions necessary for success plus "the very small number of facts known until now" generate "anomalies little compatible with the formulation of rules" to ensure progress (de Lafitte 1881, 206).

It gets worse. The varieties favored for use as grafts—especially those hybrids created just for this purpose, such as Vialla—"are, with few exceptions, quite sterile, or at least of very low fertility." What can we expect from these American roots when they are faced with trying to "nourish fifteen or twenty kilograms of Aramon? We will not force a conclusion but will wait until experiment has resolved this difficulty" (de Lafitte 1881, 206). Here, as always, de Lafitte exudes an air of reasonableness and cool objectivity, but the content of his argument is both tendentious and tenuous. No one, so far as I know, had ever suggested that there were permanent and fixed mechanisms of equilibration between the root systems and foliage systems of grapes. Nor, even worse, between sterility of the rootstock and crop load. De Lafitte here illustrates great creativity in service of his scare tactics.

He concludes the essay with a flourish. "There you have it—if I'm not deceiving myself—plenty of plausible reasons to doubt the ultimate success of the American vines. Is it therefore necessary to renounce them? No. But it is necessary to know these things in order to dispel the

FIGURE 13. Marguerite, la Duchesse de Fitz-James, 1867, by Henri Fantin-Latour. Chester Dale Collection, National Gallery of Art. Image courtesy of the Board of Trustees, National Gallery of Art, Washington.

exaggerations of a propaganda without limits, and to proceed with prudence" (de Lafitte 1881, 207). In the end, "we must never lose sight of this truth, that to save a sick vine is the ideal, and that to replace a dead or dying vine is only a costly palliative, one followed by ruin." In other words, rather than turning to the American vines, it is far better to use insecticides, and, finally, to put into play "the idea that the destruction of *the winter egg* could save a vine" (de Lafitte 1881, 208).

Clever as it was, de Lafitte's article did not go unchallenged. Hardly had its ink dried before the highly public, highly respected Mme. la Duchesse de Fitz-James, owner of a large estate, unloaded a full salvo of return fire.

Her first sentence is directly on point: "The study published in the *Revue* of 1st March by M. Prosper de Lafitte seems of a nature to destroy the hope of a just barely reborn viticulture" (Fitz-James 1881, 685). But consider his argument more carefully. In particular, do his predictions uniquely entrain disaster? Or would we face disaster even if we obeyed his suggestions? Let us see.

> In effect, the ruin that M. Prosper de Lafitte apparently predicts would follow the premature and rash plantations [of American vines] would be neither greater nor more complete than that which has already struck this vast area.... If the Midi waits for absolute certitude, it will perish, because in order to wait, not only is bread necessary, it is necessary as well that the bread not cost more dearly than the work of the man whom it must nourish. (Fitz-James 1881, 685)

To those who have studied the issue, she notes, two things are quite clear. First, in the Midi, without vines, there is no bread, which means misery, the loss of the villages, and "the demoralization of a land that love of hard work has preserved until today." Faced with this bleak future, de Lafitte's dire warnings against the American vines have no traction. Moreover, and even more pointedly, it is now well known that "certain species from America (the states of the South), species pure of all European hybridization, succeed and resist on our sunny slopes, and others prosper in our rich and healthy plains; their product is sufficiently healthful to nourish the man who cultivates them and to earn from the land what other crop cultures could not render" (Fitz-James 1881, 686). This point squarely rebuts the major thrust of de Lafitte's argument. Not enough is known yet, he had claimed, to find places where productive American vines are sufficiently adapted to provide reliable and generous supplies of healthy wine. De Lafitte buttressed this claim with a succession of horror stories and gloomy hypotheticals, all based upon the bad experiences of a number of growers, almost every one of whom had planted the *labrusca*-based vines. Fitz-James rebuts by claiming that, first, there is plenty of experience in the Midi with vines from the southern United States, non-*labrusca* vines "pure of all European hybridization," which have succeeded admirably. If she can back up her claim, Fitz-James will have effectively deflected de Lafitte's central claim. She proceeds at once to provide the requisite backup: a detailed history of America's experience with the vine, beginning with Leif Ericson!

Throughout her discussion, Fitz-James is careful to remark upon the diversity of the climates and terrains where the Americans had made their attempts to found a viticultural industry. Her reason for doing so will become clear when she finishes her tale. But for now, she takes great pains to bring one conclusion home to her readers: after three hundred years of total failure to establish European vines on the eastern part of

TABLE I   PROPOSED USES OF AMERICAN SPECIES

	Fertile, direct producers	Infertile, graft stock
South	*aestivalis,* clean taste	*riparia,* easy rooting
North	*labrusca,* foxy taste	*cordifolia,* difficult rooting

SOURCE: Fitz-James 1881.

NOTE: Mme. Fitz-James argues that only *aestivalis* should be used in France to produce wine.

the continent, the Americans began a second viticultural phase: attempts to grow the native races of grapes. This effort, she remarks, has been successful and has remained successful for more than forty years. And now, she says, the Americans have well and truly entered the third phase of developing the local grape industry by careful, patient selection and breeding of the native races. "A table," she notes, "will allow us to avoid the aridity of the details" (see table 1).

According to this table, it is the America's southern vines that should become France's sole focus. Prosper de Lafitte's arguments against the American vines in general are based upon the failures of the northern selections, the *labruscas*. But, as Fitz-James notes, "the failures of the *labruscas* confirm my faith in the *aestivales,* since in the same milieu, the one perishes while the other prospers" (Fitz-James 1881, 689). Compared to the *labruscas*, "the *aestivales* certainly resist long enough to pay us for our pains, and perhaps they will always resist. Their wine resembles French wine; its color, its degree of alcohol, varies around Narbonne from the richest, most full-bodied to the most delicate of rosés. . . . The *aestivales* from the South prosper in the region of the olive; they enrich the worker after having nourished him first" (Fitz-James 1881, 689–90). Indeed, the duchess is so confident in these vines that she has planted more than 140 hectares—principally in Jacquez—during the preceding two seasons. Obviously, based on her own experience, she would never say to her viticulturalist colleagues "Go!" but rather "Come!" and join her in her experience.

Fitz-James uses this experience to address what she perceives as Lafitte's "most salient point." She then quotes in full the passage in which he claims that those "who know these vines by what they have observed appreciate them much less than those who know them only by what they have heard. . . . And why is it that the neighbor of an apparently flourishing American-planted vineyard so rarely puts in these vines himself?" "To this," says the duchess, "I will respond by *facts,* by *domaines planted,* by *dates.*" She then lays out her program to put in 450 new hectares of grapes over the next three seasons. Her accomplices in this project will be "my own workers and sharecroppers, who care for my vines and see quite enough to form their own opinion; nonetheless, they are willing to risk their work, their time, their salaries planting these vines." Clearly her people have seen enough over the years to justify their investment in what seems to them to be a trustworthy future. By their actions they refute de Lafitte.

Fitz-James's final point is directed to de Lafitte's sarcastic indictment of adaptation, which "M. de Lafitte seems to consider as an easy excuse used solely in order to explain failures without negating the notion of resistance itself." This is a fair characterization of de Lafitte's position; Fitz-James thoroughly rejects it:

> This seems to me to be unjustly severe, for this notion expresses a truth and a traditional but unobservable property that has lately been applied to the division of French varieties according to the zones most appropriate for them. Each variety has a home region where it reaches its highest degree of perfection; from which, radiating outward, it diminishes in value proportionately to the distance: Aramon, in l'Hérault; Folle Blanche in l'Ouest, etc. It appears illogical not to expect the same thing from the American vines, a group composed of such diversity, coming from everywhere in a vast countryside. (Fitz-James 1881, 691)

The duchess's point is well-taken: if the notion of adaptation has acquired a fixed and definite technical meaning in recent French viticulture—which it obviously has, given its use by such diverse people as Millardet and Sahut—then it is illogical to either refuse to apply the notion to a group of vines with even more diversity than French vines, or, as de Lafitte does, to subvert the technical meaning into something sinister.

In the end, those who, "like me, like to believe, and desire to believe, accept the theory of a serious *viticulteur* in Texas, who has divided the *æstivales* into three groups, each prospering in a different zone, but surviving in all" (Fitz-James 1881, 692). The Texas reference is, of course, to Thomas Volney Munson, an American viticulturalist whose already well-received work with the native American vines would culminate twenty years later in the publication of the classic *Foundations of American Grape Culture*. According to Munson's theory, Norton is the *aestivalis* for the North (Missouri), Jacquez for the South (Texas), while Herbemont does fairly well in all three zones.[43] Had de Lafitte cared to pay attention to Munson's work and the testimony of the many growers themselves, he would have given the notion of adaptation the respect it was due, exactly as the duchess claims.

FIGURE 14. T.V. Munson. Courtesy Grayson
County College.

Fitz-James ends her response to de Lafitte with one final tweaking. Referring to her huge acreage of grapes, she states that "as a proprietor, I do
not partake of the fears of M. Prosper de Lafitte; they are contrary to all
that I have seen in my experiments . . . and qualifying myself only as a
grape-grower for profit. I come to combat a very shrewd agricultural tourist" (Fitz-James 1881, 692).[44] Of course de Lafitte was a bit more than an
agricultural tourist; after all, he headed up the phylloxera commission in a
very important department. Yet, unfair as her characterization might be,
she makes the necessary point that she is risking her sizable investment of
land, labor, and time based upon her own experience with the new grapes.
On the one hand, were she and her neighbors to follow de Lafitte's counsels, none of this would be risked; on the other hand, however, the Midi
would be finished as a wine-growing—indeed, as an agricultural—giant.
Without wine, the small landholder must emigrate or starve.

Prosper de Lafitte did not limit his campaign against the American
vines to the vines themselves. His was a genuine "no quarter given" war

of eradication. One newly thriving use of American vines came under his fire in mid-1881. The target was propagation of rootstock vines from wild vine seeds, a technique designed and perfected by Millardet and his colleagues at Bordeaux. Millardet first described the technique in 1877; two years later he published further on the topic and included a broadly based justification of the technique, particularly addressing the issue of whether wild seed was purely *riparia*. This was an important issue since were it not reliably *riparia*, then long-term resistance of rootstock grown from the seed could not be assured. Prosper de Lafitte's attack upon the technique focused upon precisely this point.

De Lafitte wasted no time: "As graft stock, American vines obtained from seed at first glance seem to offer seductive advantages" (de Lafitte 1883a, 535). For example, seeds don't carry the phylloxera and thus can be imported into the quarantined areas; the price is right; the graft plant is ready by the second season; and seeds from varieties that are hard to root from cuttings root well from seed. But even given these manifest advantages, he claims, "this method appears to me quite dangerous, and I wish to say why, if only to provoke more complete explanations on a most grave question" (de Lafitte 1883a, 535).

All de Lafitte's arguments turn on the notion of a "hybrid." To begin, he offers a brief run-through of grape reproduction before moving quickly to the then-known facts of inheritance. Never, he says, do offspring of different species, or even distinct varieties, fully resemble either parent to the exclusion of the other. Always one finds a blending of the characteristics in the offspring.[45] With regard to resistance, following successive hybridizations between "species or varieties, of which one is resistant and the other is not, the offspring will either be resistant or not, depending on how the state of equilibrium between these contrary influences comes out." In any case, the result "must be observed, it cannot be predicted" (de Lafitte 1883a, 536).

Certainly Prosper de Lafitte is correct about this latter point: the resistance of hybrids could not then be predicted. (Indeed, it is not a matter that can be reliably predicted even today.) Only observation would tell. But de Lafitte presents his initial point somewhat duplicitously. He speaks here of a vine being resistant or not resistant as though this were a binary property, yet he knows full well that resistance is a scalar property, appearing in all possible intensities, from "none" to "immune." Although this is not entirely crucial to his argument, treating resistance in this fashion certainly makes his point sharper and more penetrating. This is especially true since in his central argument he presents the resistance of both *vinifera* and *labrusca* as "zero"—which is itself questionable, since at least some of the *labruscas* are usefully resistant—and then infers from the presence of *labrusca* in a hybrid offspring that its resistance *could* be zero.

But before inferring this conclusion, de Lafitte must muddy the water as much as possible. Variation in offspring is not limited to hybrids, he notes. There are many reports of offspring derived from (presumably) self-fertilized seeds of Jacquez, York's Madeira, and Solonis that resemble their parent in scarcely any feature. Jacquez is the most obviously variable in this way. In part, this might be due to atavism, or genetic throwback, but it is more likely that the presumption of self-fertilization is wrong and that the American vines had been fertilized by neighboring *vinifera*. This could very well account for observed variation in France, as de Lafitte notes. In the wild, however, this shouldn't happen, according to Missouri's Dr. Engelmann. For one thing, of course, there won't be any *vinifera* present. But, more importantly, continues Engelmann, the flowering season of *riparia* is well before that of the *aestivales* and *labrusca*.

De Lafitte rejects this argument. Consider, he says, the "*minimum* interval" between the early ripening *riparia* and the other species. Among the *riparia,* some will be latest to end bloom, and that lateness will certainly provide a small overlap with the earliest blooming *aestivales* and *labruscas.* Why couldn't cross-breeding occur under these conditions? Maybe it could, he suggests. Hence, "in the presence of such incertitudes," concludes de Lafitte, "it's quite possible that we find coming from America seeds crosses between wild species made in the virgin forests of Missouri" (de Lafitte 1883a, 538). De Lafitte backs up this conclusion by an analysis of Millardet's view on the issue, which, as de Lafitte makes clear, undergoes a significant change between the two works published, respectively, in 1877 and 1879.

In the 1877 paper, Millardet clearly concludes that "it would be wise to renounce the seeds of all types of grapes from V. *aestivales* and V. *riparia* taken directly from the North American forests" (Millardet 1877, 31). In his 1879 paper, however, Millardet concludes just the opposite, due, he claims, to new data and facts produced in the interim.[46] But what about these facts, asks de Lafitte. Can they be trusted? No: "After an appreciation of these data and facts that I must regret finding myself in disaccord with the learned professor," confesses de Lafitte. But there's not much to the new data and facts: "To speak truly, everything reduces to an assertion from Dr. Engelmann," namely, an assurance "that V. *labrusca* is not to be encountered in the Mississippi Valley" (de Lafitte 1883a, 539). This assurance turns out to be crucial to de Lafitte's case against the American seedlings. As he reiterates, "Hybridization by some varieties of *V. labrusca* could, all by itself, compromise the phylloxera resistance of a vine come from an American seed." But, of course, if Engelmann is right and there are no *labrusca* vines in the Mississippi Valley, there is nothing to be feared. For his part Millardet accepts Engelmann's assurance with complete confidence, says de Lafitte. The stage is now set for de Lafitte's crushing blow: if he can find evidence of any *labrusca* vines in the Mississippi Valley—preferably by Engelmann's own admission— then the assurance will be rendered null and Millardet's confidence in the purity of *riparia* seeds from Missouri will be invalidated. And, of course, evidence—of a sort, as we shall see—is precisely what de Lafitte immediately turns up.

To begin with, de Lafitte notes that Engelmann's claim originally appeared in the Bushberg catalogue on page 17. But what inquiry, "what researches have been instituted regarding such a difficult question concerning a territory several times larger than France, where immense spaces are filled with virgin forests, full of the most diverse varieties, the most undetermined, of species without name? ... I know nothing about it, because he doesn't say a single word about it!" (de Lafitte 1883a, 540). In fact, if one reads further in the catalogue, one finds claims that *labrusca* do exist in the Mississippi Valley. Indeed, on the very next page Herman Jaeger of Neosho, Missouri, reports to Engelmann that *labruscas* in southwest Missouri, southern Illinois, Arkansas, and west Texas give two harvests of healthy grapes. We are most certainly in the Mississippi Valley, notes de Lafitte drily. On page 20 of the catalogue he notes that G. Onderdonk has reported that "our vines in Texas must be taken from the *aestivales* family because no *labrusca* has provided a sufficient and permanent satisfaction." Satisfaction or no, there are *labruscas* in the Mississippi Valley crows de Lafitte (de Lafitte 1883a, 541).

It would appear that de Lafitte has Engelmann, and thereby Millardet, on the ropes, if not hors de combat. But appearances are sometimes deceiving, and in this case they most likely are. Consider more closely the two counterexamples brought up by de Lafitte. The first thing to note is that both Jaeger and Onderdonk are grape men. Both maintain nurseries and are commercial producers. The *labruscas* they are talking about are those that are maintained commercially by growers and farmers. It is simply not the case that *labruscas* grow native in the woods of either Jaeger's southwest Missouri or Onderdonk's central Texas.[47] And that, after all, is precisely Engelmann's point: the woods of the Mississippi Valley do not have native *labruscas* silently waiting to cross-pollinate *riparias*. Millardet's argument in favor of the purity of wild *riparia* seed hangs precisely and explicitly on this point.

Certainly de Lafitte's claim that *labruscas* exist in the Mississippi Valley is correct, but it is quite irrelevant, as he knows. If the *riparia* seeds come from the virgin forests, then, unless some very extended mechanism is in operation connecting the forest with one of the isolated *labrusca* plantations, there is essentially no chance that the seed is hybridized *riparia* × *labrusca*.

Prosper de Lafitte's goal is quite clear: he hopes that by employing a clever, albeit scholastic, argument to rule out the hope that American wild vine seeds might be used by French vignerons. In the positive realm, what he wants them to do is "to persevere in the study of treatments directed against the insect," namely, the use of insecticides and the elimination of the winter egg (de Lafitte 1883a, 542).

Yet sometimes de Lafitte could not prevent his cleverness supporting the American vines. Just such an occasion came up later the same year, during the October 1881 International Phylloxera Congress in Bordeaux. At the long afternoon session on 12 October, the committee charged with examining the American vine situation presented an extremely thorough report (the central points of which have been reviewed above). In general, the report is quite optimistic about the course of developments with the American vines, although, of course, the discussion of nearly every topic ends with a plea for more investigation. One of the methods of further investigation comes from none other than Prosper de Lafitte. What he has created is a very clever methodology with which to carry out a sizable chunk of that investigation, a methodology that is based in the primary school.

According to this methodology, a collection of the best varieties known was planted "in the garden or the court of the school. One hundred and forty-nine of these establishments are already completed, and this number will rise to at least five hundred this year if enough participants are found" (Gachassin-Lafite 1882, 274). Six benefits are claimed to accrue from this arrangement. First, rather than the vigneron having to travel a long way to a nursery or demonstration plot in order to view the American vines, it is the vines that will travel to the vigneron. Second, "the youngsters, marvelously gifted for this apprenticeship of the eyes, observe the vines all year long, getting to know them and then taking the information to their families—peasant families for the most part—which would have been nearly impossible to accomplish in any other fashion." Third, by the nature of things, these collections of vines will find themselves in every terrain found in the department. They will thus be exposed to all the conditions demanded by a complete study of adaptation. Fourth, details such as the quality of each variety, responses to chemical and physical compositions of varying soils, cultural needs, and fertilizer requirements will be made available not only at each school, but across the entire range of schools in the department. In his students, "the teacher will find the elements of the research most sure, most extensive." Fifth, the Central Committee, by comparing all these observations, "will be able to establish the dossier for each species or variety in such a fashion as to resolve the question of adaptation for the whole department." Finally, the collections will put into the hands of the researcher "a scholarly corpus that will allow instruction of the pupils in all they need to know about these new vines, in particular the art of grafting them" (Gachassin-Lafite 1882, 275).

Using this method, the complete solution to the problem would be only a matter of time and communication. Conveniently enough, the Central Committee had been able "to establish its nurseries and plantations on the grounds of the Teacher's College, where the good will of the director, the devotion of the professor of agriculture, and the eagerness of the students has rendered the committee's task as agreeable as it is easy."

Although the reporter for the congress remarked that this study would probably have to endure as long as the phylloxera, it remained nonetheless that the methodology was intelligent and, if carried out carefully, would produce some excellent data. Obviously Prosper de Lafitte, no matter his own strong feeling about the American vines, was unable to refrain from designing a useful and innovative method to give the vines a chance. Of course, it must be remarked that de Lafitte's main presentation at the congress was not this one, but rather his unveiling of a new, improved method of applying CS_2 (de Lafitte 1882).

De Lafitte's campaign continued for several more years. One finds, for example, a long paean to Balbiani's winter egg theory as late as 1885, four long years after the Bordeaux conference (Magen 1885). By then, most viticulturalists, and certainly almost all scientists, had given up on Balbiani's theory. Yet de Lafitte still fought on—an excellent example of a scientist hanging on to his views no matter what the evidence to the contrary. Yet, his fight is one of the best examples of an intelligent, hardworking, and—no matter what Mme. la Duchesse de Fitz-James might say—always-thinking Frenchman. Prosper de Lafitte's campaign was already heroic by the time of the Bordeaux congress in late 1881. By then, the American vines had clearly proven themselves in many terrains to be reliable and profitable direct producers of a healthy, albeit ordinary, product. To oppose the Americans in this role was, as Fitz-James so sharply pointed out, to deny a life-saving branch held out to a people already gone under once from the phylloxera. What is more difficult to understand than his opposing the American direct producers is de Lafitte's opposition to the American role as rootstock. After all, if the Americans could successfully and generally provide roots for French tops—an issue still in major doubt at the time of the Bordeaux congress—then at least one of de Lafitte's two objectives would be satisfied: French wine from French vines would be again available. His only loss would be the cultural practices surrounding traditional wine production, but that loss seems quite minor compared to what would be gained. In the end, it simply is not clear what the basis of de Lafitte's opposition to American vines as rootstocks was. No matter what, however, he was doomed to defeat. By the end of the 1880s—six years after Prosper de Lafitte's magnum opus—essentially all the problems with American vines as rootstocks had been solved. The amount of research and experimentation necessary for this solution, however, was simply enormous. To that tale we now turn.

CHAPTER 4

The Underground Battle

Grafting on American Rootstock

The grafting of fruit, including grapevines, has been documented since Pliny. The procedure serves many purposes, but it is relatively simple in both concept and execution. Elements of two or more separate plants are physically united in such a way that root support from one is used by the other. There are nearly as many grafting techniques as there are grafting technicians, but for various reasons, mostly economic ones, only a few techniques dominate any given industry. French viticulture, however, started grafting from a knowledge base of essentially zero. When the first attempts at grafting against the phylloxera were made by Bazille in 1871, he had no experience with the technique, nor did anyone else in Montpellier or, more importantly, the SCAH. Some, of course, had experience with grafting fruit trees—apples, for example, were frequently grafted—but this experience was of only limited utility. Indeed, knowledge about grafting grapes was at such a low level that Bazille's proposal to graft grapes onto Virginia creeper or Boston ivy was taken seriously. Over the next fifteen years the knowledge base was filled in remarkably well and remarkably quickly. It is safe to say that the major issues were well understood and the appropriate techniques established by the early 1890s. But it took a lot of hard work by a lot of talented people to get the job done.

Grafting presents three distinct problems. First, the rootstock, the part of the plant that inhabits the earth, must thrive. But getting American vines to thrive in all regions of France was an extremely difficult problem to solve. Indeed, by the mid-1880s most of the research came to focus upon this problem. Second, the scion, the part of the plant that lives aboveground and produces the crop, must thrive as well. Here the issues particularly involved the problem of compatibility between the rootstock and scion, but also crucial was solving the difficulties involved in the vigor and productivity of the rootstock-scion combination. Not all compatible combinations were equally productive. Finally, there is the problem of the graft union itself: what is the best technique for putting the two vines together, given the varieties involved, the local climate, and the available labor supply? Work proceeded in all three areas at once, with progress occurring at variable rates.

Montpellier made the first move toward solving the technical problems. From 3 to 4 March 1879 a workshop on practical issues related to the grafting of American vines was held at the amphitheater of the École (Société Centrale 1879). The workshop, attended by 962 people, was cosponsored by the SCAH and the school. Presenters included both faculty members and members of the SCAH. L. Vialla presided and gave the opening address. In his talk he noted the surprising fact that it had been only ten years since grafting had first been suggested at the Beaune Congress (Barral 1869, 666–67), and only seven years since the SCAH had made its first attempts to discover techniques for successful grafting (Société Centrale 1879, 13).

Toward the end of the address, Vialla made two observations that introduced issues of an extremely serious nature. First, Vialla noted that resistance to phylloxera was not the only question about the American vines. A new and, as it turned out, devastating problem had only just been recognized: "Last year at the viticultural congress at Montpellier a question that I believe is of great importance was posed for the first time. I speak here of the adaptation of various American varieties to the diverse natures of our soils. This question is still being studied" (Société Centrale 1879, 14). The Concord disaster was now reaching its peak, and, unbeknownst to anyone at that point, the *riparia* disaster was incipient. Vialla and the other Montpellierians were only now starting to awaken to the scale and significance of the crisis of the failure of the first wave of American vine plantings. Moreover, as seen in Planchon's 1877 analysis of the Concord failure, the cause of the calamity was taken to be the climate in the "region of the olive" and not the American vines' adaptation—or lack thereof—to the soil. Thousands and thousands of young American vines, both *labrusca* and non-*labrusca*—thousands of hectares of them—were just beginning their decline toward death, even as Vialla spoke.

Vialla's second point was another harbinger of the future. Because of the American vines' bad taste (or, when the taste was acceptable, their lack of resistance), the Montpellierians believed that American vines would increasingly be used for grafting; thus "grafting will be the general rule, direct production will be only the exception" (Société Centrale 1879, 15). Vialla's words here marked a change in direction for Montpellier. Throughout the previous decade the École and the SCAH had pushed American vines as both direct producers and graft stock without discrimination. But here Vialla depreciated their direct production role and emphasized their role as graft stock. With this change of direction Montpellier chose a path that would soon set it in opposition to Bordeaux's Millardet and other researchers such as Couderc and Ganzin, who were just beginning to make some progress with hybridization. In retrospect, Vialla's speech reported events of no little consequence for the next decades of French viticulture.[1]

Following Vialla's talk, the workshop proper commenced. Planchon led off with a presentation entitled "The General Principles of Grafting and Their Application to the American Vines," the principles being demonstrated "among the vines of the École by competent men." Following this, Sahut presented "Choice and Conservation of Grafts" and Foëx "The Most Favorable Epoch [in the vine's life cycle] for Grafting," both of which included practical work. Hortolès, from the École, finished off the day with a discussion and demonstration of styles, instruments, and techniques of grafting. The second day continued the practical work, kicking off with local proprietor Pagézy's presentation "Respective Merit of the Different Varieties of Vine (*cordifolia, æstivales, labrusca*) Considered as Rootstocks for Our Diverse French Varieties." According to the advance program, if time remained after this, there would be "a rapid examination of the diverse procedures put into use at the École d'Agriculture for the multiplication of American vines—semis, bouturage, marcottage."

Multiplication, or the rapid propagation of huge numbers of copies of the rootstock varieties, was at this point the crucial bottleneck in the reconstitution on American vines. And, unfortunately, it would turn out that the best American varieties, the ones most resistant to phylloxera, were nearly impossible to multiply under ordinary conditions. It would be nearly five years before the supplies of graft stock would equal demand, and resolution of the problem involved not just individual efforts but also the direct intervention of the government in setting up nurseries (Tisserand 1885, 117).

As far as the workforce for grafting was concerned, it took a massive coordinated effort to disseminate the skills to the viticultural community: "Thus, in order to resolve this difficulty, it was commenced to organize courses for grafting, and even here the *syndicats* and professional schools played a front-rank role. Soon, master-grafter diplomas were delivered to certain vignerons, women, even adolescents who became facile at producing a great quantity of good-quality grafts" (Lachiver 1988, 432). But it took a long time to get enough grafters in the field. Immediate relief was found by importing experienced grafters from surrounding countries. Because their smallholders practiced polyculture involving interplantings of grapes, olives, almonds, and many kinds of fruit trees, Italy, Spain, and Portugal had large populations of competent grafters. Italians came to the mountainous regions east of the Rhône, Spaniards and Portuguese to Bordeaux and the Gironde in the west. Grafting paid well, and, since new candidate areas would open up as soon as their infestation became official, the grafting crews migrated constantly northward. Italian workers were reported in Beaujolais around 1885 and in Burgundy in 1890 (Garrier 1989, 128). But by this latter date, the supply of French workers had become sufficient to take over for the foreigners. In great part, the growing supply was due to a massive development of grafting education. Every department was required to participate, and "the professors of agriculture were charged to multiply the demonstrations in the trial vineyards, to give conferences, and above all to open grafting schools" (Garrier 1989, 128). Expansion was rapid and extensive; the *département* of the Rhône, for example, went from ten grafting schools with 220 students in 1885 to twenty-four schools with 778 students in 1895. From 1884 on, the central Ministry of Agriculture gave medals to participants in the grafting workshops (Tisserand 1885, 120).

For its part, Montpellier stayed at the forefront of the business. In the early spring of 1880, a year after the original workshop, another conference was held, again with an enormous crowd attending. This time the conference proceedings were published in the form of a small but very thorough handbook on the topic of grafting. Covered were choice of graft stocks and of scions, correct timing for the grafting itself, styles and techniques of grafting (including eight different techniques, with drawings and instructions), tools and machines for grafting (illustrated), tying and gluing, caring for the new grafts, a calendar, and, as an appendix, a list of local vineyards that welcome visitors. Each vineyard entry is glossed with details about which scion plus rootstock combinations were pres-· ent and their dates of planting. Those inviting visitors included Saint-pierre (director of the École), Pagézy, Vialla, Leenhardt, and many others, to a man prominent members of the SCAH. Their invitation would be taken up many times in the years to follow.

Proselytizers of the Montpellier view radiated out in every direction. Two years later, for example, Chauzit, who had been a researcher and presenter at the École, was called to the job of departmental professor of agriculture in the Gard. He immediately held his own workshop on grafting in Nîmes under the sponsorship of the departmental agricultural society (Chauzit 1884). Sahut, Vialla, and Planchon all gave talks and presentations throughout the region. In general, it is safe to say that the knowledge about grafting techniques, methods, and compatibilities between stock and scion were worked out by the Montpellierians and then disseminated as widely and as quickly as possible.[2]

Grafting is a multistep process. In concept it is simple enough: the growth zones of two pieces of wood are brought into close proximity, physically fixed together, and the join heals over. The growth zone—the cambium—occupies the major middle region of the shoot between the xylem and the phloem; bark surrounds the whole structure. After the two pieces of wood are brought into close contact, the wound calluses over: a spongy tissue exudes from each member, mixes, dries, and encases the cambium, which can then exchange vital fluids between the leaves and the roots. Although this sounds simple enough, in practice there was always controversy.

Two basic modes of grafting competed against one another: field grafting and bench grafting. Field grafting takes place in early spring in the field where the rootstock is already growing. There it is cut according to the style of graft, and the prepared scion is joined to it and secured with a piece of raffia twine. Callusing occurs in situ before the dormant stock begins to awaken from winter. Field grafting is obliged when the rootstock is grown from seed, and thus necessary in the case of varieties from species such as *aestivalis, lincecumii,* and, most significantly, *berlandieri,* which could not be propagated from cuttings. Portuguese and Spanish grafters were particularly expert at this style; a good worker could make three hundred or more grafts a day. But, since the new vineyards typically had five thousand vines per hectare, it is clear that the workforce needed for the reconstitution was huge, and their labor long. Unfortunately, in the maritime climate of west-central

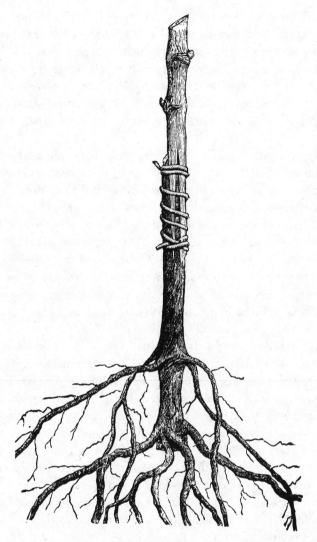

FIGURE 15. Field graft in the English, or "whip-and-tongue," style. From Barral 1883.

France—the Charentais and Nantais—the high humidity made it very difficult for field grafting to succeed since the graft join often became diseased before it had properly callused. In these regions, bench grafting was required.

In bench grafting the rootstock cutting and scion cutting are independently prepared, joined together, calloused, and then set out to root.

The work takes place during the late winter on a workbench or table inside a specially dedicated building. Callusing bench grafts requires an elevated temperature and very high humidity. For this reason, boxes are constructed in which the newly joined pieces are wrapped in wet moss and seaweed and sealed tightly; after approximately five weeks in a heated room, the vines have callused. Those whose calluses are firm and secure are immediately set out to root in special nursery rows.

By the late 1880s machines were available that could cut both stock and scion and physically join them, all in one fell swoop. But the machines were expensive, and most of the work continued to be done by hand. As noted earlier, women and adolescents were allowed into the grafting schools, with the eventual result that they took over the industry: they were "more apt, more adroit, and cost much less," as the Regional Society of the Rhône noted in 1889 (Garrier 1989, 128). Indeed, "after 1885, when productivity absolutely required improving, it was necessary to make appeal to great battalions of young men, young women, and women" in order to get the work done (Garrier 1989, 129).

Controversy also existed about which grafting style was the best. Each region seemed to develop its own style or set of styles, which it then argued was the best. In the Bordelais, for example, the "Cadillac" style involved a complicated cut on the side of the graft stock. In the Languedoc, an even more complicated chamfered graft was considered best. But in fairly short order the graft controversy settled down and the majority of workers adopted the so-called *greffe anglaise*—the English graft—which the English (and the Americans as well) call the whip-and-tongue graft. "Tongue" in this case refers to the parts produced by slits in the two cuttings that interlock; the sloping faces of the whip cut maximize the area of cambia that are exposed and touching. By 1896, the standard manual of viticulture recommended only the English graft for bench grafting (Berget 1896, 101).

The scale of the job was immense. Table 2 gives a rough estimate of the figures involved. The area to be covered with graft stock was roughly the size of the state of Massachusetts. The length of graft stock needed would go around the earth at the equator nearly eighty times. Although costs varied considerably, a recent analysis put together from a welter of data collected at the time calculates an average cost per hectare: three thousand francs (Garrier 1989, 131).[3] One well-accepted estimate has the reconstitution costing France more than the Franco-Prussian War (Convert 1900, 337).

TABLE 2 THE SCALE OF THE GRAFTING JOB, 1880

Wine-growing area	2,209,000 hectares
Vines	5,000 per hectare
Graft stocks needed	11,040,000,000[a]
Graft length	30 centimeters
Total length	3,300,000 kilometers
Total weight	290,000,000 kilograms
Cost per 1,000 graft stocks	80–125 francs
Total cost	1,104,000 francs[b]

[a]All graft values assume 100 percent efficiency.
[b]About $125,000,000 in 1880 dollars.
SOURCE: Adapted from Ordish 1972.

Unfortunately, development of grafting techniques and processes did not in itself save the day. Once the resistance of the American vines was assured, an even more complicated problem came to the fore: planting in the chalk.

THOSE DEADLY WHITE SOILS

When the Concords and other *labruscas* began to fail around 1875–76, Planchon and the others thought that it was the climate of the "land of the olive" that was at fault. Native to the cooler regions of northeastern America, these grapes grilled under the hot sun of France's south. In any case, according initially to Laliman and, finally, Planchon, many of the *labruscas* were not resistant enough to be reliable. Besides, by this time the *riparias* had been discovered and tested successfully for resistance, and thus were being planted as quickly as possible, both as seeds and from rooted vines, many of which had been sent from St. Louis. Obviously the *riparias* were the vine of the future.

But then the *riparias* themselves began to fail; not everywhere, but here and there, and with increasing frequency. Perhaps Prosper de Lafitte was right! Savants who had so wholeheartedly recommended the *riparias* immediately began to back off from their recommendations, hedging against the risk of another Concord disaster. The first warning notice was given in 1878; one year later it was repeated, more emphatically, by Louis Vialla: "Last year at the viticultural congress at Montpellier a question that I believe is of great importance was posed for the first time. I speak here of the adaptation of various American varieties to the diverse natures of our soils. This question is still being studied"

(Société Centrale 1879, 14). "Adaptation" has been mentioned before. As it came to be conceived, adaptation was a response of the vine to the properties of the environment—climatic, geographic, and agronomic—that it lived in.[4] One of the most useful definitions was that given by Sahut in the late 1880s: "One sees therefore that adaptation to the soil consists in placing each *cépage* in the terrain that suits it; or, better, in other words, to resolve the following problem: *Being given a soil of a determinate nature, how choose the cépage that could happily prosper there and be conserved for a long time in a good growing state?*" (Sahut 1888, 11; emphasis in original). The problem was that many French soils—indeed, probably the majority of French soils—do not suit the American vines familiar at the time. Vialla's warning was official recognition of the problem.

What the adaptive failure consisted in was not known; it was only known that many American vines did not do well even while others did. Systematic observation was required as the first step in understanding the problem. Observations made by the sagacious Louis Vialla were among the first to go on record in a short 1879 piece entitled "Des vignes américaines et des terrains qui leur conviennent," or "American vines and some soils that suit them" (Vialla 1879). Vialla starts out with a clear statement of the new problem: "Most important with the American vines is knowing their resistance; but nearly as important is knowing which terrains are suitable for them, which is not known" (Vialla 1879, 1). Previously, it had been believed that once resistant vines were found the phylloxera problem would be solved. Unfortunately, the world rarely matches the simplicity we project upon it. As Vialla remarked, with obvious exasperation, although it is true enough that some of our traditional vines do better in some places than in others—Folle Blanche, for example, does very well in the Charentais but badly in the Midi, whereas it is exactly the opposite for Aramon—"we have no variety that refuses obstinately, as do Norton's Virginia, Cynthiana, and Concord, to grow in a patch of good ground, or one that yellows nearly everywhere, as does Herbemont" (Vialla 1879, 2). Thus, "Concord and Clinton have succumbed miserably these last five or six years, not because of the phylloxera, as one had too often believed, but because of the terrains they grew in that did not suit them." Apparently, he argued, before they had put these vines into wide culture, the special care given to them in small gardens or around the home masked any effects of ill-suited soil. But now these effects were evident in many places.

Vialla then made a link that would soon prove to be important. He noted that most of the American vines began the process of failing by turning yellow in the spring, but then, unlike the indigenous *vinifera*, they did not regreen in early summer. Their yellowing was followed by a drying out, then the crisping of the leaves, and then the destruction of the parenchyma of the shoots, leaving "the entire vine then stunted and not far from succumbing" (Vialla 1879, 5). In this fashion the majority of the Concords and Clintons had been observed to die. Vialla concludes his piece with a careful description of the places, times, and conditions of the various failures. No theory is advanced, no generalization made. Just the bare facts are presented. It is a good start.

But the situation was only going to get worse, much worse, before it got better. Regions already hit by the phylloxera, and in which reconstitution was well apace, were located along an arc between two southern cities, Bordeaux in the west and Montpellier in the east. Although there was wide variation in the soils of the affected regions, the variation was nothing like the extremes that were waiting further north. One of the best examples was the Charentais' Armagnac and Cognac regions, which included parts of two large *départements*, Charentais-Maritime and Charentais. Here the soils were thin, infertile, and bleached white from their high chalk content; this soil was perfect for the high-quality *vinifera* grapes grown there for the production of brandy. By the late 1870s, however, the feared insect had begun to show up in the region.

In spring 1880 a conference on the American vines was held at Châteauneuf-sur-Charentais. M. le Dr. Débouchaud, the town's mayor, presided. The conference reviewed the progress of reconstitution in the areas already entrained, especially Provence and the Hérault. Optimism was the most visible emotion, especially in the conference's closing speech, given by M. le Dr. Aubert, local physician and member of all the agricultural and viticultural societies. Aubert argued, in no uncertain terms, that with the evident perfection of the use of American vines to reconstitute phylloxerated vineyards, the future of Cognac was assured. Obviously, word about the *riparia* problem and the Concord disaster had not yet penetrated the Charentais.

Three years later, the Central Committee for Study and Vigilance of Charentais decided to examine things in the Gard and Hérault rather more carefully (Carrière 1884). They send M. Carrière to do the job. The two southeastern *départements* had been selected to visit because of the rapid pace of their being replanted with American vines: there were 2,500 hectares in 1881, 5,000 in 1882, and 10,000 in 1883. Carrière noted that the amount of surface replanted increased considerably as he got closer to Montpellier. At each stop along his path, he noted each grafted variety under cultivation, when it was planted, what soil and subsoil it was planted in, and its condition. His report was very thorough, continuing for more than sixty pages. The oldest planting visited was one belonging to M. le Baron Pieyre, near Saint-Hippolyte-du-Fort (Gard), which had been planted in 1871, with part of it grafted in 1875; the planting consisted of Jacquez, Concord, and Clinton. The soil was very deep alluvial, and the vines had very well developed vegetation and abundant fruit. Because there were no problems with this particular planting, Carrière counted it as evidence of the stability and longevity of the American vines. Only a few American vines, principally Concord and Clinton, were observed in bad shape. Carrière's report was thus generally optimistic about the chances for success of American vines in the Charentais. But such misplaced optimism could not last, especially since the problem of adaptation was becoming both better known and more central.

The preceding year Foëx had published a short memoir summarizing the state of knowledge at the time and pushing it a bit further. Until then, Foëx noted, most researchers had been content to observe, record, and generalize about the conditions bringing about the yellowing—the chlorosis—"collecting the facts without making a theory about them; some, however, have searched to provide an explanation" (Foëx 1882, 6). Among the latter, Vialla, Chauzit, and Millardet had each proffered a hypothesis. Vialla argued that the successes of the American vines, since they occurred almost entirely in the iron-rich red soils, had something to do with the chemistry of the iron ions. For his part, Chauzit—still at Montpellier at this date—believed that the cause was a physical factor, especially "the depth and physical nature of the soil and subsoil." Millardet's hypothesis was similar to the Montpellier researcher Chauvin's: he "contests the action of iron oxide on the American vines; he believes that their success is due above all to an ensemble of physical conditions, among which is found the heating of the soil due to its coloration." So, according to Millardet, the red soils were important not because of their chemical powers, but rather because of their physical properties, namely, an increased absorption of heat from the sun.

After considering these three hypotheses, Foëx sided with Chauzit and Millardet. His reasoning was straightforward. He divided the problem into two possibilities: either "first, absence of some material necessary to the plant; or second, insufficient or poor functioning of the organs destined to receive the material." He argued against the first possibility, basing his argument upon a careful chemical analysis comparing the red soils, where American vines were succeeding, and the white chalky soils, where they were failing. For the elements targeted, particularly iron, the values were quite comparable. Thus, Foëx concluded, the cause must be some malfunctioning in the organs of the plants themselves.[5]

As the bad news spread northward, worries began to crop up in official publications. For example, Lot-et-Garonne's Central Committee for Study and Vigilance met on 12 March 1884. Although the reconstitution had already started in this eastern neighbor of the Gironde, the process was nowhere near as advanced as it was in the Gard and Hérault, hence the evident anxiety in the committee's tone: "The culture of American varieties, the object today of a certain favor, isn't without offering some great difficulties. Witness the Jacquez of Vauvert [Gard], the *riparias* of Plagnol [l'Herault]. The *riparias*, even the ungrafted ones, are dead. We await the information that will permit establishing, if possible, the causes of this disaster" (Comité central 1884, 6). Disaster indeed. Lot-et-Garonne had caught the fever for *riparias* no less than their eastern neighbors. Grafting upon the American species was already well started. If the *riparias* were failing now in the Gard and Hérault, then it was only a matter of time—or so the committee worried—before their own *riparias* failed as well.

The bad news brought out some extreme cases. In Carcassonne, midway between Montpellier and Bordeaux, M. Jean Izard published a tract with the excited title "Plus de cépages américaines ils meurent!!!" (More of the American vines die!!!). Izard's piece is a long, loud polemical call to arms against the American vines. In essence, his argument consists of long quotes from various authors—including Laliman—who either rail against the American vines, or cite their deaths, or both. Laliman pithily notes "that the *riparia* called Gloire de Montpellier sees its glory declining" (Izard 1885, 8).[6] Among the bad news listed is the fact that the newly widowed Mme. Saint-Pierre's eight thousand *riparia* have died; this is bad news indeed, since some authorities had only three years earlier been quite ecstatic about the glorious state of her vines (Chauzit 1884). Given all these disastrous turns, Izard concludes that the notion of "adaptation" is far "too elastic" and functions only as a way to save the claim that the Americans are resistant.

But Izard finishes his polemic on a somewhat self-indicting note. Since the Americans are dying, he says, we need to return to growing good old-fashioned *viniferas* using a secret disinfectant that has allowed him—or so he says—to grow a single *vinifera* so large that it covers an entire hectare all by itself and that bears crops of similar proportions. At this point, of course, only the most credulous were still listening.

Another disaffected writer chimed in from Aisne, up near the Belgian border. In his "La maladie de la vigne, les microbes, et la Commission supérieure du phylloxéra," M. Chavée-Leroy, an *agriculteur* from Bucy, claimed that the Pasteurians, the phylloxera-causist "microbists," obviously won the battle over etiology far too soon. As the death of the American vines shows, he claimed, it was not the phylloxera that was causing the problem, assuming that the American vines were immune. Rather, there was something wrong with the growing environment; Naudin, the respected entomologist, was right, the phylloxera was only an effect, drawn to a plant that became unhealthy as just one element in an unhealthy situation. Moreover, since Pasteur himself was now running the show on the national commission, there was no chance that the correct view, the Naudinist view, would ever be able to overthrow the microbist view. But if it were to, there would be lots of things that northern agriculturalists such as the author could advise their southern grape-growing colleagues to do. In particular, Chavée-Leroy's experience growing beets would be of great help—or so he thought—in showing them how to make the soil healthy, thereby making healthy plants that would not attract parasites (Chavée-Leroy 1885).

But most of the work being done was not of Izard's and Chavée-Leroy's ilk. For example, early the next spring the Horticultural Society of the Gironde commissioned its president, Joseph Daurel, to compile a thorough report on the successes and failures of American direct producers and graft stocks in southwestern France (Daurel 1886). Daurel, a laureate of the national Society of Agriculturalists, was the right person for the job. His report ran to over fifty pages; it was broken down by species, then by variety, each of which was exhaustively described in terms of its use, location, soil, topographical exposition, wine quality, and method of propagation. Daurel did not flinch from the truth, noting quite accurately the ultimate failure of American vines in any but the best soils.

Two *départements* to the east, the Agricultural Society of Haute-Garonne was busy preparing a detailed set of questions about the American vines to send to as many vignerons in the region as possible. Le Vicomte de Gironde, who lived in nearby Montauban (Tarn-et-Garonne), received a copy and was much impressed because, after all, "it is a question upon which depends the future of our agriculture, and thereby, one can say, a great portion of the public wealth" (de Gironde 1886, 1). Much impressed with the completeness of the questionnaire, M. le Vicomte promised that he would "try to respond, point by point, always succinctly, aiming to avoid wherever possible banalities and trivialities." It is clear from his answers that although M. le Vicomte quite favored the American vines in general, he was not prepared to hide the truth about them. Thus, when he arrived at the question "Have you observed signs of withering, either in the graft stocks or in the direct producers?" he could only be honest: "I haven't seen any signs of withering among my American vines, with only one exception; I wish to speak here of the chlorosis—a very accentuated chlorosis if the vine is grafted—that strikes without distinction among varieties all the American vines planted in white marly or schistous soil" (de Gironde 1886, 13). These were both white soils, very high in chalk. De Gironde reported here that none of the American varieties he planted in the white soils succeeded; to a vine, they become chlorotic and passed from the severe yellowing stage to withering and thence to death. Although he tried every trick he knew, nothing succeeded in saving these vines. However he was not ultimately pessimistic, since he has heard that Millardet was on his way toward finding a vine, perhaps from *V. monticola*, that would root easily and "content itself in the shallow soils of Charentais and the white soils in general."

At just about this time, Montpellier's Félix Sahut published a small monograph that would prove so useful in addressing the white soils problem that it would eventually go through several printings (Sahut 1886). Entitled "La jaunisse ou chlorose des vignes" (The yellowing or chlorosis of the vine), the tract exemplifies exactly the relentless practicality one had learned to expect from Sahut. It begins with a careful recitation of the soils in which American vines become chlorotic, with a special focus upon the *riparia*.[7] The review finishes with a set of recommendations about the treatment of old plantings and preparing new or replacement plantings. This latter set is most interesting because it includes the stern warning "Abstain absolutely from creating new vignobles in white soil, or in that which is too marly, too clayey, too impermeable, and too shallow." In other words, plant Americans only in "soils that unite those conditions that I have previously enumerated, namely only in iron-rich, sandy, deep, and well-drained soils" (Sahut 1886, 14).

Sahut concludes the short piece in a very interesting fashion, as if he were well enough aware of the skepticism his recommendations might arouse in his readers. After all, wasn't it the Montpellierians—including he himself—who had gotten everyone into this mess with the American vines in the first place? Sahut's response seems to keep all this in mind: "Just like everyone else, we desire nothing other than the return of the prosperity of our meridional region, so cruelly struck with misfortune, and we are not unaware of the fact that the reconstitution of the vignobles is the only possible means to obtain it." With this comment Sahut shows that he has not engaged in anything other than an attempt to salvage viticulture in the south. Moreover, he agrees that the only way to salvage it is the Montpellier way, by reconstitution on American vines. But that is no longer the main issue; now the question is which are the right American vines to use? Even though a huge investment had already been made in vines recommended by Montpellier, was this any reason to continue to invest? Unfortunately, no:

> But when the future of our viticulture is at stake, when one is conscious of the fact that millions have been thrown away in a dead loss, when one is persuaded that we are on a false route, courting certain disaster, it comes time to finally dissipate the illusions and put oneself resolutely and without prejudice on the search for the truth. Under these conditions it is the duty of every good citizen to sound the tocsin of alarm, to signal the danger, under what form it manifests itself, and above all, that it is still possible to evade it. (Sahut 1886, 15)

Two things are clear from this conclusion. First, the current Americans, especially the *riparias,* were a dead loss. Second, however, the danger resided in the particular American varieties recommended and not in the recommendation for reconstitution using Americans itself. What must now be done, according to Sahut, is to immediately halt planting of Americans in the terrains he proscribes, particularly the white soils, even while, at the same time, research goes forward to find new, heretofore unknown or unnoticed, Americans, vines fit for the white soils.

But in a strange turnabout, when the key to the research future was found, it was not a Montpellierian who found it. Millardet, the famed Bordelais, made the move.

NEW HYBRIDS, VINES OF THE FUTURE?

Crossing grape varieties wasn't anything new by that time. Bouschet had started crossing vines on his estate in the Hérault in the late 1820s in an attempt to deepen the color of the highly productive but lightly tinted Aramon. His Petit Bouschet was the first great success; his son produced the dark-juiced Alicante Bouschet, which went into wide use in the south by 1865. But Bouschet *et fils'* work was not, strictly speaking, hybridization. Rather, it was cross-breeding between varieties of the same species, producing what in French is called a *metis*. Yet since the mechanics of the process are the same, it was not a great stretch in practice to extend Bouschet's cross-breeding methods to work among different species, which is *hybridization* properly speaking. Conceptually, however, it was a grand stretch to understand precisely what underlay the practice, and thus what could be accomplished in hybridization. It was Millardet who made this intellectual stretch, in 1888.

Millardet's thoughts on the topic of hybridization took place within the context of his own experiments and those carried out by two quite practical men, Victor Ganzin and Georges Couderc.[8] In 1877 Ganzin, inspired by the idea of producing a phylloxera-resistant Aramon, began crossing this popular French vine on a variety of *rupestris* he had obtained from America. From the various seedlings he selected two vines to experiment with, and, ultimately, release to the public. The release was not announced until 1888, at precisely the same time—indeed, in the same issue of *La vigne française*—that Millardet's manifesto for hybridization appeared. Georges Couderc, from the central Rhône *département* of the Ardèche, had begun his own hybridization work in 1881. Both men's work would be cited by Millardet in his manifesto as providing strong evidence for the eventual success of hybrids combining American resistance with French wine quality.

Millardet's first theoretical inklings of the possibility of hybridization were signaled in a brief passage in his anthology of 1881. His remarks were direct and suggestive but at this point had very little—if any—empirical evidence: "It is a matter of nothing less than that of creating through crossing entirely new varieties, whose fruits will have the taste qualities of our indigenous varieties, to which will be joined at the same time roots with the resistance proper to American vines. Evidently, this will be the solution par excellence to the phylloxera problem" (Millardet 1881, quoted in Pouget 1990, 69). Millardet began his work on hybrids immediately and, in his typical style, methodically. Over the next several years he reported his results in various journals. Results reported in one of his first papers, published in 1882, were quite significant since they set for him the limits of what he thought he might expect from his hybridization work. Already in his possession were specimens of a fair number of what he (and some others) took to be natural hybridizations among American species, in particular the mysterious Solonis.[9] Other of his specimens came from the work of American hybridizers, both earlier (such as Charles Arnold) and, most importantly, contemporary (namely Hermann Jæger, from Neosho in far southwestern Missouri). Jæger had combed the woods and ravines of his rugged Ozark highland region searching for the best specimens of native species, such as *rupestris, aestivalis, lincecumii,* and others adapted to the hot and frequently humid climate. One of Jæger's crosses, #70, between *rupestris* and *lincecumii,* turned out to have an extremely promising future, as we shall discuss below.

In this September 1882 article Millardet set out to resolve several issues. First, he wanted to show definitively that hybridization between American species not only could occur, but indeed had occurred, of itself, naturally. This provided the "existence proof," as it were, of the complete interfertility of American species. Added to this proof, he would suggest two mechanisms (wind and insects) and show how their operation could explain hybridization even among vines with perfect (hermaphroditic) flowers. Second, he would ring the changes on what species actually had crossed in the specimens under review. Third, he wanted to analyze the specific composition of the Jæger hybrids in his possession, not only to illustrate his methods, but, just as importantly, to deny Foëx's claim that these varieties had some admixture of *labrusca*—which, if true, would of course weaken their resistance. Finally, he wanted to make clear the implications of these points for his hybridization research program.

Demonstrating the first claim was simple enough: Millardet took specimens of known natural origin, let them self-seed, sometimes for two or more generations, and then described the offspring. Many of the offspring clearly and evidently returned to the types of their parents (Millardet 1882, 473), thereby verifying one of Millardet's principles of heredity: from the seeds of the second- or third-generation hybrids, "the generation that issues from these seeds returns more or less completely to the type of the father or the type of the mother" (Millardet 1887, quoted in Pouget 1990, 69). When the "plants produced are very dissimilar" in type, they unarguably must have parents from different species (Millardet 1882, 473). As far as mechanisms of hybridization are concerned, Millardet's theories showcased his extremely fine eye for observational details. Close observations of perfect-flowered vines, he asserted, clearly show that after the capeau—the helmet-shaped covering of the sexual organs—is shed, the pollen-bearing anthers frequently stay reflexed—in no position to do any fertilizing—for hours; in some cases, he had even seen it take longer than a day for the anther to straighten out into its proper position. During this whole time, the pistil exudate was evident,

FIGURE 16. AxR1 leaves. Courtesy Regents of the University of California. Photo: Jack Kelly Clark.

signifying readiness for fertilization. If vines of other species are blooming nearby, the wind or a feeding insect might provide the carrier for errant pollen.[10] A natural hybrid would result.

Having shown a plausible mechanism for natural hybrids, Millardet then proceeded to go down the list of species evident in the crosses he had found in seeds hybridized or collected in Missouri, Kansas, Arkansas, and the Indian Territory and sent to him by various people. Parental species included *riparia, rupestris, candicans, aestivales, cordifolia,* and *cinerea.* Finding examples of each parent in some seedling or another established for Millardet the fact that all the species were interfertile among themselves. This was an extremely important fact, since Millardet believed that it was from this pool of species that the ultimate hybrid graft stocks and direct producers would emerge (Millardet 1882, 477).

Noting that *labrusca* characteristics did not show themselves in any of his Jæger specimens or their offspring, Millardet concluded that Foëx was wrong to attribute *labrusca* parentage to them. This was also true of the natural hybrids, a not unsurprising point given "the well-established absence of *V. labrusca* in the natural state in all the regions in which these vines originate" (Millardet 1882, 476).

Millardet's conclusion made several important points about his research program. First, "it is evident that hybridization is of all possibilities the most important: one can lessen or increase, within wide limits, the resistance to phylloxera and other diverse ills, the property of rooting well, and the faculty of adaptation to different soils and different climates." Among the dozen species of North American vines, at least half of them are good candidates for hybridization; thus "since each of them has its own special adaptive properties, we can hope to find suitable graft stocks for all our various climates, different soils, and different varieties." Millardet then emphasized that "graft stocks" is plural: "I must finally say it, I do not believe, and never have believed, in a universal graft stock." Even though the public "pursued for four years this utopia in the *riparia,*" it did so without understanding the impossibility of finding "a plant adaptable to the ensemble of varieties and terrains of France, Spain, Italy, etc." (Millardet 1882, 477).[11] Only by crossing could the "special aptitudes" of these American species emerge; only then would the entire range of suitabilities to scions, climates, and terrains become available: "One sees now of what importance is hybridization" (Millardet 1882, 478).

It was not until six years later that Millardet officially published notice of his first successes, although he had reported them during a speech to the Bordeaux Congress a little over a year earlier. Happily enough, Millardet's article in *La vigne française* had immediately followed an article by Ganzin that announced his own first two successes, including AxR1, the now-notorious Aramon × *rupestris* #1.[12]

Millardet begins by summing up his research program to date:

> I mentioned in 1881 the path I wished to follow in my researches on hybridization of the vine. In 1886 I gave to the congress at Bordeaux a summary of the results I had obtained until then. I insisted upon the usefulness of seedlings from seeds of first-generation hybrids, if one wished to be nearly certain of obtaining at one time, in one and the same plant, the resistance to phylloxera and to mildew of the American vine, and the quality of the product of the European vine. (Millardet 1888, 26)

Millardet insists upon the use of first-generation hybrids in order to ensure getting both the proper roots and the proper fruit. Based upon his observations, he believes that "two or three times out of a hundred, the first-generation hybrid approximates much more one of its parents in one of its organs, and the other by another organ, to such a degree that one can hope to make a direct producer having at one time the resistance to phylloxera and to mildew, and the requisite qualities of fruit" (Millardet 1888, 26). Having once secured these two or three plants, one then tests them carefully in the trial vineyards. Millardet gives as one example a cross between Pedro Ximénes and *rupestris* that had the leaves of the latter and the fruit of the former. When it was put to the test in M. de Grasset's test vineyard,[13] the hybrid proved to be *"absolutely immune to the phylloxera"* (Millardet 1888, 27; emphasis in original). Thus, here Millardet has a vine suitable for further testing as a *hybrid producteur direct* (HPD).

Millardet also is struck by something apparently new and, as it turns out, creatively important. During their ongoing research, he and de Grasset had frequently been struck by the "extraordinary development of several hybrid males, which showed us that the majority of these plants were equally immune or nearly immune to the phylloxera." Obviously, since these male vines were infertile, they were quite useless as HPDs. However, it suddenly occurred to the two workers "to try these infertile Franco-American hybrids *as graft stocks, in the bad soils where the pure American species couldn't go*" (Millardet 1888, 27; emphasis in original). This was a stroke of sheer brilliance. Pure American species tried until then either couldn't deal with the white soils *(riparia, riparia, aestivales)* or, when they could, the species couldn't be multiplied by rooting in nurseries *(cordifolia, monticola)*. But, thought Millardet and de Grasset, perhaps their hybrids could succeed where the pure species couldn't. They hoped that the tests that they had just begun would prove that hybridization "will give us the graft stocks that Texas seems to have refused to give us." The reference is prescient in two different ways: first, even as Millardet wrote, Viala was preparing his trip, his "mission viticole," to America, where he would go to Texas to find, with Munson's help, the species for the white soils; second, among the grapes that the two men had chosen to test was the head of the line that would produce 41B, ultimately the most important resource for the white soils.

As evidence of the correctness of their path, Millardet makes good use of the work of Ganzin and Couderc. After noting the release announcement of Ganzin's AxR1, he relates a story about a visit he and de Grasset had made to Couderc's institute just before the 1886 Bordeaux congress. They spent the day there, examining Couderc's productions, several of which they much admired. Millardet thus writes, "I can't recommend too highly that the public try some of the hybrids of M. Couderc; it seems to me impossible that there won't be at least one that prospers in any given terrain."

Millardet is obviously quite struck by the successes of Couderc and Ganzin, so struck, in fact, that he declares what can only be called a manifesto for the new hybrid grapes:

> Thus the year 1887 will mark a date ever remembered in the history of our devastated vineyards, our agonies and our struggles against the formidable plague that has assailed our viticulture since twenty years ago. By grace of the hybridization of our European varieties with diverse American vines, we are, from today onward, absolutely certain to obtain, in first-generation hybrids, either graft stocks of an assured resistance, and of an adaptation much

easier than those that we have possessed until now, or direct producers, resistant to phylloxera and to the most dangerous plant parasites, which are
capable of producing at the same time wine completely correct in flavor.
(Millardet 1888, 28)

As far as Millardet was concerned, the future was now open to the hybridizers: "From now on, in effect, the exactitude of the method with
which we have reached our goal must be regarded as absolutely demonstrated." Ganzin, for his part, was equally convinced; indeed, he thought
that grafting would soon be eliminated in favor of the new hybrids:
"Persuaded that grafting our non-resistant cultivated varieties on resistant species is only a transitory cultural process, destined to disappear
in the future, I have searched since 1876 to obtain by hybridization resistant *cépages* for direct production" (Ganzin 1888, 23). The editor of
La vigne française evidently agreed with the hybridizers, saying, "The
future of American vines in France, according to the most authoritative
ampelographers, is neither as rootstocks, in direct producers, but in
hybridizations, which give us a fruitful plant with French *sève*, on roots
resistant to phylloxera. This is the path taken by Millardet, de Grasset,
Couderc, and Ganzin, with some prospects for success" (Ganzin 1888,
15). Of course, it had not been easy:

> It is by the dozens of thousands are counted the hybridizations that we have
> made, M. de Grasset and I; and M. Couderc, for his part, estimates the num-
> ber of hybrid seedlings he has obtained since 1880 to be forty thousand. Fi-
> nally, it is essential not to forget that it is in great part by grace of the perfect
> concordance between the results obtained by this last observer, by M. Ganzin,
> and by ourselves that the question of hybridization can be now regarded as
> resolved, in principle, in the most favorable sense. (Millardet 1888, 29)

It would be difficult, yes, but doable; indeed, thought Millardet, it was
assuredly doable.

Unfortunately, Millardet's confidence was due in part to his—and
everyone's—ignorance of the actual complexity of inheritance. Thirty
years after its publication, Mendel's work was still unread, uncompre-
hended by hybridizers, and thus unavailable for application in the vineyard.[14] But this was perhaps a good thing; had Millardet and the other
hybridizers had Mendel at hand, they might not have tried so hard to
achieve the successful first-generation hybrid that Mendel's theory indicates to be nearly impossible, and thus they might not have done what
they did, which is very nearly achieve the nearly impossible. Luckily
for us, their theories were rough-and-ready useful even while being

scientifically quite inaccurate: "The empiricism of the breeder allowed him to rush in where the scientist feared to tread" (Paul 1996, 85).

But all this is not to say that Millardet was completely bereft of theory; to begin with, he had his "principle of heredity." Millardet often referred to his principle as an Ariadne's thread (Millardet 1877, 23). At the time when he named it, the principle was restricted to the claim that certain properties were inherent to a plant type and thus inheritable. The principle led him to his first work, isolating the pure species types of *riparia, rupestris, monticola,* and *aestivales* in order to see what properties were part of the "type" of the species. Experimentation on the pure types revealed that high phylloxera resistance was integral to each of these species (although *aestivalis* was lower than the others) and thus was available to offspring as an inheritable property. As his work developed in the late 1870s and early 1880s, Millardet came to the conclusion that whole organ systems (e.g., roots, foliage, and fruit) were the heritable elements; a vine type existed as a unity of these organic elements, which, during sexual reproduction, would be mixed and matched to produce new unified individuals in the offspring generation. One of the more explicit theories of this sort was developed by Naudin based on hybridizing experiments he had carried out during the early 1860s. Although Naudin produced no generalizations secure enough to be called "laws," he did get enough data to publish a testable hypothesis, which Darwin rejected in a letter of 13 September 1864 (Marza and Cerchez 1967). Millardet's own theory, like Naudin's, could best be described as a "mosaic" theory (Paul 1996, 83). Each individual is a mosaic, with the pieces of the mosaic corresponding to the organ systems inherited from the parents.

Mosaic theories have obvious problems, chief among them being the issue of what constitutes the inheritable object. Millardet's answer is "the organ system," but that is not enough. If the object is taken to be, for example, the root system, then one must ask, "In whole or in part?" More important for the French breeders of the time was the central question, "Does the ease of rootability necessarily accompany the roots' phylloxera resistance?" If a vine type, say *berlandieri,* is both resistant and very difficult to root, then what are the chances of getting a useful hybrid that is resistant but easy to root? It would seem that these sorts of difficult questions are in fact quite beyond the theories that even genuine scientists such as Millardet espoused. Hence, such issues do not appear in the scientific journals, nor do they concern the more practice-oriented hybridizers Couderc, Albert Seibel, and Eugene Contassot. Even

Christian Oberlin (Oberlin 1914) and his son-in-law Eugene Kuhlmann (Paul 1996, 84) paid little attention to the purely scientific issues involved in hybridization and simply got on with the job.[15] As we shall soon see, the results of these hybridizers were soon to seriously challenge the Montpellier program of grafting on American rootstock.

Montpellier's views on inheritance seriously limited its professors' hybridization experimentation. Unlike Millardet, the professors at the École, particularly Viala, Foëx, and Ravaz, did not believe that organ systems operated as units of inheritance. Thus, for example, phylloxera resistance was not an "atomic" property, as it were. Rather, the offspring's resistance was in effect considered a proportional "blend" of the resistance of the parents; by this theory, a perfectly resistant American crossed with a perfectly nonresistant European should produce offspring with just about half the original American resistance. According to this view, in all cases of Franco-American crossing, "their phylloxera resistance is contested" (Berget 1896, 53). As late as 1903 Viala was still claiming that no known hybrid had high resistance to phylloxera—an unsupportable claim, given the widely successful hybrids Couderc 7120, Oberlin 595, and Baco #1, all of which were widely disseminated at the time (Gouy 1903, 189).[16] It is evident that the official doctrine of Montpellier was that even the slightest touch of French parentage in a hybrid would automatically weaken its phylloxera resistance. As Paul notes, "this was the *'thèse classique'* of Foëx" (Paul 1996, 70). Clearly this conclusion was a direct corollary of the view from the time of Planchon and the *labrusca* disaster, the opinion that even the slightest touch of *labrusca* blood in an American vine would lessen its resistance. Belief in this sort of a "blending" theory of inheritance easily explains the always-evident reluctance of the École to push Franco-American hybridization or hybrids from the mid-1870s on. Unfortunately, they were thereby forced to concentrate exclusively upon either pure American types or Americo-American hybrids. Viala's work finding varieties suitable for the white soils was made extremely difficult by this limitation due to his blending theory. Millardet, with his mosaic theory of inheritance, was not similarly limited.

Millardet published his fundamental work on hybridization in 1891 as *Essai sur l'hybridation de la vigne* (Millardet 1891). It begins in the simplest possible way, with the sentence "One knows what a hybrid is: it is the product of the crossing of two different species." The vast majority of the book is given over "to treating nearly exclusively the techniques of hybridization" (Millardet 1891, 6). Chapter 2, for example, examines the flowers of the vine, blooming, and fertilization. Clearly the drawings are depictions from his own acute observations. All the techniques needed for hybridization are included, and even the very tricky operation of emasculation of perfect flowers is described and illustrated in detail. Concluding remarks focus on the conservation of the ripened grapes until the seeds are removed and readied for planting the next spring. Given this handbook, any reasonably proficient person could learn the fine art of hybridization. And it is clear that many did.

Of course, the exultation Millardet had felt at the time he wrote his manifesto a year earlier has not dissipated much, if at all. After remarking that in ten short years hybridization had achieved graft stock and direct producers better than any available until then, Millardet goes on to make a prediction we have seen before, but, if possible, it is made now in an even stronger fashion:

> It is extremely remarkable that so complex a result has been possible to obtain in so few years. Consequently, it seems that there is scarcely any desirable variation that it would be impossible to obtain in time, if one reflects on the wide variability of the vine, its extraordinary faculty of hybridization, and the fertility of its hybrids. Hybridization constitutes, one knows, the strongest cause of variation.[17] If by culture alone V. vinifera has given us the innumerable varieties cultivated today, what marvelous results can we await when we join hybridization to culture! (Millardet 1891, 6)

Millardet's optimism was contagious and was soon caught by many in France, even though it was resisted by his colleagues to the east, in Montpellier.

Millardet's greatest success came almost unbeknownst to him. Certainly, at the beginning, he was not looking for what he eventually found, namely, the rootstock for the white soils, his *vinifera × berlandieri* cross 41B.[18] In yet another of the odd turns between Bordeaux and Montpellier, it was Montpellier's Pierre Viala who ultimately shed the light on Millardet's discovery. And that took a trip to America.

VIALA'S *MISSION VITICOLE EN AMÉRIQUE*

By 1886 the white soil adaptation problem—with its attendant Concord and *riparia* disasters—had reached the level of crisis. Something had to be done. The issue, so far as everyone could see, could be reduced to the following argument. Reconstitution must proceed on the basis of American vines, but significant areas of France consist of white soils. American vines heretofore known cannot succeed in white soils.

FIGURE 17. Pierre Viala. Private collection of
Marie-Laure de Maurepas.

The conclusion reduces to a plain disjunction: "Either significant areas
of France cannot be reconstituted, or new American vines that can succeed in white soils must be found." Since the first disjunct is unaccept-
able, the second becomes the conclusion. But how is it to carried out?
How are these new American vines to be found? Thus began the last
great adventure of the phylloxera wars, Pierre Viala's *mission viticole
en Amérique*. It started in the Charentais.

Problems in the Charentais were staggering. By 1884 more than
160,000 hectares of vines had been attacked, and very few were replanted during the next few years, mainly because there was no known
American rootstock that could succeed in their thin, chalky white soils.
Things were so bad that in late 1886 the Central Committee of Study
and Vigilance of Charentais-Inférieure proposed "sending some dele-
gates to America in order to seek indigenous varieties of vines that grow
vigorously and without chlorosis in chalky soils identical to those of the
Charentais" (Pouget 1990, 76). SCAH supported this call, as did the
local agricultural society, which opened a bank account to provide

some financial resources. In March 1887 the minister of agriculture agreed, and by a decree of 16 March he sent Pierre Viala on "a mission to the United States of America, in order to find grape varieties able to grow in chalky and marly terrain" (Viala 1888, 5).

At that time Viala was a very young man, only twenty-eight, but he had already achieved the chair of professor of viticulture at Montpellier's École, certainly the most prestigious academic position in the entire world of wine. Viala was born in Lavérune, a small village about seven kilometers southwest of Montpellier. He went to the École, studied under Foëx, and immediately went into viticultural research. Already well known by the time he was nominated for the American mission, he would become a world-class star upon the publication, first, of his twenty-four-page report to the minister (Viala 1888), which appeared immediately upon his return, and, second, his full-length book, published two years later, which went into excruciating detail about every phase of the trip (Viala 1889). Every centimeter an École man, he carried its perspective with him to America. In certain respects, this perspective would haunt him later.

Viala was quite explicit and precise in his descriptions of the goals of his mission. First, he had to find chalky terrains, "if not identical to, then at least comparable to, the chalky marly soils in France." This wasn't easy to do, even with the help of a week spent at the USDA and the Geological Survey immediately after he arrived. What he learned was that such soils could be anywhere in a huge area in the central part of the country, between the Mississippi and the Rocky Mountains, from Montana to Texas. It was going to take some trekking, not to mention some help from the locals, to even find the terrain, let alone the grapes. Second, assuming that he could find the terrain, he would then need to find

> native vines flourishing in this milieu. It would be necessary that these vines show resistance to phylloxera, susceptibility to grafting and—a secondary but very important issue—that they could be multiplied by rooting. Finally, Monsieur le Minister, it is desired that these varieties or species already exist in France, in order to avoid onerous importations, which, because of the scientific character of this mission, couldn't be made in any case. (Viala 1888, 7)

Several very important concerns are indicated here. Resistance and graftability are taken as givens; but, even though here he calls the issue "secondary," the ease of propagation by rooting is crucial—and he treats it as such in his researches. There already have been found promising white soil vines—here one thinks of various *aestivalis* and *monticola* varieties—that were useless because they could not be propagated. Second, Viala hopes that whatever he finds already exists in France, since he certainly isn't equipped to bring anything back.[19] He doesn't seem to be worried, however, about the risk of new diseases from further importations; Sahut, on the other hand, specifically warned Viala and the industry in general that it would be dangerous to import any new material from America (Sahut 1888, 91). Nonetheless, Viala's discoveries were to precipitate a second boom in imported American material (Pouget 1990, 78).

Viala spent from 5 June until 3 December in America. During the first period of the mission he visited New Jersey, Maryland, Virginia, North Carolina, New York, and Ohio. Viala discovered "nothing of any use" to him in his travels in these states.[20] Next he visited Tennessee, Missouri, the Indian Territory, California, and Texas. He made some "important findings" in Tennessee and Missouri, findings that were confirmed in Texas. He visited the Bush and Sons and Meissner nursery near St. Louis—a necessary pilgrimage for visiting European grape men!—and then continued south and west to Neosho, where he spent some extremely valuable time with Hermann Jæger. Jæger, whose hybridizing work was already well known in France, given his tight connections to Millardet and Couderc, was especially helpful to Viala. Together the two roamed over much of that corner of Missouri where Arkansas, Kansas, and the Indian Territory met. Jæger's own work in finding and crossing the local varieties had ultimately focused upon *V. rupestris* and its many hybrids. Viala observed and reported on several of them: *rupestris × cordifolia, rupestris × riparia, rupestris × lincecumii* (for which Viala mistakenly foresaw little future),[21] and, of course, other combinations such as *cordifolia × riparia*.

One significant point needs to be made about Viala's estimation of the vines in Jæger's region. Even though he explicitly remarked the disease resistance of these varieties—for example, that of *lincecumii* and its hybrids to black rot and mildew—Viala did not find these properties of interest. In the end, it was chalk tolerance and nothing else that he was seeking. One might contrast how the hybridizers' behavior would have differed from Viala's. If Millardet, for example, had been along with Jæger on his treks through the woods and prairies, it is beyond question that Millardet would have found the disease-resistant properties of the native grapes of primary interest; after all, according to his principle of heredity, these properties were heritable, and could be passed down to offspring. Thus, for Millardet, Jæger's local vines would have been fascinating candidates for hybridization in the production of Franco-American HPDs. But Viala, in classic Montpellier fashion, simply didn't see it. He did not—in fact, could not—see the American specimens as potentially valuable progenitors in an HPD breeding program; if the vine didn't have what he needed—chalk tolerance—then he didn't need the vine. Unfortunately, his perspective on the matter was never to change.

After his visit to Jæger, Viala left for Texas. There he hit pay dirt. Guided always by Munson, either in person or by his instruction, Viala found a huge area of chalky soils extending from the panhandle in the north to the Pecos River in the south, and from the New Mexico border in the west to a region bounded north-to-south by Dallas, Austin, and San Antonio (today's I-35) in the west, which, on the basis of its fossils, he established as cretaceous in origin. Thus, this soil was geologically identical in origin to the Charentais terrains; and, from all his observations, it looked as if it were extremely similar in its chemical composition as well. Perhaps even better, he thought, looking at the entire soil structure, including the subsoil, what he had was "a terrain of a fertility inferior to the chalky terrain of Charentais" (Viala 1888, 13). If vines could make it here, they could certainly make it in the Charentais.

Luckily enough, there were native vines—lots of them—in the area. In this demanding milieu flourished "V. berlandieri, V. cinerea. V. cordifolia, V. candicans, V. monticola, and a new form considered to be a species by M.T.-V. Munson, V. novo-Mexicana, and numerous hybrids resulting from the various crossings among these species." Among these species, Viala thought that V. berlandieri, V. cinerea, and V. cordifolia would be the best because of their vigor and adaptation to the difficult soil. One of the other species, candicans, known locally as the Mustang grape,[22] although it had good phylloxera resistance and seemed to tolerate moderately chalky soils, was less valuable than his top three choices "because of the great difficulty of rooting it, even with special procedures." Obviously, even though propagation was only of "secondary" interest, it was compelling for Viala—at least when he had other choices! Luckily enough, in this case he did: berlandieri, cinerea, and cordifolia seemed to be exactly what he was looking for. All three "remain green and vigorous in the poor, dry terrains of Texas, where the soil and subsoil are formed from crumbly white chalk. . . . [These vines] do not yellow in the white chalky soils" (Viala 1888, 17–18). Moreover, given what he had seen in Belton, where some Spanish vines were grafted on *berlandieri,* at least that species, and most likely all three, would carry the graft well.

With Munson's help he also identified hybrids among the three species. He reported explicitly that these forms did not appear to be as vigorous as the "types"—the pure species—nor did their hybrids with other species appear to be worth reporting (Viala 1888, 19).[23] Here we see the formal statement of Viala's view on the proper function of *berlandieri* and the other two species: their usefulness will be exclusively as pure types, not as hybrid genitors, for graft stocks. Viala will never reject this view (Pouget 1990, 78). And, unfortunately, his refusal to consider *berlandieri* hybrids will ignite yet another battle among the *américanistes* over the best way to use American vines for grafting.

Viala's conclusions are stated with his usual vigor and in straightforward Montpellierian language: "On my return from the United States, I remain even more convinced that, with the exception of some direct producers that have proven themselves in the Midi, we must above all count on American rootstocks carrying our indigenous varieties in order to assure the reconstitution of our vignobles and the maintenance of the legitimate reputation of our French wines" (Viala 1888, 22). Further, he concludes that for the white soils, *berlandieri, cinerea,* and *cordifolia* "offer the greatest chance of success." His conclusions, he states, "are based entirely upon the observation of the milieu in which these vines flourished in the United States." For Viala, exactly as it will be for his readers, seeing is believing; if, finally, types of American vines have been observed flourishing in white soil in their natural state, then hope is fairly justified that the solution to the French chalky soils crisis has been found. Admittedly, there is a propagation problem, but Viala hopes to find either a variety of one of these types, or a specific rooting procedure, or both, which will allow a good multiplication by rooting.

Evidently, Viala takes his observation of the vines growing in their natural state to be crucial. Again the point would seem to be based in the Montpellierian view that what is important is not the possibility of hybridization in order to produce vines adapted to white soil; rather, what is important is to find natural types, naturally adapted to the white soils. Since Viala has found exactly these sorts of naturally adapted natural kinds, he feels secure in predicting their success. His readers apparently felt equally secure in accepting his conclusion.

Viala ends his report graciously, writing, "I cannot end this note without thanking M.T.-V. Munson, M. Hermann Jæger, MM. Bush and Meissner. The objective counsels of MM.T.-V. Munson and H. Jæger have allowed me to resolve the majority of the questions that I have reported to you." Perhaps even more significantly, these two American workers have stood by Viala's conclusions, "even when they might go against their own commercial interests" (Viala 1888, 23). This is a good point, given that both Americans were plant salesmen!

For his efforts, Munson was awarded the Order of Merit Agricole, the highest French honor that could be awarded to a foreigner. Viala, of course, had quietly nominated him for the award to the minister of agriculture. It is obvious that the two men, the young Frenchman and the older Texan, got along famously. In his great work *Foundations of American Grape Culture,* Munson frequently refers to Viala with evident affection, even calling him his "friend" at one point (Munson 1905, 101). Although Munson was already known in France—la Duchesse, for example, had referred to him years earlier—his fame increased mightily with the publication of Viala's report.[24] In the following years, Munson's nursery would be busy filling orders for his own hybrid creations, for cuttings of wild plants, and, by the late 1890s, for pounds and pounds of *berlandieri* seeds taken from the wild, sold at $15 or seventy-five francs per metric "pound," or five hundred grams (Pouget 1990, 79).[25]

But Viala's discoveries did not end the controversy among the *américanistes* regarding the appropriate use of the American vines. One battle remained to be fought.

THE *BERLANDIERI* WARS

Viala returned to France, committed to the *berlandieri* vines as the solution to the white soils problem. As he was soon able to demonstrate, this Texas species was quite at home in the terrible soils of the Charentais. Moreover, not only did the vines graft well, they retarded the season of the French scions—a nice bonus that provided some defense against late spring frosts—and they increased the yield of ripe grapes at the end of the season. But there was one problem, and it was a big one: the vines refused to propagate easily or efficiently from cuttings, which made them extremely costly.

Viala had known of this problem already in Texas. But he had been so taken with the sight of vines obviously flourishing amidst the bleached Texas soils—so similar to the devastating white soils of France—that his hope of finding a way out of the propagation problem swayed him into overoptimism. Viala imagined two possible escape routes. Either a new method of propagation from cuttings could be found, or a variety of wild *berlandieri* would be found that propagated easily. Neither of these possibilities was ever realized. Yet, as Prosper Gervais observed in 1896, Viala's commitment to the pure *berlandieri* species was unswerving:

> We've got to give Pierre Viala this: he has never varied. He has always pointed to the *berlandieri* as the type plant, as the ideal vine for the calcareous terrains; and without letting himself be rebutted by the difficulties of all orders that, from the start, have attended the propagation of this vine; he has succeeded, little by little, with a tenacity, a perseverance, a suppleness that one can only admire, to flesh out his doctrine, to breathe life into it, to communicate to it an appearance of life, which trumps and seduces even those most prejudiced. . . . He is the high priest of *berlandieri*. (Pouget 1990, 80)

At the congresses in 1893 at Montpellier and 1894 in Lyon, Viala was the "principal representative" of "the partisans of the pure species" (Berget 1896, 41). Viala and his colleagues argued simply and straightforwardly that the pure *berlandieri* stock was the only answer. Hence, even if these vines were expensive to propagate, the expense was worth it. Viala worked in cooperation with several horticulturalists, particularly Euryale Rességuier, whose nursery was at Alénya, just south of Perpignan in the Pyrénées-Orientales. Rességuier made several particularly useful selections from the vines grown from wild seed. R. 1 and R. 2, for example, both rapidly reach usable trunk size and have superior fructification; unfortunately, they "root only with special care" and are "costly to purchase" (Berget 1896, 51).

One somewhat more practical method for using *berlandieri*—planting the seeds of wild vines—harkened back to the first days of *riparia* and *rupestris*. When those species first came into prominence, it was under the rubric of pure species. In other words, it was argued that the purely wild vine was the only one to use, since that vine would have had no accidental admixture of either *labrusca* or *vinifera* blood, which would weaken phylloxera resistance. Hence, from American had come tons and tons of seeds of these two species taken from wild vines. These seeds were then planted, and in the second year of growth, the seedlings were grafted with European scions. Even though de Lafitte earlier had strongly attacked the premise undergirding this method, it succeeded nonetheless. Unfortunately, every batch of seedlings from wild seed exhibited a whole spectrum of variations in behavioral properties. For example, although the wild *riparia* seedlings typically showed acceptable ranges of variation in phylloxera resistance, their variation in vigor and season frequently caused problems. Variations in vigor, for example, caused problems with pruning, and variations in season caused different vines to ripen at different times. Similar to a population of wild vines, a population of "pure" species, can exhibit up to two weeks variation in ripening times, and the ripening times of their seedlings will vary as well. To have a whole vineyard of seedling-grafted vines that each ripen in its own time can cause serious problems at harvest.

In the end, problems such as these limited the method of grafting on *riparia* and *rupestris* seedlings grown from wild seed, and research focused instead upon finding and selecting the best among the seedlings. Once the best individuals had been selected, they could be propagated and used as graft stocks. By the early 1880s use of *riparia* and *rupestris* graft stocks came to be concentrated upon one or two pure selections, for example Riparia Gloire de Montpellier and Rupestris du Lot (Lafon et al. 1930, 45), plus a diverse population of hybrids. As the fear of contamination by *labrusca* or *vinifera* lessened based upon experience with grafted vines and a better theoretical understanding of the heritable features of phylloxera resistance, hybrid rootstocks came to be the main component of the *américaniste* position.

Five years later, when Viala's conclusions about *berlandieri* were first promulgated, exactly the same debate—pure species versus hybrids—was reprised.[26] It is not entirely clear why another chorus was necessary, although three points should be remarked. First, the personal views of Viala himself were central. Not only had he seen the pure species in their natural settings, but he had also had seen natural hybrids among the three chalk-adapted species; according to his observations, the latter were not flourishing as were the pure *berlandieri*. Moreover, Viala was through and through a Montpellier man, and Montpellier, although it was the leading *américaniste* institution, had never been favorable to hybrids (Paul 1996, 70).[27] Consequently, Montpellier's Viala had firm theoretical and institutional reasons for preferring pure species and disdaining hybrids. Secondly, since Viala had achieved instant fame and authority as the discoverer of *berlandieri*, his views in favor of pure species and against hybrids played a leading role. Finally, *berlandieri* hybrids were an almost totally unknown quantity, even to their "accidental" creators, Foëx, Couderc, and Millardet. Among the three, Foëx was the least aware of his creation, not least because he was a Montpellierian.[28] In the early days he had crossed a *berlandieri* with *riparia* and *rupestris*, primarily because it was there in the collection. Afterward, however, he had essentially forgotten his work, which thereby languished.[29]

Couderc and Millardet, on the other hand, in the early 1880s had quite purposefully crossed *berlandieri* and other species because of their strong interest in how the hybrids would perform in terms of their phylloxera resistance. As we have seen, both were quite satisfied with the results. Moreover, both noted the slight easing of the propagation problem in the hybrids; although not genuinely easy to root, the hybrids gave a marginally economical production of rooted cuttings. At no time during this early period, however, had either Couderc or Millardet looked at their *berlandieri* hybrids in terms of adaptation to chalky terrains.

Public notice of the potential of the *berlandieri* hybrids in white soil was not given until the Mâcon congress in 1887, when Millardet's 1883 hybrid crosses between *berlandieri* and *vinifera* were announced for the first time,[30] along with notification of a rigorous program of evaluation and testing of the vines in situ in Charentais's white soils (Pouget 1990, 82). All the interested parties participated in the on-site testing, most on their own specially dedicated properties. Couderc bought a pair of vineyards at Tout-Blanc in Charentais to test his newer hybrid rootstock selections. And although Millardet had for years cooperated with de Grasset in testing vines in the latter's Midi vineyard, for the purposes of testing his *berlandieri* hybrids Millardet found a willing partner in A. Verneuil, whose Conteneuil vineyard was in some of the Charentais' very worst, most chlorotic soils. So serious was the effort to test the new *berlandieri* vines that Montpellier's Louis Ravaz—a student of both Foëx and Viala—was chosen in 1888 by the Comité de Viticulture of Cognac to head up a dedicated research program. Four years later he was directed by the committee to set up and lead the Station Viticole de Cognac, the first viticultural research station founded entirely by a professional rather than an academic institution (Pouget 1990, 83).

Until then the Charentais' experience with phylloxera had not been unlike that in the Midi, although the phases of the disaster in the Charentais changed much more quickly since the vignerons of the two Charentais had learned much from their neighbors to the southeast who had suffered earlier. Phylloxera first showed up in the Charentais in 1872, in two widely separated vineyards near Cognac. From there the disease "expanded in a nearly explosive fashion"; the heat and drought of 1875 "favored maximally the multiplication of the insect, that is to say, the extension and intensity of its ravages" (Lafon et al. 1930, 42). Although ravaged, the two Charentaises never went completely out of wine and brandy production thanks to water and sand: vineyards in the Pays Bas de Cognac, in the valley of the River Antenne, came to be submerged and treated every winter, whereas the extensive regions of sand along the coast and the Île d'Oléron and Île Ré soon came to be planted heavily. Use of insecticides occurred on only a very small fraction of land, and it was halted as soon as grafting began to succeed.

Grafting on American vines was first tried in 1880 using the standard Americans available, including Clinton, Herbemont, Solonis, and *riparia*. In the deeper, more fertile, nonchalky soils—all too rare in the Charentais—these grafted vines succeeded. Unfortunately, and most significantly, grafting attempts in the Charentais' two *premier cru* vignobles, Cognac's Grande Champagne and Petite Champagne, failed miserably because of the extremely high level of chalk in the soils (Lafon et al. 1930, 44). But where the grafts succeeded they were convincing: they "demonstrated, in irrefutable fashion, the superiority of reconstituting the vignoble by American vines rather than by various other means of combating the insect such as insecticides or submersion" (Lafon et al. 1930, 44). The demonstration given by the pure *riparia* plantings was especially persuasive. Consequently, from this point on, there were "two great currents of opinion . . . one that recommended reconstitution by pure Americans, the other by use of hybrids—américo-×américains and franco-×américains" (Lafon et al. 1930, 44).

Over the next few years, some areas of the Charentais showed evidence favorable to the purists, but other areas confounded that view by providing evidence for the alternative. From our vantage point today, it is clear that what was at work was not whether the parentage of the graft stock was mixed or not; it was the parentage *simpliciter*. That is, where a pure *riparia* graft succeeded, so also would a *riparia*-dominated hybrid. And in a region of slightly higher chalk, where all *riparia*-based stocks went chlorotic, their yellowing leaves indicating failing nutrition, the pure *rupestris* could succeed. Of course, the *rupestris*-dominant hybrid would succeed there just as well.

But in the chalkiest terrains, the white soils around Cognac itself, nothing would succeed: "In these regions, which furnished the finest brandies of all, the situation remained critical. . . . [To solve] these difficulties, work and science will be needed in order to succeed" (Lafon Vidal, and Nayrac 1930, 45).[31] Unfortunately, success was not easy. Viala's pure *berlandieri* grew, and in fact it even flourished, in Cognac's impossible soils. But this "life raft" (as some called it) then "revealed the disagreeable surprise of extreme difficulty of rooting." Because of the rooting problems, pure *berlandieri,* no matter what Viala and the other purists claimed, could not provide the foundation for the reconstitution of the Charantais vignoble. Happily, at this point "other men of science" came to the aide of the vignerons of the Charentais: "These were the hybridizers" (Lafon et al. 1930, 45).

Couderc and Millardet both started testing their *berlandieri* hybrids in 1888. Within two years some good results were beginning to show. Two years later their results, coupled with those of the other hybridizers working in the Charentais, made the situation quite clear: *berlandieri* hybrids, plus a few other crosses, particularly Franco-American, could indeed succeed in the worst of the white soils. With that conclusion, the last major problem of the phylloxera crisis was solved. Moreover, it was solved in a remarkably short time, the roughly five years between Viala's *Mission* report and the achievement of reliable graft stocks for the white soils. Several factors explain the rapidity of this success.

First, the need for rigorous testing and evaluation had been accepted. As Berget remarks, "Experimental vineyards became veritable viticultural cemeteries where only the most robust survived" (Berget 1896, 15). Moreover, the methodology of replicated trial blocks and control groups had ultimately become standardized, a process dating from the first groping attempts at the École's Mas de las Sorres during the mid-1870s (Commission Départmentale 1877). Statistical analyses were now frequently performed upon the results. Because of these methods, individual results were now more dependable, and perceived to be so.

Second, features specific to the white soils problem had been isolated and, where problematic, solutions sought. In particular, the exact etiology of the chlorotic response had been slowly unveiled; the problem was not simply the gross amount of iron in the soil, but rather the amount present as free iron available to the plant. And this, finally, came down to the amount of active lime ($CaCO_3$) in the soil, which acted as the limiting factor. As the amount of lime increased, the free iron decreased; below a certain minimum level of free iron, chlorophyll in the leaves begins to break down, decreasing photosynthetic activity and producing the characteristic yellowing of the leaves. Professor Chauzit, who was on faculty at the École when we first saw him in 1878 and by two years later was the departmental professor of agriculture in the Gard, worked out a table indicating the suitability of graft stock types based upon active lime content of the soil (see table 3). Houdaille and Semichon further developed ionization theory, and Bernard developed his "calcimeter," which assured a relatively speedy and reliable reading of the active lime. Although much of the finer detail remained to be disclosed, the

TABLE 3   CHAUZIT'S RECOMMENDATIONS FOR ROOTSTOCK VARIETIES BASED
UPON ACTIVE LIME TOLERANCE

Soil content active lime (%)	Best-suited American vines
Less than 10	most American vines
10–20	V. riparia, Taylor, Vialla
20–30	V. rupestris, Jacquez, Solonis
30–40	Champin, Othello
40–50	V. monticola
50–60	V. cinerea, V. cordifolia
Greater than 60	V. berlandieri

SOURCE: Chauzit 1884.

overall lines of the theory of lime-induced chlorosis and its associated measurements and instruments were in place by 1892 (Pouget 1990, 84).

In addition to these two major methodological advances, it must be noted as well that a new generation of scientists had come onto the scene, this generation for the most part trained by those who had initially confronted the dreaded scourge. Planchon and Saintpierre had been the chief faculty at the École during the earliest days of the crisis. Foëx came next, appointed in 1872 after his graduation from the Swiss École d'Agriculture at Grignon. Viala and Ravaz were his students and of course also studied with Planchon. Millardet was fifteen years younger than Planchon and received most of his research training in Germany, at Heidelberg and Freiburg. All of these younger men were much better prepared than had been their teachers for the tasks at hand. When they were confronted with the second major problem of the phylloxera war, finding vines suitable for the white soils, they brought better theories, better skills, and better experience to the problem. Thus, the rapidity with which the white soils problem was solved can be credited to better methods and better scientists. In effect, the white soils problem was confronted in all respects by a more mature science than was the initial phylloxera crisis.

In 1894 Millardet and de Grasset officially announced their hybrid rootstock 41B, a vinifera × berlandieri cross, and described it as a graft stock "suitable for chalky terrains" (Millardet 1887, 513).[32] The reaction was immediate. Verneuil, who had supplied Millardet with the Charentais trial vineyard, immediately replanted with 41B his main vineyard near Cognac, which had languished for lack of suitable graft stock. Just to show his confidence in his own creation, Millardet bought a domain in the horrible white soils of Cognac–Grande Champagne, where he planted a ten-acre expanse grafted on the stock. Apparently such a demonstration was necessary, since there had been some opposition to Millardet's vine on account of its *vinifera* parentage; one suspects Montpellier doctrine at work here (Pouget 1990, 85). By two years later, Berget's authoritative viticultural text could recommend 41B as the choice for the most severely chlorotic soils, especially since it had proved to be "very vigorous and, until now, immune to the phylloxera" (Berget 1896, 51).

At this point all the major battles against the phylloxera were over and the war was won. Fittingly, a memorial marking the end of the era had just taken place a little earlier.

MEMORIALIZING PLANCHON AND THE END OF AN ERA

Planchon died at home, suddenly and without warning, at 9 o'clock on the evening of Easter Sunday, 1 April 1888. He was just sixty-six. The funeral was Tuesday afternoon. Leaving from the Jardin des Plantes, the cortege went first to the École de Pharmacie by way of the long hill down the faubourg Saint-Jaumes, into boulevard du Peyrou, left onto rue National, and thence to the place de la Préfecture. Planchon's two sons led the procession. Behind the hearse marched the several faculties of the university, led by the university's rector. At the end were various notables from the political caste, such as M. le Préfect and M. le Maire.

After a brief pause at the École de Pharmacie, the column then presented itself at the Faculté des Sciences and the Faculté de Médecine. Thus all four of Planchon's academic homes were visited in turn: botany, pharmacy, science, and medicine. Five speeches were then given, delivered by Castan, dean of medicine; Soubeiran, professor of pharmacy; Rouville, dean of sciences; Cazalis, president of the Société centrale d'agriculture de l'Hérault; and finally Sahut, vice-president of the Société d'horticulture et d'histoire naturelle, of which Planchon had been president until two days earlier. No one from the École national d'agriculture spoke since, technically speaking, Planchon was not a full-time member of that faculty, although he taught there and, as far as the students were concerned, he "belonged to the School of Agriculture" (*Annuaire* 1895, 54). No processional visit to the agriculture campus at La Gaillarde was possible, given the roughly three kilometers between it and the starting point at the Botanical Garden. But, of course, La Gaillarde was well represented by both colleagues and students of the brilliant little botanist.

Throughout the five talks, the usual boilerplate language is frequently pierced by passages that make obvious that everybody really liked Planchon and respected his incredible corpus of work. Inevitably, the focus turned toward to the phylloxera. It is evident that if Planchon was identified with any one thing, it was the war against the phylloxera, no matter the depth and breadth of his other scholarly work all around the world. Over and over the following sentiments were expressed: "He revealed the cause of the plague to us"; "It was he who uncovered the true nature of the *insecte devestateur*"; "Planchon showed us not only that the origin of the disease was America, but also that America was the source of our salvation." But one thing about these expressions is especially striking: they are all in the past tense. The war against phylloxera was over. Planchon and science had fought the invader, and Planchon and science had won. In meridional France, it was all over but the shouting.

Six years later the shouting well and truly came. The occasion was the dedication of the memorial to Planchon. Six months after his death, the SCAH organized a subscription to build a monument in his honor (des Hours 1894, 291). It took five years to collect the money, finalize the negotiations regard the monument's location with the city of Montpellier and the *département*, design the piece, and find a sculptor. Money came from throughout France and all over the world, including Brazil, Australia, Spain, Switzerland, Hungary, Russia, and, of course, Planchon's many friends and admirers in the United States (*Inauguration* 1895, 7). Ultimately, the most visible location in the city was chosen for the monument: the small park right across the road from the main entrance to the train station.[33] Baussan, a local artist, received the commission for the statue. His design is as heroic as could be expected: it is a bust of Planchon gazing solemnly into the distance, seemingly unaware of the fat bunch of grapes being offered to him by the life-size viticulturalist, clad in working clothes, over which he towers.[34] The reverse of the statue carries a large plaque emblazoned "Les viticulteurs à Planchon. 'La vigne américaine a fait revivre la vigne française et triomphé du phylloxera'" (From the viticulturalists to Planchon. "American vines have brought back to life the French vines and triumphed over phylloxera").

Dedication of the monument took place at 10:30 in the morning on 9 December 1894, a Sunday. Present were M. Viger, the minister of agriculture, who presided; M. Vincent, prefect of the Hérault; M. Castets, mayor of Montpellier; MM. les Sénateurs et Députés de l'Hérault; the city council members; Monseigneur de Cabrières, bishop of Montpellier

FIGURE 18. Planchon monument, Montpellier, with author.

(even though Planchon had been decidedly Protestant); M. Gachon, president of the consistory; and representatives of the army and the judiciary. Planchon's whole family was there, of course, plus large delegations from all the various faculties and schools of which he was a member. M.L. des Hours, SCAH president, started the ceremony by officially turning over the monument to the city. M. Castets graciously accepted the gift and then, after affording place to Planchon among the scientific greats, noted that the discoveries leading to the new viticulture—"the heritage of past experience must be controlled by a new experimentation and guided by rational principles"—had naturally taken place in Montpellier:

> If this is a city where justice is most conveniently rendered to the experimental method, this is because here is where was born the man of genius who, in spite of the exaggerations made of his doctrine, remains among the number of those who have exercised the most eminent and efficacious influence upon the determination of the conditions of scientific research in the nineteenth century. I'm speaking here of August Comte. (*Inauguration* 1895, 12)

Thus had Planchon exercised the most "patient and methodical investigation," just as recommended by Comte. And thus, "once the phylloxera was discovered, once the resistance of the American vines was demonstrated, the problem was resolved in principle." (Of course, it is worth noting that both of these milestones were attributed in largest part to Planchon.) From this point "it remained only to pass from theory to practice. It is here that the role of the École de l'agriculture of Montpellier stands out clearly" (*Inauguration* 1895, 12). Ever the booster, Mayor Castets went on to note the many ways in which Montpellier had functioned in the war against the phylloxera. In the end, "this is the way in which the French vine was saved," he claimed, and, when all due homage is received, "Montpellier has the right to glory in having been the center," where everything came together in a "perfect harmony in order to accomplish this grand work and to return to our country a husbandry upon which its prosperity depends" (*Inauguration* 1895, 14).

After his mayoral paean Castests turned the podium over to M. Cousin, the deputy from Montpellier, who provided a long and detailed interpretation of the meaning of the statue. Most important, "the monument that we inaugurate recalls a long and cruel period of sufferings and disappointments; but our viticulturalists never lost courage." In the end, Cousin took advantage of the presence of minister of agriculture to hammer home a few economic points. The Midi has suffered the most, he justly remarked, and has produced the most progress toward the new scientific agriculture. But what kind of a reward is it for the region's wines to be confronted by all sorts of false concoctions made in factories and called "wine," on the one hand, and on the other to be confronted with massive imports of wine from foreign countries, allowed in under the "temporary" relaxation of tariffs during the crisis? The deputy specifically asked the minister of agriculture for protection against these further evils plaguing the good vignerons of the south (*Inauguration* 1895, 21).

Viger ascended to the podium, uttered a few platitudes, and replied to Castets that, yes, what was happening today to the Midi's wine was not fair. He promised to try to fix things and then concluded quite sincerely by singing the praises of Planchon—"in the class of the most eminent botanists, alongside Candolle, Jussieu, Linneas"; "his renown earns him a place in the pantheon of men of science"; and "the service rendered to his country by the application of his discoveries merits him eternal recognition by all Frenchmen"—all points truly attested by the record of the energetic little botanist (*Inauguration* 1895, 28).

Pride of the day's speeches goes to M. Sabatier, "dean of the Faculty of Sciences of Montpellier, friend, compatriot, and student of Planchon, who has been charged by the monument committee to pronounce the elegy for Planchon." Sabatier's speech is beautifully composed, full of detail, laced with humor, and at all points transparently affectionate toward his departed colleague.

What becomes clear is just how talented was the young Planchon, with his amazingly wide range of interests coupled with an endless supply of nervous energy. After brilliant success as a student in Ganges, a small town about twenty kilometers northeast of Montpellier, his parents scraped together money enough to enroll him in the pharmacy program at the university. Receiving room and board at a local pharmacy, where he worked and tutored in order to send some money home to his struggling parents, his life was hard but full. It was at that point, very early in his academic life, that he discovered natural science, especially botany. When the spring came, "the season of flowers, the season when the anemones, the irises, . . . and all the beautiful flowers spread their colors in our oak thickets, the young Planchon, taken by an irresistible frenzy, deserted the pharmacy" (*Inauguration* 1895, 32). But of course he couldn't just leave the pharmacy, his work undone, so he worked out a deal with his fellow students. During the winter, when all the flowers were gone, Planchon would arise before dawn and do all the other students' tasks while they slept. Then, in return, when the flowered hillsides burst forth in spring and summer, Planchon spent his mornings among them while his friends covered for him back at the pharmacy. "When the first breath of spring was felt, he was up by three in the morning, then this small young man, of such frail and nervous body, thrust the green botanist's collection box up on his shoulder and went out into the countryside" (*Inauguration* 1895, 33).

He passed his *baccalauréat ès-sciences* on the first try and the eminent botanist Dunal immediately adopted him as a protégé. Two years later Planchon submitted his exams for a license and did his thesis for a *doctorat ès sciences.* He was only twenty-one. Still studying pharmacy, he also began the course in the Faculté de Médecine. "Was there in him at this time the presentiment that one day, in the future, he would belong as professor to three different faculties?" (*Inauguration* 1895, 34). Probably not, but at the time he already had a solid drive to learn everything he possibly could about the natural and biological sciences.

It can be suspected that only someone with Planchon's broad knowledge and experience would be capable of recognizing, as he did, the utterly novel way of life of the phylloxera; and, similarly, only someone with his energy would be capable of carrying on the battle against the establishment's mighty opposition to, first, the phylloxera-cause theory, and, second, the *américaniste* program to use the American vines—bringer of the evil—to vanquish that same evil.

Lest this brilliant savant sound too severe, too much the scholar, to be believed, we must let the students from the École de l'agriculture describe him for us in order to see his wonderfully human side.

> The alumni, and most particularly those who attended the École from 1878 to 1885, certainly still have memories of the savant botanist and indefatigable worker who was J.-E. Planchon. Who hasn't encountered him, armed with a loupe and the traditional green box of the botanists, wandering one of the pathways of the old collection of American vines, examining equally avidly the *Cissus* and the *Ampelocissus* in order to extract some secrets of their affinities, then going on to study the characteristics of budbreak and flowering of the numerous varieties of American vines that have come to be classed in the compartments of our interesting collection. Encountered by a student engaged in the more practical task of distinguishing a Jacquez from a Herbemont or a *riparia* from a *rupestris*, Planchon most often greeted him by asking about the goal of his researches, and then with a few words, put the student into the midst of his [Planchon's] own present observations. Then, after finishing his explanations with some words of encouragement addressed to those who had interrupted his research, Planchon took leave toward the École. Thanks to the editors of the *Revue de viticulture* who have allowed us to illustrate this story, our comrades will know this face, simultaneously so sympathetic, open, and always alert. (*Annuaire* 1895, 53–54)

No wonder his lectures were always filled not just with students, but also with local citizens, admirers, and people who just wanted to watch the professor in action.

During the twenty years between his first entering the École de Pharmacie at age sixteen and becoming its director in 1859, Planchon worked and researched all over Europe. He spent four years, 1844–48, assisting Hooker at the renowned Royal Botanical Gardens at Kew, where he learned English, mastered woody plants, and, perhaps, developed a sweet spot for Elisabeth, Hooker's daughter (Ordish 1972, 41). Planchon then toured many of the botanical gardens of Europe and became editor of the *Flore des serres et des jardins d'Europe* (Flora of the greenhouses and [botanical] gardens of Europe). In 1851 he returned to Montpellier, where he took a doctorate in medicine. He then went to the School of Medicine and Pharmacy in Nancy until 1853, when, summoned by the old master, botanist Dunal, he returned once and for all to Montpellier. In 1856 Dunal died and Planchon succeeded him as botanist at the university; that same year he became professor of natural history at the École supérieure de pharmacie. One year later he became professor in the Faculté des Sciences.

Ten years later Planchon, along with Bazille and Sahut, had his epochal encounter with the phylloxera.[35] And the rest, as they say, is history.

CHAPTER 5

Phylloxera Makes
the European Grand Tour

The bug respected no political boundaries. Even while it was strengthening its hold in France, advance parties were breaking out into the surrounding territory. Some invasions were slow and inevitable, proceeding at the pace of the natural expansion of the bug carried by wind and rain. Other places were hit violently, the invasion fueled by the importation and planting of infected vines from France. But whatever the pace, the expansion was unstoppable.

First and worst hit was Portugal, followed soon by Switzerland and Italy. Although the phylloxera established itself wherever it landed, the course of the individual invasions varied enormously. Italy was hit early, most likely by 1870,[1] yet over the next thirty years the disease's coverage remained spotty: one region might be devastated while the surrounding countryside remained healthy. Sicily was the exception. The bug conquered everywhere and moved across the island at the speed of a conflagration.

Portuguese winegrowers first noticed the beginnings of a decline in their vines in Régua, an important port-growing region in the southwest of the Douro Valley, during the 1862 season. But the first official identification of the malady did not occur until 1872, when a governmental committee dug up vines in the town of Santa Marta de Penaguião in the Régua (Morrow 1973, 238). By this time the bug was thoroughly established, with beachheads in all the important wine-growing regions. Victory did not take long, and Portugal succumbed totally.

Some countries, however (Germany, for example), whether by reason of climate, terrain, or social-legal discipline, did not entirely succumb to the bug, even though it made significant inroads in scattered regions. Oceans were no protection, especially as commerce based on fast ships grew. As we shall later see in greater detail, phylloxera devastated South Africa, while Australia was infected but, through an intense counterattack, managed to limit the bug's territorial acquisition. Chile was apparently uninfected, while just over the Andes to the east Argentina's vignoble rapidly fell to the invader. And in far-off California, signs of phylloxera's presence were suspected already in 1874, and by the end of the decade infection foci had been identified in most of the important wine-growing regions. Phylloxera had truly become a global epidemic.

But in July 1903 the publication in the *Revue de viticulture* of the first of nearly a dozen articles by Prosper Gervais reviewing the past thirty years of the phylloxera crisis, and attempting to predict the future of the worldwide wine economy, signified an important turning point.[2] Gervais, vice-president of the Société des agriculteurs de France and later secretary of the Permanent International Viticultural Commission (Pinney 1989, 369), was the perfect choice for the job: he was scientifically trained, had experience in the wine-growing community, and had risen quickly through the ranks of the administrative establishment, in large part because of his encyclopedic knowledge and wide vision of the crisis.[3] His summary of the state of the world of wine at that point is well worth quoting at length.

> The phylloxera crisis has been the point of departure for a profound upheaval in French viticulture, with repercussions radiating, advancing, into the viticulture of the entire world. The crisis is the origin of a gigantic evolution, of a veritable revolution, both agricultural and economic, affecting the vignoble's total situation all the way from its foundations to the sources of its life and its production, modifying cultural practices and their products, the modes of life of the producers, the consumers, and commerce itself, instituting in one way or another everywhere a new order of things, destined to expand the area of culture of the vine, the extension of its production, and, inevitably, of the consumption of wine as well. (Gervais 1903-4, 89)

Gervais' main point is that not only did the phylloxera cause a crisis, but also, as was already clear only thirty years into the struggle against the bug, the entire world of wine was undergoing a revolution both agricultural and economic engendered by the success of the scientific response to the disaster.[4] Although, in the main, scientists in all of the afflicted countries relied upon the knowledge gained in France, and particularly upon the experience of the scientists at the École and their vigneron neighbors, there were individual differences, country by country, in how the French experience was adapted and applied. Not only were the climate and terrain of each place significant, but, even more important, the culture of the place, whether it be Calabria, the heel of Italy's boot, or the Pannonian Plain of Hungary, played a crucial role in how French science was grafted into place.

ITALY

Italians everywhere, from the lowliest vigneron to the loftiest official in the Ministry of Agriculture, watched—first with interest, then with growing anxiety, and finally with horror—as the debacle unfolded in France. No one had the slightest doubt that they were next. As Marescalchi and Dalmasso noted, "In Italy, the march of the invader, by then appearing at the frontiers, was followed with intense anxiety. It is really moving to relive, by reading the newspapers of those days, the seesaw of hopes and worries that, for ten long years, kept Italian vine growers in suspense" (1931, 594). But, to its credit, the Italian government rallied quickly to legislative defense. In a series of laws, defenses against the bug were approved and promulgated. Chief among them was the Swiss method of extinction.

Reasoning that the bug needed vines to propagate itself, the Swiss decided that the steadfast application of a scorched earth policy would not only kill the present infection, but it would also cordon off the bug in previously infected areas, preventing its further spread. Carbon disulfide (CS_2) was the destructive agent. As Baron Thénard's original experiments testing the substance's insecticidal potential had shown by mishap, overdoses of the oily liquid killed not only the bugs but also the vines, the weeds—indeed, everything else in the vicinity. With typical Swiss dedication, the CS_2 was applied ruthlessly at the first sign of an infection in and around a large area surrounding the focus. Done the way the Swiss did it, it was a successful method.

But it didn't work in France because the French were incapable of doing it the Swiss way. As Gervais later noted, when one compared results in Swiss and French regions that were closely analogous to one another in terrain and climate, one was "forced one to conclude that the system of *protection* imposed in Switzerland has better safeguarded viticultural interests than the *liberty* accepted and tolerated in France" (Gervais 1903–4, 635; emphasis in original). Although this method was strongly supported by the quarantine laws enacted by the government, Italy was even less successful than France using the Swiss method. This is especially true in Sicily, where the method was so strongly disliked that application of the extinction method was usually evaded and the official reports falsified (Morrow 1961, 14). Sicily's invasion was first officially reported in the summer of 1882, and Sardinia's just one year later. In both places the outbreaks were massive and along a wide front. The enormous size of the regions needing treatment was a major factor mitigating against the process. Another important factor, and most likely what drove the enormous dislike of the process of extinction in Sicily, was the type of agriculture followed everywhere in Italy but most devotedly in Sicily: polyculture.

A typical Italian or Sicilian farmer would be a smallholder with only a few hectares to call his own. Somehow a living would have to be wrested from that small plot, through true subsistence farming. The farmer would thus have to diversify his crops to satisfy the needs of his family over the year. In the genuinely Mediterranean regions of Italy, a typical farm would have fruits, nuts, cereals, and vegetables for human consumption, plus fodder for the livestock, which would certainly include some goats, a cow for milk, an ox for working the fields, probably a pig or two, and chickens for both meat and eggs. And, inevitably, grapes would be grown to provide wine for the family. Wine was essential to the diet, functioning as a major source of storable carbohydrates, which provided 60 percent of the daily caloric intake in a place like Calabria (de Lorenzo et al. 1999, 25). All the rest of the family needs would be provided by the other diverse plants and beasts on the homestead. And therein lies the problem.

The method of extinction did not kill just the phylloxera and the vines; it killed absolutely everything, including trees, flowers, grain, fruit, and berries. Following the treatment there would be no subsistence for the subsistence farmer. The government offered subsidies, of course, but these were not a viable option for a population depending on polyculture. Everyone had to have a few olive trees and some almonds, but reestablishing either of these types of trees was a long-term task: it could take a decade to bring an olive or an almond tree into bearing. Even grapevines require half a dozen years to produce a useful amount. It is thus not surprising that there was vigorous and sometimes violent resistance to the extinction process: "These methods actually sometimes led the vine growers involved to engage in violent revolts, like the one in 1898 that culminated in a bloody riot at Valmadonna, in the province of

Alessandria, with five people killed and thirty injured!" (Marescalchi and Dalmasso 1931, 595). In the face of such resistance, the government soon ceased its efforts to promulgate extinction and turned its hand to other methods.

Just as had the French during the La Défense phase of the struggle with phylloxera, the Italians tried their hand at sand, sulfide, and submersion. The latter was tried first, in early September 1879, near Valmadrera, the site of the first officially identified phylloxera focus. After obtaining permission from the local government officials, an agreement was reached among local vignerons, a steam-powered pump was set up to draw water from Lake Como to the vineyards, and the submersion process started. As in France, the technique was successful—so long as Faucon's intervals under water were followed scrupulously. But the technique was limited in Italy, even more so than in France, by the vineyard terrain: Italian vineyards tend to be found on hillsides, which are difficult to submerge. By 1905, the year of the greatest extent of the flooding, barely four thousand hectares were submerged in the whole of the country (Morrow 1961, 25).

Sand planting, which the French had demonstrated to be effective against the insect (and just as effective against wine quality!), was considered a possibility in Italy. But again, as in France, it was discovered that not all sands were equal. Tests were conducted at an experiment station in Rome, and, in results reported by Paulsen, it was suggested that the sands of Caseta and Messina provinces would be the most likely candidates (Paulsen 1933, 164). The test plantings were successful, but at no time did sand planting become an important element of the struggle against the bug.

Italian experience with sulfiding was similar to that of the French. Yes, the technique worked, but it was clearly only a palliative. Many vineyards required more than one treatment in a year; treatment techniques were fussy and demanded careful, well-trained applicators; and effectiveness varied according to soil type (heavy, damp soils were relatively impermeable to the gas). All of these factors summed up to one undeniable fact: sulfiding was a very expensive proposition. Moreover, it was an operating cost and not an investment. By 1905, near the peak of sulfide's use, Paulsen and the other leading Italian scientists had persuaded the relevant members of the wine industry—bureaucrats and growers—that Foëx was right. Sulfide must be viewed only as a temporary measure, used to buy time while the system geared up for the only long-term solution: use of American vines, either as direct producers, graft stock, or, as they were rapidly coming on line, one of the new hybrid direct producers (Paulsen 1933, 174–75). Unfortunately, putting American vines to work in Italy was no simple thing, especially since the Italians seemed determined to repeat some of the mistakes—even the "Concord catastrophe"—of the French.

The first activity undertaken by the government was a rather odd one: years before the first official admission of phylloxera in Italy, local government agencies distributed many seeds from American vines and proposed to hold a contest four to five years down the road to choose the best among the grown-out seedlings (Paulsen 1933, 156). Unfortunately, no records of the winner and his fate are available. This contest clearly reveals two underlying psychological-sociological factors, factors that in the long run determined the course of events with the American vines. First, the Italians were extremely leery of the American vines as foci of infection, which mitigated against importation of the vines themselves. This apprehension lasted for nearly a generation and had some extremely unwelcome results. For example, in 1882, two years after a thoroughly modern American vine nursery had been set up on the island of Montecristo, the discovery of some infected American vines at a nearby site unconnected with the nursery led fearful local authorities to pull out the established American mother vines (Morrow 1961, 16). Second, the expert sources were both confused and insecure about the proper route to take with the American vines.

Their confusion is evident in the recommendations—published as late as the mid-1880s—that resistant vines suitable for planting in the Italian vignoble included various *labrusca*-based varieties. Although it had already been ten years since such a dangerous recommendation had been made in France, the French mentors of Italian scientists and authorities were unable to dissuade the Italians from recommending such vines as Clinton, Concord, and other *labrusca* varieties for use in Italy, both as direct producers and as rootstocks.[5] Moreover, just as the French had been enthusiastic in turn about each new American species—*riparia, rupestris, berlandieri*—when each was discovered, only to find that each of these pure species had its own problems, the Italians followed the same pattern, with the same unfortunate results: each pure species had its own virtues and vices, although they were localized to Italian conditions.

A major factor in the disarray among expert opinions about which American vines would be best in the various Italian regions was the wildly diverse conditions of terrain and climate in the country. Grape growing in Italy stretches from the mountains in the north, through the central plains to the vast arid stretches of the south, and then into Sicily, with its own varied terrain. As Gervais noted, "The extreme variety of soils and climates found there make necessary, as in France, usage of a great number of rootstocks, to which most likely will be added some creations of Paulsen, Grimaldi, and Cavazza" (Gervais 1903–4, 635). The reference to the creations of Paulsen, Grimaldi, and Cavazza is telling. Even though there had been a large number of rootstocks—either found naturally or created—that had undergone trials and been in France, these had been successful under French conditions of soil and climate. When these selections proved unsuitable for certain Italian regions, the Italian scientists were forced to create and perform trials on their own hybrids.[6] Several of the stocks bred by Paulsen and his colleague Ruggeri were quite successful and are still in wide use in the Mediterranean vignobles of wine-growing countries in Europe and North Africa (Galet 1979, 197–98). Paulsen 1103, for example, a *berlandieri × rupestris* cross, is the eighth most popular rootstock in Italy. Ruggeri 140, another *berlandieri × rupestris* hybrid, has become extremely important not only in the calcareous Italian vignoble, but, since 1960, in France as well.

The wide diversity of the Italian conditions is also revealed by the geographical patterns of destruction (or the lack of same) inflicted by the invading bug. In the far south, Sicily was conquered early and quickly, and the same was true for Sardinia. In the far southwest, Calabria was stricken in 1883, but Apulia, directly opposite Calabria in the southeast, was not invaded until 1906. This patchwork pattern was repeated all over Italy. In the 1930 Phylloxera Commission report, forty-one of ninety-two provinces were totally infested, forty-eight partially infested, and three—Napoli, Rieti, and Frosinone, the latter two to the northwest and southwest of Rome, respectively—were by some miracle still clean (Paulsen 1933, 178–88). As these statistics show, the bug's coverage of the Italian vignoble was not nearly as complete as it was in France. In this the Italian wine industry was lucky: there was still time for responses to be crafted, frequently in conjunction with the scientists at Montpellier.

Cavazza was the first Italian scientist to study at the École, in the late 1870s; in 1885 six advanced enology and viticulture students left Italian universities to study in Montpellier. Among the six was Paulsen himself (Morrow 1961, 17). After his course at Montpellier, Paulsen was immediately appointed to direct the Royal Nursery of American Vines at Palermo. Once Paulsen took control, the situation in both Sicily and the mainland settled down, at least in the sense that research into discovering the appropriate rootstocks for each region began energetically.

But, once the reconstitution began in earnest, it was not a smooth or rapid process. Here the differences between Italy's situation and France's is illustrative.

While polyculture marked much of the Mediterranean region in both France and Italy, the parcels of French freeholders were, on average, much larger than those of the crowded and relatively tiny plots in Italy. This meant that reconstitution on American rootstocks proceeded by larger units in France, where fewer vignerons needed to be persuaded to replant per unit area reconstituted. Things went much more slowly in Italy.

A second major difference relates directly to the spotty coverage of the bug in Italy as compared to France. In France, when an area got infected it soon perished. This did not happen in Italy. Even in 1930 the vineyards of ancient Latium, a large area surrounding Rome, and the huge region of Napoli were essentially free of the disease. Given the well-known "it can't happen here" attitude, it proved to be essentially impossible to persuade the growers to apply preventative reconstitution to these huge areas of bug-free vignoble. Vignerons simply resisted pulling out perfectly healthy vines based on the prediction that someday they would become infected. Only after the Great War, when the bug had done its damage, could the reconstitution begin seriously in these large vignobles.

Moreover, it cannot be discounted that in many regions of Italy—the Veneto and Alto Adige, for example—significant amounts of wine were being made from those Old Americans—Clinton and Isabella chiefly— that had successfully taken hold there a generation or so earlier, as will be discussed more fully below. These successes also mitigated against reconstitution on American rootstocks.

By the early 1930s, however, the end was in sight. Once appropriate rootstocks had been found for each region, reconstitution proceeded with all due speed. From that period on, the pace of replanting on American rootstocks was not constrained by scientific limitation, but rather by social, cultural, and—it is perhaps not going too far to suggest— psychological resistance to pulling up the traditional or Old American vines and planting traditional vines anew on American roots. It became a case of persuading villages and individuals to make the change. In the end, everyone reconstituted. Marescalchi cogently sums up the final results: "We can say that this is how—in any place where the implacable aphis was killing the vineyards—modern viticulture was born. It was an actual new viticulture, as opposed to the pre-phylloxeric one, not only because it was based upon new operations of culturing technique, but above all because it was guided and supported at each step by science rather than by empiricism or tradition" (Marescalchi and Dalmasso 1931, 595). Just as in France, the intervention of the professors was crucial not only for defeating the bug, but also in the formation of a totally new, scientific, viticulture.

IBERIA

Together, Spain and Portugal comprise huge vineyard areas producing enormous quantities of wine. While most of their production is distributed locally, a significant amount is exported. For centuries, Spain's sherries and Portugal's ports have been imported in great volumes into Britain and the other areas of the United Kingdom. Similarly, since prerevolutionary times Portugal's Madeira wines have slaked the thirst of many Americans, including, most notably, Thomas Jefferson. During the phylloxera infestation these exports were seriously diminished, with obvious economic consequences.

Suspicions about phylloxera invasion first arose in Spain in 1872, when vines in the Ribera del Duero region began showing signs of sickness. The suspicions were confirmed by 1875, when the bug was found in the Baleares, Mallorca, and several places in the Málaga district (Sobrado 1990, 38–39). Various local agricultural societies went on the alert, and the government, via the Consejo Superior de Agricultura, sent out bulletins about the presence of the bug. Yet not much was officially done; in particular, no embargoes on the importation or transport of vines, either from abroad or within Spain, were enacted (Sobrado 1990, 38—39).

One of the most important phases of the reaction to the invasion, which was now evident, was the convening of a large international congress in Zaragoza in October 1880. Organized by the Junta de Agricultura, Industria y Comercio, the provincial court, and the county of Aragon, the congress invited the most important scientists from Europe, including both Planchon and Saintpierre from Montpellier (Sobrado 1990, 26). The Junta de Agricultura, Industria y Comercio took the task of the congress with the utmost seriousness: "Losing the grapevine would be a certain calamity for our country, not just due to the immense loss to the public treasury, but for its effect on the existence and daily sustenance of poorer classes in the cities and, especially, in rural areas" (La Comisión Organizadora 1880, 6). In their view, the congress would allow the authorities in Aragon "to strengthen their efforts even more to save (if possible) our number-one agricultural treasure from the menace that threatens it" (La Comisión Organizadora 1880, 6).

What is surprising in this manifesto is the caution expressed by the "if possible." Clearly, by this time the French were already beginning to beat back the bug by applying insecticides, flooding, planting in sand, and, most important, using American vines as both direct producers and grafted rootstock. The commission made their reason for this caution clear in their next statements:

> Several measures have been proposed or planned by the two entities mentioned above, but there is fear that they may not suffice, because in Aragon little is known about the history of the disease, which would be the base for any subsequent work; it would be convenient to gather next month of October all experts from all nations in an international congress to study the disease and to discuss the means to combat it or to reduce its calamitous results. (La Comisión Organizadora 1880, 7)

Surprisingly, then, what we find in Spain generally, and Aragon in particular, is a genuine paucity of knowledge about the phylloxera, its nature and history, and, importantly, the most efficacious ways to combat the plague. Whatever its source, the knowledge gap was real, and the Spanish were explicitly aware of their ignorance. Their hope was that by convening the international congress, foreign experts would advise them about the bug and the best means of combating it. Precisely these points were covered by the French experts. By the end of the conference, the Spanish were "unequivocally convinced of the priority of American vines in the war against the phylloxera" (Sobrado 1990, 26).

This realization came later to Portugal. Despite the geographical continuity of Portugal and Spain, the progress of the invasion and the responses to it took dissimilar courses, in part because of differences in both physical terrain and social culture. At the time of the invasion, roughly 300,000 hectares of vines were cultivated in Portugal, with an annual output of between four and five million hectoliters (Gervais 1903–4, 657). While wine production is found in every region of the country, the premier districts are to be found along the Douro River, in the highlands extending from fifty or so miles above the Atlantic port of Porto to the Spanish border. In these vignobles are grown the vines that produce the world-famous port wines, and it was in these very same vignobles that the first signs of the disease were noticed in the early 1860s. But official identification of the phylloxera dates only from 1872, when a government commission dug up vines in the town of Santa Marta de Penaguião, near Régua (Morrow 1961, 4). Curiously enough, the bug languished here, causing its typical damage, but not moving at all rapidly in its conquest of adjoining territory. Due to the phylloxera's apparent isolation in the northeastern part of the country, the government response was lethargic: the first "permanent and official Antiphylloxera Service" was not instituted until 1878, six years after the official identification (Periera 1913, 4).

All this changed in 1879–80 when the plague broke out of the Douro Valley, rapidly moving south and west approximately eighty-five miles to vineyards surrounding the old university town of Coimbra. In the following year the front moved another fifty miles south to the important market town of Leiria. Within two years after that, it was knocking at Lisbon's doors; east of Lisbon, Setúbal, the region of luscious sweet wines, was attacked in 1884, and "in 1886 the famous Colares red wine vineyards near Cintra were found to be affected" (Morrow 1961, 8).

Colares provides an interesting case. Many of the vines in the district are planted in the landward side of the Atlantic Ocean dunes, dug into sheltering cubbies in the sand, which protect them from the wind and salt spray coming off the beach. Because of their sand planting, the Colares vines were never devastated as were the vines further inland. Again, as in France, planting in sand insulated the vine against the ravages of the insect.

While the government was going through a series of phylloxera commissions (three separate bodies between 1878 and 1886), with all the indetermination entailed by such turmoil, some progress was, in fact, made on the ground. In August 1877 Portugal sent a three-man delegation, led by Dr. Manuel de Oliveira of the University of Coimbra, to the Lausanne International Phylloxera Conference. On their way home from the conference the delegation spent some time in the Midi, visiting with the Montpellier scientists and observing the various responses then being made to the bug. Upon their return to Portugal, the delegation published a full report of their activities (Demole-Ador 1877, 4–11).

At this point, the French were fully engaged along three fronts: the defense via chemicals, sand, and water; the reconstitution via Old American vines; and reconstitution via grafting on rootstock. In the next few years the Portuguese became involved with the first and third of these measures. For whatever reason, however, unlike the rest of Europe, they never committed much time or energy to reconstituting via the Old American vines.

Sand planting, of course, was already envisaged along the coast, for example at Colares. Flooding was strongly encouraged, but, due to the hilly terrain throughout much of the Portuguese vignoble, it was never a widespread option. Ultimately, no more than two thousand hectares were ever protected in this way, less than 1 percent of the total extent of vines (Pereira 1913, 4).

Protection via CS_2, however, was widely adopted. By 1884, as Francisco d'Almeida e Brito, inspector general of viticulture and the Phylloxera Service, wrote to the editor of *La vigne americaine,* the chemical was being used in all regions of the country. Interestingly enough, just as in France, the railway system was not only used for the transport of canisters of the chemical; as well, many of the depots served as distribution points for the gas, the injectors, and information about their employment (Morrow 1961, 11).

American vines made a slow start in Portugal. From the beginning, both the government and the vine growers were cautious about importing American vines, which were viewed as a source of contamination. In 1880, Laliman's lecture on the American vines to the Central Agricultural Society of Portugal met with a hostile reception (Morrow 1961, 15). Around 1880 the Phylloxera Commission, along with providing information on the proper employment of CS_2, also promulgated strongly the idea of growing American vines from seeds. Indeed, it even provided the seeds. Evidently it was believed (correctly) that the seeds would not be a source of contamination. Vines grown from these seeds were used to stock nurseries from 1882 onward. All this changed in 1885, when the government, noting the successes with reconstitution on American vines in France and Italy, relaxed the prohibition. From that point on, both cuttings and rooted vines were allowed to be imported (Morrow 1961, 15).

But the vast majority of vignerons still hesitated to reconstitute on American vines. What finally turned the tide derived from a peculiarity of the organization of viticulture in Portugal, what Gervais calls "the industrialization of viticulture," which was more advanced in Portugal than any other place in the world (Gervais 1903–4, 658). This small country was home to some of the largest vineyards anywhere: the grand plantation of the dos Santos family comprised some four thousand hectares, and was certainly the largest single vineyard in the world. By the early 1880s, dos Santos and the other large holders had seen that reconstitution on American rootstock was the only possible way to save Portuguese viticulture. So, rapidly and thoroughly, dos Santos and the other owners had done precisely that, reconstituted their huge holdings on American roots. Once the remaining vignerons saw the success of the "industrial" reconstitution, they went the same way.

One factor aided the Portuguese reconstitution enormously, especially as compared to their French, Italian, and Spanish confreres. The Portuguese vignoble terrain is, for the most part, schistous soil, "very favorable to the American vines: the *riparia* × *rupestris* [e.g., 101–14Mgt and 3309C] and Rupestris du Lot are perfectly adapted there" (Gervais 1903–4, 657). Unlike the rare rootstock vines required by the chalk soils of France and Italy, these other rootstock vines root easily and are extremely compatible with essentially all the European vines.[7] Thus the Portuguese reconstitution required no time-consuming experimentation to discover—or perhaps even to produce by hybridization—vines that would root, be compatible with the European scion, and, finally, thrive in the local terrains. From that point on, things went fairly smoothly in Portugal's reconstitution.

Back in Spain, however, things were not going so smoothly. Even after the conference in 1880, and another in Madrid in 1891, the Spanish vignerons were still stubbornly dragging their feet. Indeed, even as late as his 1903 report, Gervais worried about Rioja and sherry:

> But the question is one of knowing whether the replantation of the vignoble will take place in Spain in a fashion to reconstitute, to maintain in their integrity, these precious types of wine in more than name only? It is not out of line to ask this question. From the start the reconstitution properly so called has not been without suffering some serious difficulties due to the nature of the soil, the aridity, and the climate. (Gervais 1903–4, 636)

Obviously Spain's climate is different in many ways from that of France, and thereby would need to accumulate its own research and experience with rootstock varieties, some of which would doubtless need to be new. However, it is clear that, even given these difficulties of terrain and climate, Spain's twenty-five-year delay in substantive response had other reasons, ones not prevalent in France or even Portugal. One of those reasons was that, in spite of the "prodigious encouragement of the national and provincial governments," some vignerons hesitated to replant, and some "abandoned the culture of the vine, replacing it with other crops (beets, potatoes, and forage), the success of which is guaranteed by the richness of the soil" (Gervais 1903–4, 636–37). Additionally, there was the straightforward reluctance of the vignerons themselves to make the investment of time and money necessary for reconstitution: "Elsewhere the progress of reconstitution is slow, and in spite of the efforts of publicists, of savants and eminent professionals producing encouragements and counsel, it appears that there is a manifest hesitation. The question is the cost of reconstitution versus the benefits" (Gervais 1903–4, 637).

This hesitation was widespread and, in the end, extremely costly to the Spanish vignoble. The government made an exhaustive survey in 1909, with the results published in 1911 (Ministerio de Fomento 1911). Data collected by the survey included vine area planted, by region, prior to the invasion; area lost to the bug; area reconstituted; area remaining but not reconstituted; and area replanted in other crops, including olives.[8] Zaragoza, site of the first international phylloxera conference hosted in Spain, is a typical example. According to the data, prior to the official date of phylloxeration in January 1900 (although infestation was first admitted in 1897), 85,500 hectares were under viticultural production. Infestation was noticed first on the border with Navarra, and "propagation was very rapid in the northwest and southwest directions." After the invasion, "the surface of vines totally destroyed is valued at 75,823 hectares . . . with 4,548 free of the plague and 5,129 in normal conditions of production" (Ministerio de Fomento 1911, 39–40). But the reconstruction was difficult: "The reconstitution of the vignoble is being effected in this province exceptionally slowly." It must be kept in mind that Zaragoza, prior to the phylloxera, was an important wine province. As it went, so went much of Spain. By 1909 only 10,375 hectares had been replanted. Of the nonreconstituted hectares, 342 had been replanted in olives, while 68,780 hectares were destined to be planted with other crops. It took decades to bring Zaragoza back to anything like what it was before the plague.

Spain, unlike Portugal, took a long time to recover from the phylloxera. The process was painfully slow and very expensive, but, in the end, it was successful. Today, Spanish wines are fully returned to their just prominence on the world wine scene.

GERMANY AND CENTRAL EUROPE

Of all the countries invaded by the bug, Germany incurred the least disastrous effects. This was not the case, however, in countries further south and east: Austria, Croatia, Dalmatia, and particularly Hungary were devastated. Recovery and reconstitution happened differently in each country according to the same limits we have seen already: the nature of the terrain, the climate, the local wine-growing culture, and the involvement of the government.

The Germans were nothing if not rigorously thorough in their response to the invasion. From the very first infection site, at Annaberg Gardens near Bonn in 1874, every focus was identified and investigated and the probable source of infection determined. Annaberg Gardens, for example, had been infected by wood arriving directly from America (Kessler 1892, 16). In the end, nearly two hundred distinct sites were identified and investigated. A major finding of the thorough investigation was the "evident certainty that the major part of the infections were by flying insects coming from neighboring foci" (Kessler 1892, 20). Thus, noted Kessler in his official review of the invasion, "It seemed to me distinctly that the flying form of the insect thereby played a major role in the infections" (Kessler 1892, 20).

This had a major implication for the spread of the insect in Germany. Kessler and others investigated the effect of climate and weather upon the flying form of the insect. The investigators found a large decrease in the number of flyers caught in spider webs as the days shortened and got colder after September (Kessler 1892, 23–24). From this it followed that the rate of expansion of the invasion was dependent upon temperature, and, in the long run, on the German climate.

These results were echoed by those reported by de Horváth at the Hungarian Phylloxera Research Station. After noting that the invasion's expansion rate was noticeably slower in Hungary than in the Midi, de Horváth goes on to say:

> The cause of this slowed expansion is explained by the climate of Hungary. Our researches have demonstrated that *the phylloxera disease expands proportionally more slowly in Hungary than in the Midi of France* because here the winter dormancy of the parasite is of longer duration. The phylloxera emerges here later in the spring and ceases multiplying earlier in the autumn than in the Mediterranean climate, with the result that the number of its annual generations is smaller. (de Horváth 1882, 2; emphasis in original)

A major problem in the Midi was that, with the exponential expansion of offspring every generation, the four or five generations of phylloxera born every season provided a veritable conquering army of "soldiers" all across the expanding invasion front. Central Europe's considerably shorter growing season diminished the numbers of offspring by appreciable amounts, thereby slowing the invasion. Thus, although the bug's victory was inevitable, the course of the war was considerably longer, which provided more time for counterattacks.

Of course the Germans didn't rely upon their climate to defend their vineyards against the foreign invaders. They mounted an overwhelming defense, swarming against the bug as much as it swarmed against them. As Gervais, in almost awed tones, remarks, "The insect was combated by the method of extinction so rapidly, so energetically, and with such success that its propagation had nearly always been stopped as soon as it started" (Gervais 1903–4, 260).

Germany's practice of the method of extinction, unlike Italy's and France's, was rigorous and unrelenting. Once a focus of infection had been identified, a large quarantined area *(Sicherheitsgürtel)*, or cordon sanitaire, was declared around the site. The entire area was then mowed to the ground and disinfected with CS_2. A narrower area, tight around the infection focus, was then spread with petroleum, which remained until late January. During this period the CS_2 continued its work, "in which the previously infecting insects were killed" (Kessler 1892, 28). Beginning in the early spring, the vine roots were ripped up to a depth of 1.25 to 1.5 meters and then burned. Then a lethal dose of CS_2 was administered at the rate of two hundred grams per square meter, which killed every living thing in the soil. For the next six days and nights a military cordon surrounded the site, guarding it against any reentry. "Generally a detachment of twenty-four men was necessary" for the guard work, notes Kessler laconically (Kessler 1892, 28). Needless to say, the method of extinction was nowhere else so successfully carried out. In the end, the bug scarcely pained Germany: by 1903 it had "little or no influence on the general viticultural situation, on the German vignobles' production, or on the sale of wine" (Gervais 1903–4, 260).

Austria was not so lucky. Its warmer climate and hospitable soils led to a much more severe infestation. Lower Austria, with its very important Burgenland vignoble, together with neighboring Styria, Croatia, Tyrol, and Dalmatia, comprised 247,000 hectares of vines. By 1903, more than 85,000 of those hectares had been contaminated by the bug, even though serious attempts had been made to beat back the infestation by sulfiding (Gervais 1903–4, 606).

The region's reconstitution on American roots "was effected nearly everywhere without encountering any serious obstacles" (Gervais 1903–4, 606).[9] One feature of the region was both a plus and a minus: the soil. On the one hand, the stony soil offered a hospitable environment to the phylloxera. On the other, it was also hospitable to *riparia* and *rupestris* rootstocks, which, as we have seen, provided excellent affinity to most of the virus-infected European *vinifera* vines.

One very sad note about the region dates from the early 1880s. In a short news article, *La vigne française* remarked that "several French firms have successfully made plantings of vines in Dalmatia." It went on to say that "others are preparing to do the same in Croatia and Slavonia, countries completely immune to phylloxera" (La vigne 1880, 112). This was "the Golden Age of the Croatian Adriatic," as Gamulin notes.

He continues, "As French vineyards were devastated by phylloxera in this period, the prices of Dalmatian wines rose sharply. Production and export of Dalmatian wine was stimulated; vast portions of land were converted into vineyards on the terraces sloping down island and mainland hills" (Gamulin 1996, 1).

During the next decade, the vineyard surface planted in these countries by French companies increased tenfold. None of the vines were grafted. And, of course, contrary to *La vigne française*'s claim, none of the countries in the region were "immune" to phylloxera. During the period 1900–1905 the bug arrived, the vineyards crashed, and the great Slavic diaspora began. On the Dalmatian coast, for example, declining land use, caused in great part by the phylloxera, "has been, among other things, a potent force in fostering migration overseas during the last hundred years, especially to North America" (Johnston and Crkvenčič 1954, 356). Among the many Slavic immigrants were several hundred thousand Croatians who ended up in Chicago and now form a significant population in that Midwestern metropolis ("Croatians" 2008). New Zealand was a destination for many of the Dalmatian winegrowers, who, once settled in their new land, founded what has now become a world-renowned wine-producing area (Trlin 1979, 82). It was not until after the Second World War that many of these Slavic vineyards began to be reconstituted on American rootstock.

Hungary's story was much more typical, and certainly much less tragic, than that of the Slavic lands. Phylloxera was noted in Hungarian vineyards around 1870 (de Horváth 1882, 9).[10] By 1881, 2,200 hectares had been invaded, and the rate of infection was rapidly increasing. In 1880 the Phylloxera National Committee was established, and in 1881 the government consolidated its interventions against the pest with the Hungarian Phylloxera Station, with Dr. G. de Horváth, noted botanist and member of the Hungarian Academy of Sciences, at its head.[11] De Horváth immediately began action on several fronts. He set up test vineyards at Farkasd and Szendró, where various viticultural experiments were established, including submersion, the testing of Old American varieties, grafting experiments, and CS_2 trials.

In one respect Hungary was extremely lucky: it "possesses a considerable area of a terrain composed of a very fine and moving sand. . . . Many thousands of hectares of these dunes are found in the vast central plain of the country, extending from the foot of the Carpathians" to the southern border with Serbia (de Horváth 1882, 10). As the French had shown, sand of the right sort blocked phylloxera colonization, a datum immediately grasped by the Hungarians. Even before the phylloxera invasion, these sands had important vine plantings, but once the invasion began to seriously encroach upon wine production, new planting in the dunes began in earnest. These new plantings were responsible for much of the gain in Hungarian wine production during the 1890s (Gervais 1903–4, 607).

Submersion, which was tested at the Hungarian Phylloxera Station and shown there to be effective in the light soils that covered most of the country, was not used extensively because of the region's topography and especially because of a lack of sufficient water (de Horváth 1882, 12). Sulfiding, which was also tested at the station and shown to be effective, was recommended for whatever sites were amenable. The method of treatment "employed was that of the Compagnie des chemins de fer de Paris à Lyon et à la Méditerranée [PLM]" (de Horváth 1882, 15). A complete system of transportation and product distribution was set up, again modeled on the scheme designed by the PLM, and by the late 1890s was actively in use. In 1903, just over eighteen thousand hectares were being sulfided (Gervais 1903–4, 607).

Interestingly enough, when the Hungarian Phylloxera Station first began considering insecticidal treatments, they announced themselves open to suggestions from the public concerning methods and materials. Eighteen "inventors" replied that they had a procedure ready to be tried. But the station had a rule that must be followed, namely, "submitting to experiment only those procedures [for] which the inventors communicated the constitution and mode of application"; "secret remedies were passed over in silence." In the end, only six inventors "exposed their processes." Unhappily, the station "had the regret to encounter nothing new," and, indeed, "the authors of these processes gave witness, except in some rare exceptions, of an absolute ignorance of the subject, proposing . . . processes as impossible as they were ridiculous" (de Horváth 1882, 14). Shades of the Montpellier experience with garlic, shrimp bouillon,[12] and goat urine!

De Horváth's major effort during the station's first years was to conduct trials of the Old American vines and of rootstocks, both old species and new hybrids. His goal was straightforward enough: "The most important and serious question that the Hungarian viticulturalist can ask about the American vines is if the exotic varieties can be acclimated to our climate and if they will conserve their resistance against the attacks of the phylloxera in the soil of Hungary" (de Horváth 1882, 17). Both questions were soon answered in the affirmative.

Prior to the beginning of the trials, the minister of agriculture had sent the "distinguished viticulturalist M.E. Nedeczky" to America to study the target vines in their home situation, to assess their state, and to ensure that, when the initial cuttings were shipped, that they would be "perfectly authentic." Nedeczky traveled to several important vignobles in the United States, but "above all in Missouri, where he had procured from several viticulturalists six thousand cuttings," from whence he continued on to Bush and Sons and Meissner at Bushberg, near St. Louis, where he ordered 108,250 cuttings (de Horváth 1882, 20).

Other vines were ordered from France and Styria. In the end, enormous trials of fifteen different cultivars[13] and rootstocks were set up at the three locations, with the station receiving 60,450 cuttings, Farkasd 57,200, and Szendró 24,440, for a grand total of 142,090 cuttings (de Horváth 1882, 21). Although at the time of his first report de Horváth's trial plantings were new, he was confident that the behavior of most of the new vines would parallel those of the small collections of Old Americans that had been planted at Farkasd a dozen years earlier. These vines were flourishing "amidst fields of European vines destroyed by the phylloxera" (de Horváth 1882, 17).

State activities by the Hungarian government were extensive. By 1899 forty-six state nurseries had been set up. During that year six million grafted plants, 36,500,000 American vines, and forty-five million European plants were sold to Hungarian viticulturalists. Clearly, these numbers reveal a recovered, not to mention reconstituted, national wine industry. The only "weak point that can oppose the rapid expansion" of the industry into "so happily placed particular conditions of climate and soil" is the state's "lack of capital . . . paralyzing the development of Hungarian viticulture." But, other things being equal, "one can envisage the more or less near eventuality of serious growth in the vignoble and wine production of this country" (Gervais 1903–4, 608).

Hungary's neighbors Romania and Bulgaria illustrate two quite different outcomes in the phylloxera wars. Romania's phylloxera crisis began in 1884; by 1898 29,000 hectares (out of a total about 163,000) had been phylloxerated. By the end of the year following the initial invasion, a forceful state response had begun: laws setting up local committees of surveillance and defense were put on the books and a national viticultural service was begun (Gervais 1903–4, 696). In the next decade nurseries for American rootstocks were established in conjunction with a multiple-site trial of various rootstocks. Although Romania's soil and climate vary considerably, it was soon found that the methods of experimental testing used by Montpellier—where many of the officers of the Romanian viticultural service visited and studied—revealed a range of choices suitable for each of the regions (Gervais 1903–4, 697). By the mid-1890s, the concerted national effort was beginning to pay off not only in the reconstitution of much of the original Romanian vignoble, but also in the addition of nearly forty thousand hectares of new plantings. This is strong evidence, as Gervais notes, that the viticultural service "has opened to Romanian viticulture a sure path, exempt from the failures and costly errors that have elsewhere marked the debut of the reconstitution" (Gervais 1903–4, 697).

Bulgaria was not so fortunate. By 1881 phylloxera had been observed in the country. Of its roughly ninety thousand hectares of vineyard surface, nearly one third was thoroughly infested by the turn of the century, despite the early warnings given of the onslaught. A major problem, notes Gervais, was "the lack of serious [state] organization, which has not permitted fighting efficiently against the insect, whose progress has been rapid." In addition was "an invasion of cryptogamic diseases, which, badly combated, have caused in some places the ruin of the vignoble" (Gervais 1903–4, 699). Many of the early battles against the insect (and the diseases) were fought by individual growers or small regional consortia. It was not until early 1896 that there was established an official plant protection agency, whose first task was organizing the phylloxera defense (Plovdiv 2008). A major move was inviting Montpellier's Pierre Viala (of Mission Viticole fame) to visit Bulgaria, observe the problems, and suggest solutions. As could be expected, Viala recommended setting up of trials of American rootstocks at various places in the county, creating nurseries for the accepted rootstocks, and establishing locations where grafting could be both put into production and taught.[14]

Unfortunately, Viala's recommendations could only be partially carried out, mostly due to the impoverishment of the national government. Nursery output in particular was highly restricted. Each viticulturalist could receive annually from the state nurseries only four thousand cuttings that were forty centimeters in length, enough to produce—with luck and skill—around eight thousand plants, sufficient for not more than three hectares. Moreover, each recipient had to sign a certificate that he, indeed, owned a vineyard that needed replanting! At this pace, Gervais notes, replanting could only just keep pace with the progress of the phylloxera (Gervais 1903–4, 699).

EUROPEAN LESSONS

If there is any single point made clear by the course of the phylloxera invasion in Europe and the battles against it, it is that the level and quality of state intervention was key to success. In places where the state organized too little and too late, such as Bulgaria, the phylloxera completely outpaced reconstitution. But in other places, such as Germany and Romania, where state response was fast and furious, there was a decided decrease in pain and suffering—at least as compared to what might have been.

A second compelling point that must be noticed is Montpellier's presence everywhere in the story. Wherever the decision was made to reconstitute large portions of the vignoble in *vinifera,* Montpellier was appealed to: either the locals traveled to Montpellier to observe and study, or, as in the case of Bulgaria, Montpellier came to them in order to observe, study, and recommend.

It must be noted that in many European countries it was not only the Montpellier solution that was followed. With the possible exception of Germany, all of the invaded countries of Europe planted, first, Old American vines and, later, as they became available, new *hybrid producteurs directs* (HPDs), the so-called "Franco-American hybrid" vines alongside their reconstituted *vinifera* vineyards. Montpellier was always and everywhere against these two solutions to the problem. But they were widely, and successfully, adopted nonetheless. Even today, the living results of all three solutions—reconstitution by grafting, by planting Old Americans, and by planting HPDs—remain in place throughout the European countryside.

CHAPTER 6

The Bug Goes South

New Venues, Same Story

Grapes and wine making came early to the southern hemisphere, on the heels of the colonizers. By the time of the phylloxera each of the major southern wine-producing regions—Australia, South Africa, and Argentina—had developed local industries of significance, and, in the case of South Africa, it had achieved considerable importance on the international scene.[1] But when the bug arrived in each of these places in turn, abrupt change was the immediate result. As we have seen before, many of the changes phylloxera produced were the same everywhere: loss of production, disastrous effects on the local economy, anxiety and even panic among the growers, and desperate searching for solutions. Luckily for these three regions, the Europeans, and particularly the French, had preceded them into and out of this desperate state and were available to provide some help.[2] But in addition to the typical effects, each region had its own peculiarities, both biophysically and culturally. We begin with Australia.

AUSTRALIA

In 1788 the first thousand Europeans arrived in Australia to set up a penal colony. With them came cuttings from Rio de Janeiro and the Cape of Good Hope, brought by Governor Phillip and planted at Cove Farm that same year (Walsh 1979, 1).[3] Over the next two decades various attempts to set up vineyards were made by both penal officers and free settlers—most of the latter from wine-growing areas in Europe, particularly Rhine-Hesse and central France—without much success. Diseases were a serious problem, particularly anthracnose, produced by a canker-causing fungus. Moreover, the soil-climate interaction, so different from anything ever seen in Europe, required serious study and experimentation before successful methods, such as irrigating vine roots during high-temperature periods to keep them cool, were developed and deployed (Walsh 1979, 1).

Among the most important viticultural pioneers was Gregory Blaxland, a native of Kent who arrived in the country in 1806 and immediately began experimenting with vines on his new 450-acre Brush Farm near Eastwood, New South Wales. Blaxland realized that the vine types planted previously had been chosen haphazardly at best, and that careful observation and selection from among a great number of varieties was the only method that might improve the prospects of the Australian vignoble. He began importing vines from France and Spain and planting them out on his farm. By 1816 he had succeeded in producing a significant amount of decent wine from his selections, which he kept healthy using various techniques he had developed, especially those preventing anthracnose (Walsh 1979, 2). In 1822 Blaxland exported twenty-six gallons of a light red wine to London, where it was apparently enjoyed considerably.

James Busby, another important early viticulturalist, arrived in Australia in 1824 from England by way of viticultural school in France. He brought with him a large number of cuttings with which to furnish his new two-thousand-acre land grant in the Hunter Valley. Seven years later Busby returned to Europe to collect more vines; he returned with 614 different varieties collected in Montpellier, Luxembourg, and Kew Gardens. It was from this stock that most future selection and experimentation in Australia derived (Walsh 1979, 2). During the increased immigration of the 1840s and '50s, successful vineyard expansion in New South Wales and Victoria brought viticulture into Australia's mainstream agricultural economy. From that point onward there was no doubt about the ultimate significance of wine in Australia's future.

Phylloxera appeared in 1877 in a vineyard at Geelong, Victoria, just across the bay from Melbourne (Whiting and Buchanan 1992, 15). By 1910 outbreaks had been recorded near Sydney N.S.W. and Brisbane in Queensland, but it "was more widely spread across southern, central and northeastern areas of Victoria" (Whiting and Buchanan 1992, 15). Victoria was indeed devastated: "Australia is often regarded as fortunate in having only relatively small vineyard areas infested by phylloxera. In fact, these areas were large and important grape areas until they were infested by phylloxera" (Buchanan 1992, 15). But the devastation was in part willful, owing to official Victoria's disinclination to accept expert advice from François de Castella (1867–1953), the Sydney-born son of an immigrant Swiss vigneron.

The teenaged de Castella left Australia in 1883 in order to study natural science in Lausanne, and viticulture and enology in France. Upon his return in 1886 he managed a large vineyard at Yering, east of Melbourne. In 1890 he joined the Victoria Department of Agriculture as a viticultural expert. In the next few years he traveled all around the state, gained hands-on experience with farmers, and engaged in writing his influential *Handbook on Viticulture for Victoria,* which appeared in 1891. For some reason he was soon thereafter let go by the Department of Agriculture, after which he went back into private vineyard management. But his involvement in the growing phylloxera disaster was far from over. After phylloxera was discovered in Bendigo in 1893, de Castella a year later criticized the government's policy of "death by extinction" and opted instead for regional quarantine "and the introduction of phylloxera-resistant, American rootstocks, as had been done in Europe. Although he was aware of the latest methods adopted by the French in combating the pest, his advice was ignored" (Dunstan 1993, 604).

By 1907, the Victorian wine industry verged on total collapse. Hans Irvine, one of Victoria's most powerful wine industry members, convinced the Department of Agriculture to rehire de Castella, send him back to Europe to survey the most recent methods, and, this time, to listen to what he had to say (Dunstan 1983, 438). His journey, "fully documented (1907–09) in the Victorian *Journal of the Department of Agriculture,* was the basis for the reconstitution of affected vineyards in central and northern Victoria" (Dunstan 1993, 605).

Victoria's disaster led many to predict that Australia's viticultural trajectory would parallel those of other regions, which had sadly been observed so many times before. As de Castella himself noted, "The ultimate reconstitution on resistant stocks of existing South Australian vineyards is unquestionably but a matter of time." Indeed, he went on to say, "it is as inevitable as it has proved to be in other wine countries" (de Castella 1942, 5). This sense of inevitability pervaded many expert judgments over the years. Early on Michele Blunno, the first director of the phylloxera research station, noting that "the onward triumphal march of phylloxera over the French territory filled with dismay the viticulturalists of other States from the very initiation of the campaign," argued that the war against the pest must be conducted "on different strategic lines" because "the most energetic and even drastic measures will but delay a general contamination" (Blunno 1914, 3). In other words, phylloxera would inevitably conquer the whole Australian continent. More recently, noted German wine scientist Helmut Becker confirmed this view in no uncertain terms: "You have no alternative but to prepare for grafting onto resistant rootstocks by establishing a foundation nursery with stocks suited to your conditions; this is because phylloxera will, one day, be introduced into South Australia (in spite of your efforts to keep it out)" (Becker 1960, 83). According to these views, phylloxera would inevitably win the battle. The dissenters, however, had points to make as well.

Viticulturist B. G. Coombe, in an important report for the South Australia Department of Agriculture, first remarks upon the "inevitability" chorus: "Dalmasso states an often-quoted opinion that phylloxera must inevitably invade all grape-growing areas of the world." Yet, he cautions, with respect to Australia, "this may be an over-pessimistic view which is dangerous if it tends to weaken the care and enthusiasm with which quarantine measure are implemented" (Coombe 1963, 21). And, as we shall soon see, over the years the quarantine measures have had much success. G. A. Buchanan, Victorian entomologist, makes a similar point: "It surprised overseas scientists to learn that in Australia, phylloxera was confined to small areas. 'Expert' opinion was that phylloxera would eventually invade the areas that are now uninfested" (Buchanan 1992, 21). But the extremely strict quarantine measures, coupled with inspection and the three-stage infection classification, have so far been quite successful: "Quarantine restrictions have minimized the spread of phylloxera in Australia for about 80 years while the wine grape industry was relatively static" (Whiting and Buchanan 1992, 20).[4]

Yet the national antiphylloxera regime is not without its critics. As Buchanan notes, "In spite of its apparent success in containing phylloxera within the 'vine disease districts' since 1900, some viticulturalists are skeptical of the value of such quarantine regulations" (Buchanan 1992, 21).[5] Moreover, phylloxera outbreaks have accelerated recently, and there are worries about the future, especially due to the fact that the industry is expanding so rapidly.

One of the major reasons for the success of the quarantine has been climactic: dispersal of phylloxera is climatically affected, "especially in hot, dry grape areas where spread by winged insects is limited or non-existent" (Coombe 1963, 21).[6] Moreover, as shown at the French

Mediterranean stations Vassal and Uvale, other forms of the phylloxera have limited ability to spread as well: "windblown gallicole nymphs are responsible for the infection" on the leaves of American vines in the sand at Vassal, one kilometer from the nearest vineyard, while Uvale, twelve kilometers from the nearest vineyard, shows no infection (Buchanan 1992, 19).[7] Given these limits on the spread of the bug, it is clear why "there is little doubt that the major factor in this success has been the relative isolation of vineyard areas" (Buchanan 1992, 21). But times are changing.

Over the last twenty years the growth of viticultural area has been dramatic; many traditional areas are being planted more densely, and new areas (Queensland's Granite Belt, for example) are being exploited. The results are predictable: "The recent upsurge in planting wine grapes in cooler areas close to phylloxera-infested vineyards has dramatically increased the risk of further spread of phylloxera" (Whiting and Buchanan 1992, 20). Because of this risk, "the trend now is to plant with grafted vines" (Whiting and Buchanan 1992, 20). Luckily enough, because of the "inevitability chorus," an ongoing research program into rootstock testing and selection dates back well over a century.

In 1899 the first viticultural research station established specifically to test rootstocks was set up at Howlong N.S.W. under the direction of Professor Michele Blunno. A large number of stocks, especially those proven under European conditions, were planted in trial blocks alongside ungrafted *vinifera*.[8] According to Blunno, the three primary study areas were "resistance of the stocks to phylloxera, their adaptation to the various soils, and the affinity that must exist between them and the European vines" (Blunno 1914, 13). This last question was of special importance to Australia, since, for a great number of table grapes and not a few wine grapes "which are either little grown outside this continent or are grown in some country where reconstruction is still in its infancy," there "are no data about their affinity with the various phylloxera-resistant stocks" (Blunno 1914, 12).

Unfortunately, even while the experimentation at Howlong was being set up and conducted, many New South Wales growers, frightened by what was happening next door in Victoria, went ahead and grafted over their existing vineyards, or made new plantings with grafted vines, based upon the information from Europe and South Africa. Although the results of this were not entirely unforeseeable, they were nonetheless dispiriting: "The fact is that after more than eighteen years of experiments [at

Howlong] and more than fourteen or fifteen since reconstruction was undertaken on a large scale, vignerons feel that some of the resistant stocks with which many thousands of French vineyards have been reconstructed, are worthless in their country" (Blunno 1914, 14). But such problems were to be expected. After a lengthy and detailed review of the course of death and resurrection in Europe, Blunno concludes that "phylloxera and the subsequent reconstruction of vineyards in the most important vine-growing countries is fraught with difficulties of all kinds, financial, technical, and even affecting public health and public peace" (Blunno 1914, 48).

However, the results were not always negative: "In other cases vines grafted on some other sorts of stocks which grew rather indifferently at first gradually improved in vigour and yield, and now they leave nothing to be desired" (Blunno 1914, 15). In the end, since "the oldest vines on phylloxera-resistant stocks in this country are but ten or eleven years old . . . a few years must elapse before each grower will be able to say for certain that they have proved satisfactory" (Blunno 1914, 24). Ultimately, with the pursuit of an intense and dedicated experimentation program at Howlong, Blunno hoped "to be able, after a sufficiently long period, to find a suitable stock for every variety of grape" (Blunno 1914, 62).[9]

Blunno was not alone in his foresight. Both Victoria and South Australia took measures to examine the phylloxera situation with an eye toward establishing rootstock trials. In Victoria, the indefatigable Hans Irvine, who had championed the rehiring of de Castella by the Victoria Department of Agriculture, argued for "the systematic introduction of phylloxera-resistant American root-stocks at a time when government policy opposed it" (Dunstan 1983, 438). He then set up and maintained a rootstock trial nursery at his own vineyard until his death in 1922, at which point a proper experiment station had been established at Merbein to carry on his work.

Unlike next-door Victoria, South Australia has remained phylloxera-free, certainly in part due to its rigorous prevention campaign, which was initiated on a statewide level by a consortium of government and growers in 1899. But, unsatisfied with a purely defensive campaign, the consortium in 1942 engaged de Castella to review the entire antiphylloxera situation and make suggestions, especially regarding rootstocks. De Castella believed in the inevitability of the arrival of phylloxera, probably because of his long experience in France. Thus, although he agreed that "strict quarantine, that has long been the basic plank of

South Australian anti-phylloxera policy, should be continued, its stringency being in no way relaxed," he also believed that the state should prepare for the inevitable by beginning rootstock plantings and organizing grafting education (de Castella 1942, 5). On this latter point, de Castella's ideas were quite practical, not to mention entertaining: "In order to train the large number of skilled grafters who will be needed, I would suggest the institution of classes and the giving of prizes and certificates to those who qualify, as is already done in South Australia for the training of skilled pruners. . . . Points could be awarded for operational speed and skill, and percentage of successful unions" (de Castella 1942, 7). Certainly grafting would flourish under such a regime!

As far as the rootstocks themselves were concerned, there were important questions to be answered about adaptation, "the suiting of the stock to conditions of Climate and Soil," the most important of which were soil penetrability and the lime content of South Australian soils. Settling the lime question, "in my opinion is the most urgent problem confronting your Board" (de Castella 1942, 9).

Finally, there was the issue of supply. De Castella sharply warned that everything must be done to make sure that there was an adequate supply of rootstocks available when the need presented itself. "Such is the main lesson of Victoria's experience," he argued. "Tardy availability of 'resistants' was the factor which impeded, far more than any other, the progress of reconstitution." Avoiding this impediment "demands the timely establishment of adequate plantations of 'MOTHER VINES' to produce the 'resistant wood'" (de Castella 1942, 7).

As fate would have it, de Castella's belief in the inevitability of infestation in South Australia turned out to be wrong. The state remains a phylloxera exclusion zone today, as does Western Australia, Tasmania, and both the Northern and Australian Capital Territories. But, needless to say, the quarantine measures remain strictly enforced, and, like all other grape-growing states, South Australia "has rootstock source blocks and an industry-run body to distribute cuttings" (Whiting and Buchanan 1992, 20).

One issue that faced each of the states was the selection of rootstock varieties on the basis of their adaptation to the local environment. Becker states the problem well: "Imported stocks will not give you satisfactory results in the long run because in most cases they only perform best in their country of origin" (Becker 1960, 83). This is to say nothing more than that selection pressures brought by both the environment and the researcher integrate in such a strong fashion that rootstock selections are rather precisely tuned.[10] South Australian researcher Coombe described the problem clearly: "A stock which performs well in one country," he said, "may be quite unsatisfactory in another. This is mainly a reflection of the array of growing conditions in different countries" (Coombe 1963, 31). For example, AxR1 and 1202C[11] failed in South Africa, while Jacquez, an Old American, succeeded admirably against all expectation. Consequently, "these facts stress the importance of local testing rootstocks. South Australia's growing conditions are undoubtedly unique as are those of any country, and there can be no substitute for stock trials to gauge our own needs" (Coombe 1963, 32). If this conclusion needed any further proof, the California phylloxera disaster of the 1980s provided it.[12]

SOUTH AFRICA

In the early 1600s, Dutch ships bound for the East Indies would stop at the Cape of Good Hope to resupply with water and food traded with the indigenous people. In 1652, Commander Jan van Riebeeck set up a permanent colony at Table Bay on the Cape. Van Riebeeck "was keen to establish viticulture at the Cape and made energetic efforts to achieve this end" (National Library 2003, 1). Vine cuttings from France were sent to him to be planted in the company's garden in 1655. On 2 February 1659, his labors rewarded, van Riebeeck wrote in his diary, "Today, praise be to God, wine was made for the first time from Cape grapes" (National Library 2003, 1). In 1695 van Riebeeck's successor, Simon van der Stel, constructed the model Constantia Estate, south of Cape Town on the Cape Peninsula. Van der Stel did everything right on the estate, "planting the vineyards according to the established norms of Europe as an example for other wineries at the Cape to follow" (National Library 2003, 1).

Constantia's vineyard was planted, so far as we know, with a variety of muscat grapes, which would have made wines that were very aromatic, and, if they were made properly, with a residual sweetness. The wine of Constantia very rapidly became an international success. Exports grew, and Constantia became the wine of choice in many European royal courts.

Van der Stel was no less a grape enthusiast than his predecessor. Soon after his arrival he toured the surrounding country looking for prime vineyard territory. He found it in the "long verdant valley known as the

Wildebeest," which he promptly renamed Stellenbosch (National Library 2003, 1). Van der Stel was right: Stellenbosch was ideal for vines, and it soon became the center of the Cape wine industry, a role it still plays.

Over the next two centuries the South African wine industry had some terrible ups and downs. Britain took the Cape from the Dutch in 1806, and production boomed, especially when Britain reduced customs duties on wines from the colony. The year 1822 was the Cape's best, but then exports crashed three years later when Britain reinstituted full duties. Growth resumed in the 1840s and 1850s, crashed in the 1860s, and resumed in the 1870s. From 1819 until 1875, the number of vines in Cape earth climbed from thirteen million to seventy million (Giliomee 1987, 39). Unfortunately, success brought with it great peril. As the Vine Diseases Committee reported in 1880, "In no country in the world is there so large a district proportionally, wholly and entirely dependent on viticulture, and which, if destroyed, would involve in abject poverty so large a body of men relatively" (quoted in Giliomee 1987, 39). In Stellenbosch, for example, three-quarters of the cultivated land was in vines, with the remainder in wheat. The Cape Colony lived by the vine, and it nearly died by the vine as the vines died by phylloxera.

Initial reports of the European disaster were not worrisome: "The Cape wine farmers and the government were originally not very concerned about phylloxera because they regarded it as a 'European problem'" (van Zyl 1984, 26). Their attitude changed when the news arrived from Australia: "However, when this plague broke out in Australia in 1875 [sic],[13] causing much damage, the Cape government and the wine farmers took various measures to prevent the importation and spreading of the insect" (van Zyl 1984, 26). Chief among these measures were an embargo on imported vine materials and a rough system of inspection. Unfortunately, these measures were not successful for long, and the dreaded bug was identified in a vineyard in Mowbray, near Cape Town, in early January 1886. This event was soon followed by the discovery of the insect "in vineyards in the Moddergat district, near Stellenbosch—an important viticultural area" (van Zyl 1984, 26). The worst fear had been realized.

When the Vine Diseases Committee made its inquiries in 1880, it had been all too aware of the Australian infestation. With this in mind, it queried its witnesses about the probable consequences of an invasion in the Cape: "The question has again and again been put by the commission to wine farmers, 'What could you do if phylloxera were to appear in your vineyards?' The invariable answer has been, 'Our ruin is certain; our farms are too small and unsuitable for other purposes to such an extent as we could subsist on, and we must trek'" (quoted in van Zyl 1984, 26). This dangerous possibility was clearly in the government's mind when the disease was identified in 1886. Several actions were taken immediately. First, a strict system of inspection, eradication, and quarantine was put into law, and then into action. Follow-up on this was executed in October 1892, when a phylloxera commission was instituted to examine the effectiveness of the scheme. Second, the government immediately founded vineyards for the propagation of American vines—Old Americans—and American-based rootstocks. As van Zyl describes it, "To provide farmers with American vines and grafted vines, the government established nurseries in the Western Cape, among others at Stellenbosch and Paarl, and in distant areas like East London and Fort Cunynghame" (van Zyl 1984, 27). Eight nurseries were established between 1887 and 1889, with plantings mostly of *riparia* and *rupestris* varieties. In the nurseries, research into rootstock affinity and adaptation were carried out, as well as instruction in grafting and new methods of vine growing (Phylloxera Commission 1893b, 6).

Because of an initial lack of American vine material, selections were made from *riparia* and *rupestris* seeds, but they were "unsatisfactory" (Southey 1992, 28). When they became available early in the twentieth century, stocks such as AxR1, AxR2, and 1202C, which were by then standard, were propagated widely and grafted in the field. Unfortunately, AxR1 and AxR2, "in particular -1, were found to be susceptible to phylloxera and their use rapidly declined" in the 1910s and 1920s (Southey 1992, 28).[14] One oddity concerned the Old American Jacquez. Although this variety had been widely planted in France, its phylloxera resistance was site-specific: in dryer soils, it was not fully resistant. Yet, "ironically, despite its poor reputation with respect to phylloxera resistance, Jacquez was the most widely planted rootstock in South Africa until 1960" (Southey 1992, 28). Here again we see the curious effects of site-specific adaptation. A vine that was not well known for its phylloxera resistance here succeeded widely in a phylloxera-rich environment. Another factor encouraging the use of Jacquez was its ease of use, "principally the result of its good affinity and graftability."[15] Essentially all the principal varieties grown in the Cape were easily grafted onto Jacquez.

But while affairs were gearing up for eventual success on the American vine reconstitution, things were not going at all well in the inspection, eradication, and quarantine scheme. The bug was spreading fast, and in the end "millions of vines had to be destroyed and wine farmers suffered great financial loss" (van Zyl 1984, 26). In an effort to assess and, he hoped, make the scheme more effective, the administrator of the Colony of the Cape of Good Hope, Lieutenant General William Gordon Cameron, set up the Phylloxera Commission in October 1892. The commissioners were J.S. Marais, a legislator; Wynand Hofmeyr; and Peter MacOwan. The brief of the commission was wide and exacting: first, "enquiry should be made as to the operation of the Vineyards Protection Acts and as to the scope and effect of the Regulations presently existing for the Quarantine" (Phylloxera Commission 1893a, 5). According to the acts, very intense and invasive inspections—involving the digging up of a significant number of vines on each property, since no other discovery method could be used prior to a vine's observed sickening—were to be made through the colony. When phylloxera were found, a wide swath of surrounding vineyard would be eradicated and the property quarantined. The quarantine was especially onerous: in an infected area "no vegetable produce shall be grown within twenty yards of any vine or vines on any farm or portion of a farm in such proclaimed area, unless previous notice thereof shall be sent, in writing . . . to the Inspector of Vineyards" (Phylloxera Commission 1893a, 9). This ban focused especially upon potatoes and other root produce. Nurserymen were banned from transporting horticultural material into or out of an infested area, and, in general they were severely constrained in what they were allowed to propagate, grow, and sell. Additionally, importation of horticultural material from offshore was forbidden.

Second, the commission's brief held that "enquiry should also be made as to the extent to which *Phylloxera Vastatrix* has spread through the Vine Districts, and as to the damage resulting to the Vinegrowing and Wine-making industry, and further as to what extent the American Vine has been cultivated as a disease-resisting stock, and the best methods for promoting the further cultivation of the American vine" (Phylloxera Commission 1893a, 5). Here the charge to the commission was again wide-ranging and exacting. Clearly, what the government was seeking in this second charge was a thoroughgoing national report of the offensive/positive aspects—as opposed to the first charge's defensive/negative aspects, that is, quarantine—of reconstitution based on American vines. With regard to quarantine and the spread of the insect, the *Preliminary Report* contained nothing but sharply worded criticism of all aspects of the quarantine scheme and fatalistic pessimism regarding the spread of the insect.

Commissioners traveled to every important wine district: Constantia, Stellenbosch, Paarl, Wellington, Worcester, and Ashton. They took evidence from "a number of witnesses—experts and officers superintending the Government American vine plantations, and carrying out the present Quarantine and Import Regulations," and talked also to "several farmers and others affected by these Regulations," including vegetable merchants and nurserymen (Phylloxera Commission 1893a, 7). The findings were disquieting.

First, avoiding a wonderful irony, the commissioners rejected the claims of several witnesses "that the operations carried on in the work of inspection have been the chief cause of the further spread of the phylloxera" (Phylloxera Commission 1893a, 8). Second, they rejected the quarantine as effectively providing any "probable retardation of the phylloxera," which thereby "does in no wise justify the annoyance caused to proprietors, the damage done to the vines, and, above all, the heavy expenditure entailed by the inspection" (Phylloxera Commission 1893a, 8). Indeed, they went on to say, "With reference to the evidence adduced, it appears . . . that the existing quarantine regulations have nowhere been effectually enforced. . . . [In] the infected Moddergat area, all the witnesses examined were agreed that the quarantine regulations were vexatious and detrimental to their interests, and that the manner in which these regulations were carried out offered no protection whatever to their vineyards." But the most depressing conclusion was saved for last:

> Even if all possible precautions were taken, and the strictest enforcement of all quarantine regulations was to take place, the winged form of the phylloxera—the Coloniser—is nevertheless carried from place to place by the wind. In comparison with this means of further infection, against which no quarantine regulations can give any security, the possible spread of the insect by the conveyance of fruit and other farm produce from infected into non-infected areas, sinks into insignificance. (Phylloxera Commission 1893a, 10)

Here is the reason why, unlike Australia, the Cape vignoble was inevitably doomed to destruction by the bug. In Australia the winged form is relatively rare and, more importantly, sexually immature. Moreover, until recently the Australian wine districts were quite separated, making spread by crawlers very slow. But in the Cape, all the suitable wine land is planted, sometimes densely, from one end of the territory to the other, thereby offering solace and comfort to the phylloxera drifting in on the breeze. No wonder that in the space of a decade and a half all the own-rooted *vinifera* vines were gone. There was no hiding place. In the end the committee was forced to conclude that "the extirpation of the phylloxera in this country has become an impossibility" (Phylloxera Commission 1893a, 13).

Within a very short time the battle was over: "The American and grafted vines saved the Cape's entire wine industry from total destruction," and by the first decade of the twentieth century, "the wine industry recovered to such an extent that it already began to suffer from the evils of overproduction." This is to say that, "although phylloxera caused great financial loss to the wine farmers during the period 1886 to 1900, it also had positive results." Just as in France, reconstitution meant rationalization and the greater application of science and technology to the growing of grapes. Plantings were denser, vines were healthier, and the production of better grapes led to better wines, which led to higher prices and rising incomes. Again, just as in France, "contrary to expectation, phylloxera at the Cape was thus not a catastrophe, but a mixed blessing" (van Zyl 1984, 27).

ARGENTINA

Legend has it that the first vine in Argentina was planted in Santiago del Estero in 1557 by a monk who brought cuttings over from the Spanish military settlements in Chile. Although it is difficult to sort out the exact details, the Spaniards brought vines along with them as they conquered vast areas of North, Central, and South America. It is most likely that the path began with the "common black grape" that Cortés brought to Mexico in 1520. The grape continued on to Peru in the 1530s and '40s, and in the 1550s it ended up in Santiago, Chile, from which it traveled over the mountain to Santiago del Estero (MacNeil 2001, 836). The grapes were most certainly of the variety Listan Prieto, now identified as Spain's Palomino Negro, also called "Mission" in California and Mexico, "Pais" in Chile, and "Criolla Chica" in Argentina (Alley 2007). Although the grape's thick skin allows it to flourish in the hot, dry desert climates of these countries, the wine quality leaves much to be desired. Still, it sufficed to supply needs of the clergy and the military alike.

The Argentine vineyard surface expanded over the centuries, principally to meet local needs, since, until very recently, nearly all wine made in Argentina stayed in Argentina. While the Criolla Chica remained the mainstay variety well into the nineteenth century, other immigrant groups brought their own vines with them. French settlers, for example, brought

Malbec from Bordeaux and Syrah from the Rhône; Italian immigrants—
of which Argentina has many—brought the red Bonarda;[16] Torrontés, the
luscious white increasingly grown in San Juan and Mendoza provinces,
came originally from the Spanish Rioja. These special successes of Argentinean winemakers are rounded out by all the usual suspects, including
Cabernet, Chardonnay, and Merlot.

Phylloxera came to the Argentine vignoble early: "The phylloxera was
introduced into the Republic of Argentina in the year 1878 on vines coming from Marseilles, and were not recognized by the customs." By 1895,
"the insect inhabited roots in Belgrano, San Martin, Moreno, Chacarity,
Caballito, and around La Plata and for certain in Bahia blanca" (Comisión de Estudio 1938, 5). These areas are all within the Buenos Aires region,
far from the major vine-growing areas of Mendoza and San Juan. The
spread of the bug was at first quite slow, taking nearly three decades to
move from the Federal District into the major vineyard area. Underlying
this initial slow spread was the fact "that no winged forms have been
observed in Argentina, and that infestation spreads only in the partheno-
genic and wingless forms" (Cesar 1948, 80). As we have seen in Australia, lack of winged forms restricts long-distance travel by the bug to some
sort of inadvertent "hitchhiking," usually on agricultural products or
equipment.

Against this sort of transport, the government instituted the usual quarantine laws, beginning with Law 2384, of October 1888. But "through
experience," "the inefficacy of legal means to prohibit the introduction
of vines from the infested zones" was demonstrated (Comisión de Estudio 1938, 10). Sooner or later, infested material would bring the plague
to the vignoble.

"In the year 1929 the first focus in the province of San Juan was observed," which would begin the flood. Provincial authorities estimated
that 1,497 hectares were infected; the area doubled by the 1930–31 season, and by 1937 the area "the commission considers phylloxerated at
the present in the province is 15,000 hectares" (Comisión de Estudio
1938, 86). Obviously, the insect was moving with breathtaking speed.
But since there was no winged form, how was it traveling?

Argentina's desert climate turns out to be a curse as well as a blessing.
Because of the dry air and the lack of rainfall during the growing season,
most of the pests—both fungal and faunal—do not subsist in numbers
great enough to threaten the health of the vines in any way. But, on the
other hand, because of the dry air and the lack of rainfall during the

growing season, the vines can only survive via irrigation. The Cuyo region, comprising the provinces of San Juan and Mendoza, constitutes 88 percent of the Argentinean vignoble, and it is 98 percent irrigated. Unfortunately, "Irrigation is a most important carrier of Phylloxera in Argentina," and this importance "is increased by the methods of irrigation used, and by the fact that many vineyard soils crack readily, thus facilitating the movement of the insect" (Vega 1956, 72). Under these conditions, Argentina turned out to be even more hospitable to the bug's colonization than France!

Faced with an exploding problem, the Ministry of Agriculture instituted the Comisión de Estudio del Problema Filoxerico (Phylloxera Study Commission) on 31 July 1937; the commission gave its formal report on 21 October 1937, in San Juan. In many ways, the report seems otherworldly, invoking déjà vu in the strongest possible way. The commission is tasked to "advise the minister of agriculture of which means to adopt in the province of San Juan to rationally combat the phylloxera of the vine, and to establish the basis on which to found a national organization for this purpose" (Comisión de Estudio 1938, 1). What is most peculiar is that the commission was formed fifty years after the first sightings of the bug in the country. Then, in its proceedings, the commission looked at every method tried—successfully or not—in Europe, Australia, California, and South America. In addition to strict quarantine, it examined insecticides (CS_2), submersion, sand, and, of course, grafting.[17]

Insecticides were out: "The modern orientation of the antiphylloxera battle *discards as unuseful and onerous* the employment of carbon disulfide . . . and counsels its use only when the indications are this use will be optimal for the viticulturalist" (Comisión de Estudio 1938, 8). Submersion was tried in both Mendoza and San Juan. After fifty-five to sixty days of twenty-centimeter inundation, 95 percent of the bugs were killed. However, "in the province of San Juan it is not easy to obtain irrigation water in the great quantities needed for the submersion treatment" (Comisión de Estudio 1938, 9). Even worse, "submersion is difficult to implement in Cuyo because of the slope of the land and the soil's permeability (with a consequent leaching of nutrients)" (Vega 1956, 74). So, even if there were enough water, it could not be used. Sand planting was just plain impossible: "There are no antiphylloxeric sands in Argentine vineyards" (Vega 1956, 77).

So, as always, the solution was clear: "We advise the grafting of vines of European origin on stocks of American species or their hybrids *as the unique solution* for assuring the new vineyards from the consequences of this frightening plague" (Comisión de Estudio 1938, 9; emphasis in original). Immediately entrained was the entire apparatus of the grafting scheme: trials, nurseries, local testing, and experiments with affinity and adaptability. The commission was clear-eyed about it: "It is indispensable that numerous trial vineyards be installed in order to establish the adaptability of those rootstocks to the provinces and territories that are dedicated to the cultivation of the vine and discover which are most convenient for grafting on the principal varieties of *vinifera* cultivated there" (Comisión de Estudio 1938, 88). In addition, as Cesar later pointed out, the stocks should be grown, tested, and selected directly, right in the middle of the infected areas, in order to ensure that the selections would be "adapted types" (Cesar 1948, 80).

After some years suitable stocks were found, and the affected areas reconstituted. But, in an extremely lucky twist of fate, some of the more important districts in Mendoza, cut off from the infected areas and using their own systems of irrigation, have remained free of the bug until the present day.

SOUTHERN LESSONS

In many ways, the phylloxera invasion in the southern hemisphere led to the same consequences that it did everywhere else. Growers panicked, governments blundered around, and the scientific establishment slowly but surely took the necessary steps. The European experience was found to be a useful guide for those organizing combat against the bug. Yet, for all the similarity between the experiences of the different regions, there were crucial differences. In South Africa the bug luxuriated in a full life cycle, flourishing both above and below the ground. As a consequence, the insect dispersed itself rapidly and successfully. Australia and Argentina, on the other hand, confronted a bug that languished in place, its mobility dependent upon man's helping hand. In Argentina the irrigation system provided a widespread and easy system for the hitchhiking bug. Down Under, once the threat was recognized—and, more importantly, the bug's mobility limitations understood—scrupulously applied quarantine and cordon sanitaire regulations allowed Australia to become (and remain) the only infested vignoble to successfully manage phylloxera's expansion by exclusion. Whether, in a time of rapidly expanding vineyard surface, this regime can be successfully maintained is an empirical question that can only be settled as time goes by.

Once again, even though the invasion differs, the phylloxera experience—the culture of reaction—is just the same in each of the distinct vignobles that it invades.

CHAPTER 7

The Old Americans, or How the Fox Conquered Europe

"But their wine is undrinkable," or so said Leo Laliman when he first alerted the wine world to the phylloxera resistance of the American vines. Yet the thirsty French vignerons learned to drink these undrinkable new wines soon enough. Within a few years the original American vines—the Old Americans—had spread throughout the south of France and were rapidly expanding their territory to the south and east.[1] In the end, the Old American vines could and still may be found all the way from France's Atlantic coast in the north, to Greece's Aegean coast in the south.

These vines were originally selected—or, in some cases, bred—for adaptation to local growing areas stretching from America's Northeast to the trans-Mississippian west, from Texas to the Carolinas. This is a vast expanse of land, with enormous variation in climate and terrain. The Old Americans exhibit a similarly enormous variation in their adaptations. Yet one characteristic predominates: nearly every one of these vines includes in its makeup features derived from the American species *labrusca*. This species originates in the forested areas of the Northeast—Pennsylvania, New York, and Massachusetts, among others—where the winters are cold, the summers warm, and humidity is frequently a problem during the growing season. Unlike most American species, *labrusca* grapes and clusters can be of a good size, and although their sugar levels when fully ripened are low compared to that of European grapes, their acid levels are quite manageable, unlike other American natives.

But coupled to these generally positive traits is one huge negative: that *labrusca* taste! Variously described as "foxy," "musky," "raspberry" (by the French), or "strawberry" (by the Austrians and Italians), it is without question strong and unmistakable; the taste that North Americans know from grape juice, grape gum, and grape jelly is the *labrusca* flavor, the taste of the fox.[2] Needless to say, this flavor was unfamiliar to the French and other Europeans; indeed, not only was it unfamiliar, but in wine they found it distasteful. It was that aspect of the *labrusca*-based grapes that made their wine, as Laliman declared, undrinkable. But, faced with the prospect of either drinking Old American wine or none at all, the French and other Europeans adapted quickly.

Many of the Old Americans—although this was unknown to anyone at the time—were in fact accidental hybrids of American natives with European *vinifera* varieties. Even though the centuries-long effort to grow European vines in America inevitably failed, many of the efforts lasted long enough for local native vines to be pollinated by the Europeans. Such hybrids, when they were successful, were obviously attractive to locals and were selected and tried extensively.[3] In this way, many of the Old Americans' labruscoid properties were attenuated, in one way or another, by their European "blood."

One extremely important pair of vines, Jacquez[4] and Herbemont, were not *labrusca*-based hybrids. Indeed their origin was—and remains—controversial.[5] Of all the Old Americans, these two vines most resembled traditional European vines in the flavor and quality of their wine, although the appearance of the vines themselves was clearly foreign.[6] Since both of these vines were fairly well adapted to the climate and soils of many of Europe's Mediterranean vignobles, they were extensively planted throughout those regions.

Some of the Old Americans, including, most importantly, Concord and Catawba, did not have much phylloxera resistance. Only in certain soils could these vines prosper in the face of the insects' attacks. This weakness was the negative aspect of the European "blood" that the two varieties contained, since it was their European genetic complement that weakened them in the face of phylloxera attack.

In the end, a suite of six Old American vines proved adaptable and productive enough to satisfy the needs of a vast region stretching from the French Atlantic to southern Greece. Among the whites, Noah was the most widely accepted; Clinton and Jacquez were the most widely planted reds. Isabella, another red, was very popular as a landscaping item as well as a wine producer, although its strong *labrusca* taste took some getting used to. All of these vines, in addition to sufficient phylloxera resistance, showed useful resistance to the other imported American scourges as well. All, for example, could thrive without spraying against the dreaded powdery mildew. In many ways, except for their taste, the Old Americans were ideal replacements for the traditional European vines in that dangerous new environment emerging throughout the continent in the 1870s.

Dealing with the *labrusca* taste was a complicated matter. In the case of the whites, it was often a question of picking the grapes before they were ripe, since the peculiar muskiness did not become overpowering until the final stages of ripeness. Once it was discovered that sugar could be added to a fermentation to increase its strength, the alcohol levels of these white wines could be adjusted to desirable limits. Reds were a different matter, however, and more difficult to handle. While the Old American red *labrusca* grapes had plenty of color when fully ripe, they also had plenty of foxiness at that point, too. If the reds were treated as the whites were and picked before reaching full ripeness and full flavor, then their color would be unsatisfactory. Most vignerons evidently attempted to control the flavor intensity by cellar manipulations, especially the use of excess oxygenation via frequent rackings, during which the wine was moved from one container to a new, clean one full of air (Galet 1988, 377–79). This is a dangerous procedure, since most oxidized red wine has off colors and funky flavors.[7] However, if the rackings were performed early in the life of the new wine, the dangers were lessened.

But, for the most part, people simply got used to the strange new flavors. And by the time that the second-generation winemakers came along, many vignobles considered the Old American flavors to be entirely appropriate, indeed, as part of their patrimony. As we will see below, this shift in attitude is today producing the ultimate irony in many wine regions of Europe: the Old American varieties have become part of the terroir, part of the tradition, and attempts are being made to resurrect and recognize their production and sale against the prohibitions of the European Union.

The Old Americans became important contributors to wine production all over Europe. At their peak in France, in 1935, more than sixty thousand hectares of vines produced about three million hectoliters (Galet 1988, 377). In Italy, the Veneto had hundreds of acres of Clinton (Prial 1993), while Alto Adige produced thousands of gallons of "strawberry" wine from the *"uva fragola"*—an Old American *labrusca* (Mori 2008). In the Trentino region, the Isabella—both black and white versions—were widely planted, both for the production of strawberry wine and for arbors, which adorned nearly all homes in the region (Fontanari 2004, 45).[8] In Central Europe, principally Austria, Croatia, and Hungary, thousands of acres of the Old Americans were planted. In Austria, in particular, a wine analogous to, and as famous as, Italy's Fragolino came to be prized: the Uhudler,[9] made from Isabella, Noah, and Othello, among others (Kellerviertel Heiligenbrunn 2008). Finally, when the Old Americans reached Greece, they were adopted immediately for their wine but were even more prized for their use in arbors. No home, farmhouse, or taverna was complete without a splendid arbor of Isabella, Clinton, or Othello.

In the end, regardless of their supposed "undrinkability," the Old Americans were cultivated from one end of Europe to another, supplying millions of gallons of wine, and uncountable acres of cool shade, to a grateful population. But the European governments evidently were not so grateful.

ASSAULT ON THE OLD AMERICANS

From the very start, the official French position was that if a vine had any American "blood," its wine would be of a lower quality. Indeed, the reckoning was that there was a direct relationship between American "blood" and poor wine quality. American vines brought the phylloxera disaster to France, and they would forever be guilty of this sin. In the mind of official France, American vines remained suspect, and their wines inferior. This idea persists in Europe, where it has been given the force of law.

For its part, Germany outlawed American vines—and, indeed, any hybrid vines containing American "blood"—during the *Nazizeit*, the Nazi era, beginning in 1934 with laws against miscegenation—the mixing of species—in grapes. Hybrid grapes, including the Old Americans, were totally excluded from German vineyards, with severe penalties for transgressions.[10] Needless to say, the American vines disappeared instantly from the vineyards of Germany.

Of course, these political tendencies were aided and abetted by economic forces. By the late 1920s, France was suffering from a wine glut of enormous proportions. In part, the glut was due to the accelerating production and importation of Algerian wine, which had nearly doubled during the decade. But equally disastrous was a wine glut in mainland France due to the increase in production brought about by the scientific improvements during the reconstitution after the phylloxera disaster. The government in 1927 suspended new plantings in the high-quality Appellation d'origine contrôlée areas, but it had no means to enforce the edict (Galet 1988, 377). After finally putting into place regional inspectors operating under the auspices of the Institut National des Appellations d'Origine, a decree was published in January 1934 outlawing certain varieties—Noah, Isabelle, Clinton, Othello, Jacquez, and Herbemont, the principal Old Americans—demanding that they be uprooted.[11] Since it was impossible for the government to enforce the edict, given its laws and workforce at the time, it was further required that all vignerons report annually the amount of wine made from the prohibited varieties.[12]

Of course the vignerons had no interest in ripping up their vines. Inputs and the costs of production were minimal: the Old Americans didn't need to be sprayed, they didn't need to be fertilized, and they were completely comfortable being left alone to produce a decent crop year in and year out. And after all these years, wine drinkers had not only gotten used to the foxy-raspberry taste, but in many cases they preferred it. Over the next several decades there was a struggle back and forth between the huge population of small vignerons—usually freeholders practicing poly-cultural agriculture—and the government, with the government always prohibiting and sometime enforcing the prohibition, and the vignerons always hanging onto their vines for dear life.

In fact, by 1953 there were more Old Americans planted in France than there had been at the time of the original prohibition in 1934. At that point, the government once again girded its loins for battle, this time using the carrot: 1,500 francs were offered to vignerons for each hectare uprooted. But after three years, when only 30 percent of the remaining vines had been pulled out, the carrot was exchanged for a stick: a fine of 3,000 francs per hectare, plus, in cases of recidivism, imprisonment from ten days to three months, depending on the severity of the case. Back and forth things went. By 31 December 1963, the original 62,000 hectares of prohibited vines had been reduced to 17,837 hectares, at least according to the official numbers. But this was still too many hectares as far as the government was concerned, so a new law was promulgated. The new law of 1964 quite strictly prohibited traffic in wine from the cépages prohibés anywhere but to the distillery. However, there was an exception made "for viticulturalists aged sixty-five or older, who possessed fewer than twenty-five acres of vines, and who commercialized no part of their harvest" (Galet 1988, 380). This exception expired with the viticulturalist in question.

By the time of the great vine census of 1968, there remained only 8,585 hectares of prohibited vines.[13] Galet ends his discussion of this fascinating chapter of the phylloxera story with the remark that "at present [1988] only a few isolated vines in gardens or in arbors standing before homes must still exist" (Galet 1988, 380).

His pious thought could not be more wrong.

FROM OUTLAW TO IRONY

In many areas of France, particularly the fiercely independent Vendée (in the west, between the Loire and La Rochelle) and the isolated mountain country of the Cévennes-Ardèche (west of the Rhône, between Lyons and Montpellier), the outlaw Old Americans not only exist, but they flourish under loving care. One of the underground delights of the Vendée is its *pineau de Noah,* an eau-de-vie, or brandy, distilled from Noah wine. Noah, of course, is one of the *cépages prohibés.*

American financier Bill Bonner describes his introduction to this beverage, in the cellar of his friend Damien in the Vendée:

> "Ah . . . here, you have to try this . . ." he said, turning to a smaller barrel, covered in cobwebs. "I've been saving this for more than 10 years. This is made with the 'noah' grape. You know, it's illegal to grow the vines in France and illegal to make the wine. But it's really good."
>
> We tried it.
>
> "Hmmm . . . Damien, this is very good," we replied.
>
> We drank our glass . . . continued our conversation . . . and then thought to ask.
>
> "Why is this illegal?"
>
> "Oh . . . they say it drives people crazy. I don't believe it. I know plenty of people who drink it. They're not crazy. Well . . . come to think of it . . . they are crazy. Every one of them. But it could have been something else that made them that way." (Bonner 2007)

References such as this are not at all uncommon. On a British website offering upscale vacation homes with pools in the Vendée, the author speaks about the wines of Noah, the white, and Oberlin, the red:[14] "It is illegal to grow and sell these grapes so I'm not sure where you can get to try them out other than to quietly ask the older men in the small villages. If they like you you may be lucky" (Gites with Pools 2007). The famous *pineau* is also remarked: "I have never heard of these being available commercially though if you go to a dinner party with the locals you will often be given a glass as an aperitif. These Pineaus are exceptionally fruity and the locals claim the finest is made from the Noah Grape. It seems this method does not work so well with traditional grape varieties" (Gites with Pools 2007). Here we see clear evidence of a distinct preference for the "undrinkable" Old American over the traditional French *vinifera* varieties. Those foxy grapes can do some jobs better than their subtler European cousins.

In the Ardèche, it is not just the foxy Old Americans that have hung on; it is also Jacquez. "Jacquet," as the variety is called locally, is perfectly matched to the region and terroir: "In the sunny valleys of the Cévennes region, the stalks and broad, decorative leaves of the Jacquet vines, over 100 years old, blend beautifully with the dry stone walls that overhang the rivers. The optimal conditions of sunlight and sloping exposure allow for the development of a colorful, rich, flavorful wine" (Vignes d'Antan 2008).

The situation in the Ardèche is fascinating. In this landscape of rocky escarpments, clear-running streams, and terraced plantations, the Jacquez grape has come to occupy a central, indeed crucial, role. Here this vine is uniquely adapted to the terrain and climate and dependably produces a sturdy, rustic, and correct table wine. For the Cévennes, Jacquez has become the major provider of that lifeblood of French culture, wine.

But, of course, since 1934 the government has more or less seriously attempted to tear up the Jacquez plantings. Yet the folks of the Cévennes have never submitted to the officials from Paris. Their countryside is hard to reach and difficult to inspect. It is much easier for officials to leave it alone, which Paris, for the most part, has done. So there Jacquez still flourishes, alongside a movement to restore the Old American to glory.

In 1970 the French government created the Parc National de Cévennes in the region, located chiefly in the departments of Lozère and Gard. Occupying a total area of 910 square kilometers, the park was instituted to bring ecological and cultural stability to a region of unique environments, both natural and human. French national parks are singular in that they not only focus on unique terrains, but they attempt as well to capture unique ways of life, and, in the end, to preserve both. But this brings conflict with the wider world. Certain of the folkways preserved in the parks are not consistent with regulations promulgated from Strasbourg and Brussels, headquarters of the European Union. This is painfully evident in the case of the Old Americans, and especially in the

Cévennes, for Jacquez. According to EU dogma, American "blood" in a vine necessarily lowers the wine quality. Consequently, Old American vines are, from the start, considered to be of low quality[15] and thus must be extirpated, no matter their history and place in the local culture. But, in a splendid irony, with the founding of the new national park in the Cévennes, cultivation of Jacquez has become a nationally protected folkway. Park community members were quick to seize upon this opportunity to preserve their local vines and wines. An informal group was formed soon after the opening of the park, and in 1993 it became the official association called Mémoire de la Vigne (Vignes d'Antan 2008).[16] Since that time, this group has been very active, not only in the vineyard and winery, but also in politics, lobbying for the reversal of the EU laws against all American hybrids and the 1934 French prohibition of the Old Americans.

There have been some useful results of their activities. First, and most interesting, although commercialization of the Old Americans is strictly illegal, because of the patrimonial status of the association, members of the association are allowed to purchase cases of the wine. This has stabilized production, which is a benefit, even though production remains at a much lower level than it was before (Galet, quoted in Vignes d'Antan 2008). Additionally, this more or less official recognition of the wine and its production has led to some influential exposure of the situation. Writing recently in the *International Herald Tribune*, Thomas Fuller described the history and plight of the Ardéchoise, spoke about the lobbying efforts, and even talked about the wine as if he had drunk some.[17] Most important, as Fuller reports, Pierre Galet, "perhaps the world's leading expert on grape varieties," "believes the ban on American vines is anachronistic" (Fuller 2004).

Adding to the weight against the ban are two other works, one popular, the other scientific. The first is Freddy Couderc's *Les vins mythiques* (The vines of myth), a paean to the wines of the grandfathers: "Clinton, Jacquez, Isabelle, Concord, to cite only these, are above all names which resonate in a quasi-magical fashion in the collective imagination of the Cévenne ardéchoise et du Bas-Vivaris" (Couderc 2000, back cover).[18] The first printing of this beautifully written book sold out almost immediately, and it has gone into a second printing. Its theme, that the Old Americans are now as much a part of the patrimony of the Cévennes as any vine is or could be, is a powerful one, and, indeed this theme is also a major premise in another work supporting the old Americans, Herminie Piernavieja's master's dissertation at the Université du Vin, Suze-la-Rousse.

The dissertation is entitled "Le Jacquez, un cépage charge d'histoire: son adaptation au terroir cévenol et les enjeux de son maintien" (Jacquez, a history-laden grape variety, its adaptation to the Cévennes terroir, and what is at stake in its survival there) (Piernavieja 2005). In this work she provides a detailed history of just how Jacquez came to the Cévennes, and how it turned out to be exceptionally well adapted, both to the terroir and to the aesthetic sensibilities of the locals. Then, in the scientific parts of the dissertation, Piernavieja directly confronts the official canard used to denigrate the Old Americans since the time of the 1934 ban: that wine from these vines carries a toxic dose of methanol.[19] The author's laboratory tests show definitively that Jacquez is not only relatively low in methanol, it is also quite a bit lower in methanol than some of the revered traditional *vinifera* red varieties (Piernavieja 2005, 98).

Piernavieja's concluding argument is strong and pointed. Even if there are general rules against the Old Americans in the European Union, the terroir of the Cévennes has been specifically chosen as a special place via the founding of its national park. Since, as her dissertation demonstrates conclusively, the history and adaptation of Jacquez to the terroir inexorably leads to the conclusion that this unique *cépage* is part and parcel of this unique place and its folkways, for this reason alone, the French bureaucrats should exempt Jacquez from the laws made in faraway offices in Strasbourg and Brussels. It is clear that the dispensation on sales to members of the Mémoire de la Vigne, no matter how grudging, is the first significant step in that direction.

The Cévennes is not alone in this battle. In Austria, the traditional wine made from Old American grapes, Uhudler, has recently been granted a similar dispensation. Although this unique wine is not allowed to be trafficked in the open market, several wine-making districts, particularly the Südburgenland, have moved toward the same solution as in the Cévennes. In 1992 was formed the Verein der Freunde des Uhudler (Association of the Friends of Uhudler), among whose members production and sales of the Old American wines is allowed (Uhudler 2008). Since advertising is permitted, and membership is open, it is clear that the future of these wines is not only safe, it is promising.[20]

In Italy, *fragola* wine has not yet achieved such protected status, although it is likely such a day is not far away. In a recent issue of the glossy official government magazine of Italy's Trentino region, the authors strongly encouraged the home planting of Isabella in order to make fresh juice, dishes for the table, and wine (Fontanari 2004, 46).

And finally, as French vine expert Galet notes, many of the newly admitted members of the European Union have large plantings of the Old Americans. It is quite impossible that these countries can be compelled to rip out their plantings just because France had an overproduction problem in 1934: "You can't tell the Hungarians, Bulgarians, and Romanians to uproot their vines," Galet avers (Fuller 2004). Old Americans, despite seventy-five years of official persecution, are in Europe to stay. Certainly this particular consequence of the phylloxera disaster is by now a permanent feature of the European wine patrimony.

CHAPTER 8

Phylloxera Breaks Out (Twice) in California

Grapes came to California early: Spanish padres brought vines with them from Baja California when they founded the first Alta California mission, Mission San Diego, in 1789 (Davidson and Nougaret 1921, 3). Undoubtedly the grape variety would have been the Mission, since it was the only variety planted at missions that were founded earlier and later.[1] This variety is enormously adaptable, flourishing equally well in the Central Valley, the Foothills, or along the coast where the missions were founded. Its wine is low in acid and alcohol, but there is lots and lots of it, due to Mission's rugged fruitfulness. Very shortly after its introduction, vineyards of Mission grapes were spread throughout the territory. But not all Mission wines were equal. In the 1840s, General Vallejo claimed that "the Mission grapes grown at the Sonoma Mission were of a better quality than those grown at the other missions in California"; moreover, and more importantly, "a recognized superior quality of wine was made from them" (Davidson and Nougaret 1921, 3).

Davidson speculates that "it was probably because of the reputation that the first commercial vineyards of wine grapes were established in the vicinity of the town of Sonoma" (Davidson and Nougaret 1921, 3). But what was lacking in Sonoma—indeed, throughout California—were better varieties of wine grapes. A solution to this problem came about in an interesting fashion: competition between Vallejo and Agoston Haraszthy to plant the finest vines and make the finest wine in the Sonoma Valley (Robinson 2006, 123). Haraszthy himself is the stuff of legend, as is his involvement in California's nascent wine industry. A Hungarian-born noble, Haraszthy immigrated to Wisconsin in the late 1840s.[2] He was enticed to California by the Gold Rush and led a wagon train to San Diego, where he immediately took up real estate development, politics, grape growing, manufacturing, and who knows what else. Several years later he moved to the San Francisco Bay Area, where he tried growing grapes in the city and San Mateo County (both locations failing because they were too cool). Finally he moved to Sonoma, where he reprised his San Diego burst of activity, eventually focusing much of his effort on developing vineyards of high quality on the hillsides around the town. Haraszthy, along with his partner Henry Appleton, planted three hundred thousand vines in 1857–58,[3] fifty acres in 1860, and, in 1862, "seventy thousand European vines" (Davidson and Nougaret 1921, 4–5). These vines, from all available evidence, were taken from among the one hundred thousand cuttings of 315 different varieties Haraszthy had brought back from Europe the preceding year at the behest of California governor John Downey (Robinson 2006, 337).[4] It was in this Sonoma vineyard that phylloxera first showed signs of life in California (Davidson and Nougaret 1921, 4).

PHYLLOXERA APPEARS

Phylloxera probably infested some California vineyards in the late 1850s, but its ravages were identified only as "a disease of vines from unknown causes" (Davidson and Nougaret 1921, 4). That the bug's presence and responsibility remained unknown in this period is completely unsurprising: after all, Planchon and his colleagues didn't locate and characterize the culprit until 1867. The first definitive identification of the insect in California occurred in 1873, when H. Appleton and O.W. Craig found, on the roots of a vine of the latter, "the insect, or louse, known in Europe by the title of phylloxera vastatrix, and in the United States as pemphygus vitifoliae" (Appleton, quoted in Davidson and Nougaret 1921, 4). Once the identification had been made, earlier problems in Haraszthy's vineyards were explained. During the period of massive expansion of his vineyard beginning in 1860, Haraszthy had had problems with dead and dying vines. During 1868 about three acres of vines were uprooted and the area replanted. Over the next four years even these newly planted vines sickened and died, and, as Appleton later noted, in retrospect, "all the symptoms were observed of vines dying from the vine pest" (Appleton, quoted in Davidson and Nougaret 1921, 4). In 1873 it was thus clear to Appleton, Craig, and, of course, Haraszthy himself that California in general, and Sonoma in particular, was bound to have big problems with the bug.

One question, of course, was how did the bug get to California in the first place? At that time it was easy to associate the arrival of the bug with the arrival in 1861 of Haraszthy's shipload of vines and cuttings, particularly those from France (Bioletti 1901, 3). But this association did not make sense, as several workers noticed. In the first place, there were clear signs of the bug's destructive work already by the late 1850s in Sonoma vineyards. Second, had the bugs come in on Haraszthy's imports, there would have been a diffusion of the disease wherever the imports had been distributed, which was not the case. Finally, there was a simpler explanation available: phylloxera had come to California from the same place that had spawned the French phylloxera—the Eastern United States, on imported Eastern vine material.

Davidson found clear evidence that there was a planting boom of Eastern varieties—Old Americans—during the 1850s, propelled by such blandishments as this one, for Catawba, published in the weekly *California Farmer* on 23 January 1855: "We sincerely esteem the Catawba grape, one of the very best varieties for cultivation in California. Longworth of Ohio, whose famous Catawba Champagne is now esteemed equal to any wine imported, says it is the very finest wine grape known. Will be found far superior to our California Grape, We earnestly urge our cultivators to give the Catawba a careful trial" (cited in Davidson and Nougaret 1921, 5). Walker notes another source, namely, settlers coming to California from the East, bringing rooted, and perhaps even potted, vines along with them as they traveled by wagon train across the country (Walker 2000, 210). In the end, the presence of Eastern vines was significant in wine-growing areas. For example, by 1875 a vine census in the Gold Country counties (including Amador, Placer, and Calaveras) counted no less than 23 percent American varieties among the vines counted (Davidson and Nougaret 1921, 7).[5] There can be no doubt that such an influx of Old American vines would have introduced the dread bug.[6]

But no matter where the phylloxera came from, California was stuck with the bug, then and forever. The next issue, then, was what to do next.

At first, not much was done at all. For whatever reason, the spread of the bug seemed to be fairly slow beyond its initial infection sites. Only later was it discovered that the winged form—the "colonizer," as Bioletti terms it[7]—did not occur in California, which meant that long-distance infestation via wind-assisted flying insects was not to be a problem (Bio-letti 1901, 4; Pinney 1989, 343). The slow pace of expansion lulled growers into a dreamlike false security. As Pinney describes it, "for years growers in the afflicted regions had pretended that the threat was not serious, or that it was under control, or that it did not exist, or that it would go away by itself" (Pinney 1989, 344). "A strong contributing factor was the enduring belief that growing conditions in California were different enough from those in Europe" that the course of the phylloxera would be different as well (Wolpert 1992, 7).[8] Exactly as their colleagues in France believed, Californians evidently also believed "It can't happen here." But, exactly as their French colleagues learned to their dismay, not only *could* it happen here, but it *would* happen here.

In Sonoma, although the winged form of the bug was rare,[9] the terres-trial form was perfectly adapted to accepting help from humans, especially by inadvertently hitching rides on vineyard equipment, vine materials, and, as Davidson shows conclusively, picking boxes. These contain-ers, fashioned from wood, featured many joints, breaks, and gouges; in short, they were perfect hiding places for the nearly microscopic devasta-tor. Since filled boxes were carted to the wineries from vineyards during harvest, and then, once emptied, parceled out indiscriminately to other vineyards for the return, it was easy to transfer a hidden horde of insects from an infected vineyard to an uninfected one: "Certain grape growers have noticed that the first signs of phylloxera in their vineyards appear at places where they have been in the habit of dumping boxes for the conve-nience of grape pickers" (Davidson and Nougaret 1921, 10). Such disper-sal, although slower than that via the winged colonizers, was effective enough to build up a sizable infected area in a few years. Thus by 1880 Sonoma had more than six hundred destroyed acres, and "growers had to face the fact that they were in serious trouble from phylloxera" (Pin-ney 1989, 343). At that point, the state decided to act.

During that year the legislature set up the State Viticultural Commission, established the Phylloxera Board, and instituted the Department of Viti-culture at the University of California in Berkeley (Walker 2000, 210). The Phylloxera Board got busy immediately: "It surveyed the infested areas; it made and published translations of the standard French treatises on recon-stituting vineyards after phylloxera attack; it tested the innumerable 'reme-dies' that had been hopefully proposed since the outbreak of the disease in France in 1863" (Pinney 1989, 344). The university's first professor of agri-culture, Eugene Hilgard, already busy on viticultural projects, now became point man for the newly authorized Department of Viticulture, making a series of appearances around the state at various venues. In early April he appeared before the Sonoma Vinicultural Association, and, after he advised "the wine-growers of the State to turn their attention to the production of wine of a superior quality, rather than to the manufacture of great quantities," the meeting turned its attention "to inquire into the best plan for ridding the vines of the phylloxera" ("War on the Phylloxera" 1880, 3). A committee composed of Hilgard and Charles A. Wetmore was appointed, and "Mr. Hilgard was requested to secure the services of one of the students at the Agricultural College, who is to travel over the State examining all vineyards in the search for this destructive pest, and report the results of his examination" ("War on the Phylloxera" 1880, 3). The association gave Hilgard $150 toward the project and commissioned him to petition the regents of the university to do the same. Thus was born "the first systematic attempt to rid the vines of this destroying insect" ("War on the Phylloxera" 1880, 3).

And phylloxera truly was a destroying insect in California. What happened in the Santa Clara Valley is a classic cautionary tale. This lovely valley, which runs from the south end of San Francisco Bay up to the hills of the Coast Range halfway to Santa Cruz, and today is home to Silicon Valley, was a paradise for fruit trees and grapevines. From 1885 to 1895, which marked the period of most rapid expansion of its vineyards, Santa Clara produced almost one third of the dry wine in the state. But the vines began to die in 1893, and the decline was so rapid that it outpaced the rate of earlier growth. According to Davidson, "phylloxera was responsible for a far greater share of the destruction of the Santa Clara Valley vineyards than has been ascribed to it" (Davidson and Nougaret 1921, 15). Similarly bad things were happening in Napa: between 1889 and 1892 roughly 10,000 acres were destroyed; by 1900 only 2,000 bearing acres remained in the county (Pinney 1989, 34). Over the whole territory of the state during the forty-eight-year period 1866–1914, 90,000,000 vines were planted—approximately 150,000 acres—and estimates by the USDA's George C. Husmann put the loss to phylloxera over that time as nearing 75,000 acres. Professor Bioletti of the University of California's Department of Viticulture makes a similar estimate, while Charles Wetmore, of the Board of Viticultural Commissioners. "considers this estimate conservative" (Davidson and Nougaret 1921, 15). Thus, according to these estimates, California lost nearly half of its vineyard planted during this period to the devastating bug. There was no question that something needed to be done.

FIGURE 19. George C. Husmann.
Courtesy Gail Unzelman.

COMBATING THE BUG

The four remedies—insecticides, flooding, planting in sandy soil, and the use of resistant rootstock—ultimately played a large part in focusing the efforts of the state and the university in combating the bug. Yet, although Wetmore's 1882 initial report of the State Viticultural Commission mentioned the four methods, no serious testing followed (Wolpert 1992, 4). Then, a full nineteen years later, Bioletti reprised the four with some care, revealing the freshness retained by the subject over the years. Bioletti's report starts with a very interesting motivational paragraph, well worth quoting in full.

> The late discovery of the phylloxera of the vine in several important grape-growing districts which have hitherto been regarded as exempt has awakened widespread interest, and even alarm, among both wine- and raisin-producers.[10] In order to satisfy the demand for information regarding this serious enemy of the vine, to allay the alarm of those who exaggerate its menace to the industry, and at the same time to rouse up to prompt and intelligent action

FIGURE 20. Professor Bioletti demonstrates pruning a vine in 1923. Courtesy Special Collections, University of California, Davis.

those who are inclined to minimize the danger, the following brief account is issued. (Bioletti 1901, 3)

The first and most amazing point about this introduction is that it is even necessary! Twenty-eight years after the first discovery of the bug in Sonoma, after thousands of acres in Santa Clara have been destroyed, and after Napa has almost been wiped out, Bioletti still finds it necessary "to rouse up to prompt and intelligent action" those involved. Somehow, it is as if it were all new again.

Bioletti's discussion is straightforward. He begins by dividing the infestation zones into various categories, and then he makes recommendations for behavior in each of the zones. First is the uninfested zone. Excluded from this zone is not only all material related to the vine, but also "all other plants, such as nursery stock, potatoes, etc. which are taken from below the ground" of infested areas (Bioletti 1901, 7). This regime is exactly the same as that already in practice in South Africa and more severe than most quarantine exclusions seen in other parts of the phylloxerated world.[11]

In the second type of zone, "a few small infested spots are known in the district" (Bioletti 1901, 7). In this zone the Swiss-German "death treatment" is followed: first the infested areas are surveyed and the vines uprooted and burned, this to be followed by the construction of an embankment "around the whole vineyard and then running water on to it until it is converted into a lake" (Bioletti 1901, 7). Disinfection with lethal doses of CS_2 are used as well in areas where treatment by flooding isn't possible. The goal here is not only to kill the bugs but also to kill any and all root fragments that might offer sanctuary to the bugs at a later time.

But sooner or later, either the second type of zone will develop further infestations, or perhaps the zone was from the start so thoroughly infested that "it is practically hopeless to attempt to eradicate it" (Bioletti 1901, 8). In this case, the entire zone is to be thoroughly surveyed and such quarantine and eradication methods as are possible should be used. In the end, however, it must be realized that these are only delaying tactics. Bioletti's analysis and description of the future course of action mirror precisely those we have seen earlier in South Africa:

> However conscientiously and completely these measures are enforced, a time will arrive sooner or later when the cost of inspection and eradication will be greater than any benefit to be derived from them. We are then face to face with the third set of conditions: we must accept the phylloxera as a permanent inhabitant of the district and simply consider the best method of growing our vines in spite of its presence. (Bioletti 1901, 9)

It is at this point that Bioletti introduces his discussion of the four remedies.

"Of the many thousands of methods proposed and tested," he remarks, "only four are at present used to any important extent," namely,

1. Injection of carbon disulfide;
2. Flooding or submersion;
3. Planting in sand;
4. Planting resistant vines. (Bioletti 1901, 9)

The first two aim to kill the insect, but neither approach is permanent and both must be repeated annually, which is expensive. Planting in sand is "of very limited and local applicability" in California. Thus, "the use of resistant vines" is "the only method that need concern grape-growers in California" (Bioletti 1901, 9). Bioletti divides resistant vine use into two categories, as direct producers (both Old American and new hybrids) and as rootstocks. He ends up conclusively recommending the latter, but only after dismissing both sorts of direct producers in a surprisingly cursory, ill-informed, and off-handed way.

After describing the initial efforts to replant the devastated French vineyards with Old Americans, Bioletti concludes that "these varieties, however, all proved unsatisfactory." This is an odd thing to say given that, at the very time he was writing, more than 125,000 acres—only 3 percent of the French vignoble but nearly 100 percent of the California vignoble—were planted in these very vines, and the number was growing rapidly. By 1939 750,000 acres were in direct producers, and at their peak, in 1958, 42 percent of the French vignoble was planted in these vines (Galet 1988, 373–74).[12] Bioletti, however, offers three reasons why these vines were all "unsatisfactory": many, especially the *labrusca*-based varieties, were sensitive to phylloxera under French conditions; second, "most of them were poor bearers compared with the prolific European vines"; and, finally, "the character of the fruit" differed so widely from the *vinifera* that "there was little sale for the fruit" (Bioletti 1901, 12). Of these reasons, only the first would seem to hold. Both the second and the third are completely belied by the numbers Galet relates. Even more incriminating is the fact that, contrary to his third reason, there is now, and has been ever since the first generation after the phylloxera came to Europe, a strong demand for "strawberry" wine.

The new generation of *hybrid* direct producers, vines such as were coming from Baco, Seibel, and other hybridizers, is mentioned and dismissed by Bioletti in almost the same breath: "The merit of these new varieties, however, is chiefly their resistance to Peronospora and Black Rot," which is unimportant to the California grape grower, since, given "the dryness of the climate, there is no likelihood of trouble from these serious fungus diseases" (Bioletti 1901, 13). And with this comment, Bioletti dismisses all thought of direct producers and turns to his discussion of resistant rootstock, a discussion that was to set the pace and tone for the next generation and beyond of California researchers and grape growers both.

PUTTING DOWN NEW ROOTS

Bioletti, of course, wasn't the first Californian to recommend planting on resistant rootstocks. As early as 1878 Hilgard had told a meeting of Sonoma winegrowers that "the vineyards of the future" would have to be planted on resistant rootstocks, just as was happening at that very moment in France (Pinney 1989, 345). Unfortunately, "Many, perhaps most, vineyardists in California were slow or negligent in acting on the advice given by board and university alike. Fewer than 2,000 acres, it was estimated, had been replanted to resistant rootstock by 1888, all this while serious depredation was occurring" (Pinney 1989, 345). But the reconstitution problem, as Pinney notes, wasn't due entirely to "human stubbornness or parsimony." Rather, there was no scientific consensus about what rootstocks should be used (Pinney 1989, 345). This caused some serious problems. In any case, following the recommendations from the university (and, soon after, the board) to make new plantings with grafted vines—even in the absence of scientific recommendations as to which stocks to use—a number of growers did in fact begin to make the effort to reconstitute their vineyards on resistant rootstocks.

Unfortunately, in their eagerness they repeated some of the mistakes lately committed by their French brethren. In 1877 Millardet had published his studies on *riparia* raised from seed collected in the wild. He announced that pure wild *riparia* seedlings were much more resistant than the Old American direct producers and should be used to replace them as rootstock (Millardet 1877). There was soon a strong demand for wild *riparia* seeds from the Missouri Valley: in 1881, Missouri's Dr. Engelmann sent between five hundred and a thousand kilograms of wild *riparia* seed to nurseries in France. Much the same thing happened in California: "Early California viticulturists, like their European counterparts, initially used wild grapevines indiscriminately. It was generally believed that any native grapevine could be used as a rootstock and would be resistant to phylloxera. Huge shipments of cuttings of wild vines were made from the Mississippi River Valley" (Wolpert 2002, 4). Following the French, *riparia* was a Golden State favorite; when the French soon went through a vogue of seeking *rupestris,* Californians followed suit. Wetmore, "without much evidence," recommended *V. californica*. His recommendation was duly followed, and, unfortunately, thousands of grafted vines soon failed (Pinney 1989, 345).[13] Conditions stayed like this through the 1890s, with only haphazard results. For example, the Old American Lenoir (known as Jacquez in France) enjoyed a Californian vogue during which it was indiscriminately planted, but it most certainly did not succeed everywhere (Pinney 1989, 345).[14]

Bioletti sums up these nearly two decades of chaos with his analysis of the properties necessary for a suitable rootstock. First, he say, we must begin by rejecting "all unselected and unnamed varieties, such as the ordinary Rupestris and Riparia, which have caused so much disappointment and loss." This rules out the use of randomly procured wild seeds, cuttings, and vines. Second, we must reject "all insufficiently resistant varieties, such as Lenoir" and other direct producers. Finally, then, "our choice of a resistant for a particular soil, climate, and scion must depend on its qualifications as regards *affinity* and *adaptation*" (Bioletti 1901, 14). *Affinity* refers to the interaction between scion and stock: not all stocks graft well to all scions; moreover, while some combinations graft suitably, other problems—for example, mismatch in the diameters of mature stock and mature scion—rule out these partnerships. For its part, *adaptation* refers to the stock-soil interaction. *Riparia*, for example, tolerates wet and shallow soils better than does *rupestris*, while *berlandieri* handles droughty and limey soils better than any other species. As regards climate, some stocks speed up the growth cycle of the scion, which is useful in shorter season areas; other stocks retard the growth cycle, which is adaptive in longer season areas.

One thing that cannot be overemphasized about Bioletti's criteria here is that he is speaking about finding a resistant stock for a *particular* soil, climate, and scion. He is in no way seeking a type of universal resistant stock: every distinct soil, climate, and scion-stock combination must be tested carefully, over the long term, in order to determine what works, and where. Because of this need for specific information, Bioletti in his concluding remarks is quite simply unable to make specific recommendations beyond the sort of thing seen above regarding, for example, vines that are tolerant of wet, shallow, or limey soils. He mentions the promise of some of the *riparia × rupestris* hybrids (3309C and 101–14Mgt), but then he says that he must forgo further analyses. Research needs to be done, it is obvious, before the serious reconstitution of California's phylloxerated vignoble.

Bioletti began the desperately needed rootstock research program in earnest shortly after this bulletin was published. He appointed Frederic Flossfeder to begin trials at the University Farm (now U.C. Davis) and the Kearney farm station near Fresno. Selecting from what was available in California already, and filling in gaps through importation, Flossfeder ultimately began experimenting with twelve rootstocks: three named pure species selections (Riparia Gloire, Rupestris St. George, and Rupestris Martin), eight hybrids, including 420A *(berlandieri × vinifera)*, 3309C, and 101–14Mgt (both *riparia × rupestris*); and one Old American, Lenoir/ Jacquez (Walker 2000, 211).

But the Board of Viticultural Commissioners recognized the immense scope of the project and asked the USDA to do some of the work as well (Pinney 1989, 345). This was accomplished in 1904, when George C.

Husmann, son of the Missouri wine scientist and immigrant to California, began a project of wide scope: 102 different rootstock selections were to be tested at twelve different locations around the state (Lider 1957, 59). All future rootstock trials and recommendations would follow from the work initiated by Flossfeder and Husmann.

TRIALS OVER THE YEARS

Flossfeder and Bioletti published their results in 1921 (Bioletti et al. 1921).[15] From the start, their report acknowledges its dependence upon the French work, but they remark that "owing to the great differences in soil and climate, however, we can accept the French conclusions only in a general way and as a guide for our own tests" (Bioletti et al. 1921, 83). The choice of stocks to be tested had Prosper Gervais as "the principal guide," "both because of his acknowledged competence in the matter and because most of his work was done in southern France, where the conditions are in many ways similar to those of California" (Bioletti et al. 1921, 84). The twelve stocks selected, as noted above, comprised pure American species, American hybrids, and *vinifera*-American hybrids, including AxR1. Bioletti's general conclusion regarding these latter is revealing: "Although a little less resistant than varieties of exclusively American origin, all that are used largely are sufficiently resistant *where the soil and other conditions are favorable*" (Bioletti et al. 1921, 85; emphasis added). With regard to AxR1 specifically, the report notes that "like most hybrids with *vinifera*, it is attractive to the Phylloxera, *which may weaken it in poor or dry soil*" (Bioletti et al. 1921, 88; emphasis added).

Several general conclusions are reached. First, because of variations in affinity and adaptation, "a stock which is excellent in one set of conditions may be a failure in another," and thus "no stock can be found which is best in all cases" (Bioletti et al. 1921, 90). Second, although "good results may be expected with any of the stocks listed," "perhaps the best are 3309, 1202, 101–14 and A.×R. No. 1" (Bioletti et al. 1921, 124). Third, the report does not provide information on either "the relative resistance of the various stocks to Phylloxera, [or] their adaptation to various soils" (Bioletti et al. 1921, 91). On the former issue, Bioletti remarks that it "may be considered in most cases as settled by the work and experience of European investigators." Thus, the resistance of the stocks chosen, according to European "work and experience," is "sufficient," except "perhaps under unfavorable conditions for the . . . vinifera hybrids," including AxR1 (Bioletti et al. 1921, 91).

On the issue of adaptation, decisions cannot be made on the basis of a single location; rather, they will await the results of "observations in many locations." Meanwhile, however, "most of the stocks used . . . have proved to be well adapted to a wide range of soil conditions" (Bioletti et al. 1921, 91). These results are summed up by Bioletti later in the report: "The questions of resistance to Phylloxera and of adaptation to various climatic and soil conditions did not come within the scope of these investigations for reasons already given. The principal points studied were the quantity and quality of the crop and the perfection and permanence of the unions" (Bioletti et al. 1921, 97). Clearly, from the point of view of scientific thoroughness, Bioletti and Flossfeder's results can only, and at best, be called "preliminary." Phylloxera resistance, the overriding goal of the program, has not been tested at all at this stage. Rather, it is only *assumed* on the basis of the French and other European results, and it is a tragically flawed assumption at that. By the time of Bioletti's report the French had already disavowed most *vinifera*-American hybrids, AxR1 specifically, as being deficient in resistance, and AxR1 had failed in 1908 in Sicily[16] and in 1915 in Spain (Galet 1979, 200; Granett et al. 2001, 399). But these failures are not noted in Bioletti's report. Given the widespread professional literature regarding the various failures, it is hard to believe that Bioletti had missed the news. We can only speculate about the reasons for his silence. But it is certainly odd. Thus, while on the one hand it is claimed that French and European experience will be relied upon in judging phylloxera resistance, on the other hand, it is not relied upon in the specific case of the *vinifera*-American hybrids. The consequences of this will be disastrous.

Husmann's results were not unlike Bioletti's. After issuing a preliminary report in 1910 (Husmann 1910), Husmann focused his efforts upon AxR1, AxR2, 1202C, Lenoir, St. George, Riparia Gloire, 420A Mgt, 161–49C, 101–14Mgt, 3306C, 3309C, 1616C, 1613C, Dogridge, and Ramsey (Walker 2000, 211). Work continued through the 1930s, with a final report being issued in 1939 (Husmann et al. 1939). The stocks 1202C, 1613C, St. George, Dogridge, and AxR1 were picked out as the generally most adaptable stocks. These conclusions, according to Lider, "listed the more vigorous stocks and those which showed the widest adaptability and best resistance to phylloxera" (Lider 1957, 59).[17]

Wolpert's analysis of the Bioletti and Husmann results is direct and to the point: "The work of both Husmann and Bioletti can best be described as screening trials in that they attempted to sort through a large number of European hybrids. They also utilized a wide variety of scions which had commercial potential at that time. Both researchers, by the nature of their screening efforts, were able to draw only general conclusions" (Wolpert 2002, 5). At this point, nearly forty years into the rootstock evaluation program, it was still the case that only general recommendations were available. Further work would be needed before particular recommendations regarding specific soils and specific stock-scion combinations would be forthcoming. Harry Jacob, a viticulturalist at Davis, would begin to provide the further work, starting in 1929; upon his untimely death in 1949, Lloyd Lider would take over. Lider's work, published in 1957, together with its almost accidental 1978 follow-up, would constitute the core of rootstock investigations in California into the twenty-first century.

MORE TRIALS: JACOB AND LIDER

Jacob was hired by the university in 1921 after completing a master's degree in plant pathology there. During his career at Davis he worked on many viticultural projects—low temperature and grape storage, girdling and grape size—in addition to his long-term rootstock trials ("University of California" 2007). As Wolpert notes, Jacob "learned from Bioletti and Husmann, and sited the trials in production regions" (Wolpert 1992, 52). It was Jacob's "aim to take the most promising of the list of experimental rootstocks available from previous investigations, and to test them with commercial scion varieties under actual vineyard condition" (Lider 1957, 59). More specifically, Jacob tried twenty-one different stocks, including AxR1 and all the other usual suspects, at ninety-nine sites in seventeen counties (Walker 2000, 211). In 1943, in some preliminary results, Jacob "acknowledged that AXR#1 was less resistant to phylloxera than St. George and even failed in dry sites 'where phylloxera attacks are likely to be most severe'" (Wolpert 1992, 52).

Unfortunately, Jacob died before announcing any final results. Luckily, Lloyd Lider was available to take up Jacob's work. Selecting several of the "most promising" of Jacob's trial stocks—AxR1, 1202C, 3306C, 3309C, 99R, 420A, and Dogridge—and using the most complete of Jacob's sites, Lider began his work.[18] In the end, fourteen of the trials, "only a portion of the total number of locations," provided secure data (Lider 1957, 61).

Lider published a full report in 1957, emphasizing that "the past fifty years" of work by Bioletti, Husmann, and Jacob "have provided a *basis for selection* of a limited number of rootstocks for the cooperative field trials which have been considered in this report" (Lider 1957, 58; emphasis added). Lider's distinction here is an important one. As Wolpert noted earlier, Husmann's and Bioletti's (and, evidently, Jacob's) work "can best be described as screening trials" (Wolpert 2002, 5). Lider obviously agrees: their work provides a preliminary "basis for selection" for his own work, not final results. It is only with his own work that extensive comparative field trials will have finally been done, more than fifty years since Bioletti's first explicit call for such work to be done, and nearly seventy years after the State Viticultural Commission remarked the need for California's own experimentation with resistant rootstocks.

Lider describes the goal of the program as "to evaluate the performance of a few chosen experimental rootstocks with a number of fruiting varieties over as wide a range of environmental conditions as possible in the north coastal vineyards of California" (Lider 1957, 58). According to this, only a few rootstocks, but a large number of scions, will be evaluated in terms of their performance. Phylloxera resistance is not mentioned as a specific criterion for evaluation. Rather, it is performance—vigor and production—that will be the targets of the trials. While it is true that vigor and production would presumably suffer in the face of phylloxera damage, etiology in this case would have to be presumptive. This is especially true in the face of other pests, particularly nematodes, which could (and indeed would) complicate the results. Thus, phylloxera resistance is only a derivative measure, inferred from overall performance, including, evidently, survival itself.

Lider's methods were straightforward. In a series of tables he presented data comparing various measures of vine anatomy (scion circumference), performance (estimated crop production), and grape anatomy (cluster and berry size) plotted against stock and location. One measure was especially noted: there is "a rather distinct correlation between the trunk circumference of the vines on the various rootstocks and the average amount of fruit they produced" (Lider 1957, 61). From this Lider inferred that the trunk measurements "reflect the vigor of the different stock-scion combinations," and this, in turn, is a measure of "their capacity for the production of fruit" (Lider 1957, 61). Vigor and cluster size also show significant correlations. It must be noted that these observations speak to the grape growers, who, at that time, would have been paid almost entirely on the basis of the weight of their crop. Measures of interest to the winemaker—berry color, juice total, soluble solids (including fermentable sugar), and total acidity—showed "no detectable difference between the fruit produced on the different root-stocks" (Lider 1957, 64).

In the end, Lider's observations show "a rather consistent relative order of performance when one is compared to another," even taking into account differences in location and scion (Lider 1957, 64). Since this data can be taken as evidence in favor of a "one stock fits all" hypothesis, it is a rather significant finding, one that would reverberate throughout California for years.

WHAT ABOUT PHYLLOXERA?

Lider makes no measures concerning phylloxera, nor is the topic even mentioned in the section of the report on results. Only at the beginning of the section entitled "Discussion" do we confront this dread issue. Lider's argument is direct, explicit, and unmistakable:

1. "The north coastal counties of California currently are generally infested with grape phylloxera."

2. Own-rooted vines, both original and replanted, "have been destroyed."

3. "The use of phylloxera-resistant rootstock is a necessary requirement for new vineyard plantings."

4. Thus "the judicious choice of which rootstock to employ in any given vineyard soil is an important decision to be made by the vineyard grower" (Lider 1957, 64).

Lider then goes on to introduce the seven most important paragraphs in California rootstock literature: "From the experience of growers in California, the published findings of research workers in the past, and from the data presented above in this report the following summary statements can be made concerning the stocks of greatest interest to the growers of this area" (Lider 1957, 64). Only two stocks are recommended: Rupestris St. George and AxR1. He writes, "A survey of the data collected has shown that the most popular phylloxera rootstock used commercially, Rupestris St. George, is not the most suited under the conditions to which it was tested, whereas, the experimental stock, $A \times R^{\#}1$, is generally the most vigorous and productive of the array of stocks under consideration" (Lider 1957, 58).[19] None of the other stocks in the trial—99R, 3306C, 3309C, 5A, 420A, Dogridge, and 1613C—are recommended.[20] Three of these summary discussions demand further analysis.

Until the time of Lider's report, Rupestris St. George was the most popular rootstock in California, despite no solid data or recommendations by any researcher, an extreme oddity worthy of a thorough examination in itself (Morton 1994). Lider uses no uncertain terms in his analysis of Rupestris St. George's virtues and vices: it "roots very well from cuttings, is compatible with practically all V. *vinifera* varieties, and exhibits a high degree of phylloxera resistance" (Lider 1957, 64). It is not at all clear where Lider has gotten his data on Rupestris St. George's phylloxera resistance, unless he is relying upon French results. On the negative side, the stock tends to produce shot berries (small, hard, unripened grapes), straggly clusters, and, worst of all, low yields, which, of course, is Lider's major desideratum.

Yet, even in face of these deficiencies, "there is still a place for the use of Rupestris St. George in California vineyards" (Lider 1957, 65). In shallow, unirrigated valley soils that are "usually low in moisture late in summer," Rupestris St. George can be used with "heavy producing wine varieties" (Lider 1957, 65).[21] Underlying this recommendation is a pointed bit of reasoning: "In such shallow, dry soils, the phylloxera resistance of St. George may be a prime criterion in its choice as a stock" since these soils present the greatest phylloxera risk (Lider 1957, 65).

With this highly selective recommendation, Lider attempts to end decades of entrenched practice, namely, treating Rupestris St. George as "the universal rootstock" (Morton 1994). This near-prohibition of St. George, coupled with Lider's strong endorsement of AxR1, which we will examine in a moment, would ultimately expose north coastal California vineyards to a repeat of the phylloxera disaster it experienced in the 1880s and '90s.

It also must be emphasized that Lider links the need for phylloxera resistance to a particular soil type: shallow, dry, unirrigated valley soils, typically those found on the valley hillsides. These soils are most certainly very hospitable to phylloxera (Galet 1988, 311). But in France and most other regions of the world, phylloxera risk is certainly not limited to these soils; in France, Portugal, Italy—and nearly everywhere— even deep, humid, and irrigated soils get infested with phylloxera. Yet Lider, just like his predecessors, seems to have endorsed quite strongly "the enduring belief" that California was special, or at least that the phylloxera threat in this special place was limited to shallow, dry, unirrigated soils (Wolpert 2002, 7).

As for AxR1, "this variety appears to be the nearest approach to an all purpose rootstock for the coastal counties of California" (Lider 1957,

65). Indeed, "in the more fertile soils, adequately supplied with moisture, that is, one *[sic]* the valley floors of these counties this stock is at present the best choice" (Lider 1957, 65).[22] Lider bases this recommendation on several factors, especially "its great vigor" and the fact that "the yields of varieties grafted on it have been consistently the highest of any of the stocks used" (Lider 1957, 65). Although Lider doesn't mention it here, it was widely known that AxR1 was a delight in the nursery: it rooted well from cuttings, mother vines were highly productive, and (something of great importance) it grafted easily on all of the commercially important scions ("University of California" 2007; Walker 2007).

But what about AxR1's phylloxera resistance? Here Lider's discussion is puzzling. Right off the bat Lider notes that the "phylloxera resistance of these stocks[23] is not high, as experiments in other viticultural areas of the world have demonstrated."[24] Certainly the failure of AxR1 in so many countries could not go unremarked. Indeed, in the face of this overwhelming evidence, why even bother to perform trials using AxR1? "It is understood that in very dry, shallow soils, or in areas where phylloxera can be quite serious, they may do poorly or even fail." Here again the special exclusion by soil type raises its head: in dry, shallow soils or other areas "where phylloxera can be serious" (wherever these might be), AxR1 "may do poorly or even fail." But why is there no mention here of the areas of rich soil where AxR1 had also failed, for example in France, Spain, and South Africa? Again, there is something special about California: "In the extensive trials that they have been subjected to in California, however, they have done *remarkably* well" (Lider 1957, 65; emphasis added). Lider's use of the term *remarkably* is a warning to us, although it was likely not intentional on Lider's part. It *is* remarkable, indeed; not just striking, it ventures into the realm of the miraculous that, among all the vignobles of the world, AxR1 should find itself resistant to phylloxera in California.

Yet, there we have it: based upon its yields, its performance in the nursery, and its observed success in phylloxerated soils, AxR1 is "the best choice" for planting on the fertile valley floors of north coastal California. Lider published in 1957. Within a very short time the great planting boom of the 1960s and '70s would be upon California, with millions of new vines planted on the highly recommended AxR1; very shortly after that it would begin to fail, as it had in France, Italy, and South Africa.

The last possible escape path is blocked by Lider's discussion of 3306C and 3309C. These two rootstocks, developed by Georges Couderc long before from a cross of two resistant American species, *riparia* and *rupestris,* had been successful wherever they were tried. In fact, at the time of Lider's report, 3309C was the second most popular rootstock in France (Galet 1988, 210). Lider's conclusion about these stocks is that, based on "the data gathered in these trials and from the numerous observations made upon their performance in California over the past fifty years, *they cannot be recommended* for further planting in this state" (Lider 1957, 66; emphasis added). France's second most popular rootstock cannot be recommended in California? Why not? Because of its performance. But performance in what sense? These two stocks "have been thoroughly tested under a wide range of viticultural conditions in the grape-growing regions of the world." And obviously their performance, that of 3309C especially, has been satisfactory or it wouldn't be so popular. But what about California? "These two stocks have performed satisfactorily in California but neither have been outstanding in the field trials when compared with other stocks." This is the only premise that Lider offers for his conclusion that neither of the Coudercs can be recommended. A look at his data reveals the basis of his conclusion: on the vigor/productivity scale, neither of the Coudercs came in first anywhere, with any scion, while AxR1 repeatedly did. And on this basis alone the Coudercs are ruled out, and AxR1 is ruled in.

The word *phylloxera* does not appear in the paragraph.

LIDER'S FINAL SHOT

In 1974 Lider returned to several of the test plots he had last observed nearly twenty years earlier. Although the reason for his revisit at that particular time is not clear, he relates his long-term motivation quite straightforwardly: "Finally, of special interest to modern California viticulture, was the question of the longevity of the vines grafted onto Ganzin 1 [AxR1], since that stock, with its questionable level of phylloxera resistance in Europe, was the rootstock chosen for a majority of the new plantings in the coastal districts" (Lider et al. 1978, 20). In 1953, when data gathering was terminated and final publication envisaged, the oldest vines in the Jacob-Lider trials were only eighteen years old. Lider indicates in his 1957 publication that he was acutely conscious of the brevity of this period; thus it made sense in 1974 to return to the original plots—if any still existed. Luckily, six were still relatively undisturbed and available for a final analysis.[25]

But there was also a more pressing need, created by Lider himself, for information on the long-term performance of AxR1. In his 1957 paper

Lider had strongly recommended AxR1 as the sole candidate for planting on fertile, irrigated valley floors, casting St. George in what was at best a very minor role on the dry valley hillsides. As Lider later observed. "During the last few years, extensive new plantings of wine grapes have been made in these coastal districts, with a major portion grafted onto Ganzin 1" (Lider et al. 1978, 18). "Extensive" is a good term to use here, for the numbers are very large indeed: "The five counties in California which contain practically the entire phylloxerated vineyard acreage in this north coastal district are Mendocino, Sonoma, Napa, Solano, and Alameda. Of their total grape-planted area (nearly 60,000 acres) more than one-third has been set in the past five years" (Lider et al. 1978, 18). Later estimates would note that during the great north coast planting boom of the 1960s and '70s, "between 30 and 40 thousand acres of vines were set using this rootstock" (Lider et al. 1995, 14). If, in 1957, Lider had been uneasy about the dearth of information concerning the long-term health of AxR1 grafts, his uneasiness could only have magnified greatly by 1974, when thousands upon thousands of acres of new plantings had gone in, all stocked exactly according to his 1957 recommendation. The time had come to revisit the earlier trials.

In his 1974 follow-up Lider reprised his earlier methodology, with one important exception. Each of six varieties was compared on four or five stocks using trunk circumference and productivity measurements.[26] Fruit was sampled for sugar and total acidity. Both of these procedures had been followed in the initial observations. New this time was a direct measure of phylloxera infestation. A one-half- to one-inch sample was taken from primary roots, the pieces then being "examined for phylloxera infestation after being cultured for 30 days at 24°C in a moist, aerated plastic container" (Lider et al. 1978, 20). Given the similar methodology plus the new phylloxera data, Lider thinks that his results are probative since "the six trials reported here are quite representative of the vineyard sites upon which vines are planted in these coastal districts" (Lider et al. 1978, 23). Identical methods plus representative sites equals trustworthy data.

His data are presented in a series of figures and tables. In the figures, which exhibit trunk circumference over time versus stock type, the new data are simply used to extend in a straight line the previous results.[27] Lider's interpretation of the results reveals no surprises, except for the poor performance of AxR1 in trial six, which shows both 420A and 1202C outperforming the Ganzin stock using a Zinfandel scion. In his discussion, Lider notes that the site was infested with phylloxera pre-trial, the soil was of low fertility, and it was unirrigated. Moreover, in the phylloxera examination of the roots from this trial, the bug "was found on the roots of Ganzin 1"; hence, "this probably accounts for the reduction in relative size of the scions" (Lider et al. 1978, 24).

Indeed, "phylloxera larvae were present on the roots of Ganzin 1 in each of the five trials established on prior infested sites, a significant finding" (Lider 1978, p. 22). Only trial one had "deep, fertile valley floor soil," and "although precise information is lacking, it most likely was not a replant site" (Lider et al. 1978, 20). This last point supports an inference that the site most likely wasn't phylloxerated.[28]

Overall, the data from his follow-up "support the [1957–58] California recommendations[29] that Ganzin 1 be restricted to less stressful sites and sites that would tend to limit the impact of phylloxera" (Lider et al. 1978, 22). Yet, even in light of this restriction, AxR1 remains the sole recommended stock: "the outstanding performance of Ganzin 1 in trial 1, on a site having minimum growth stress and no prior infestation of phylloxera, bears out the recommendation for Ganzin 1 on such sites" (Lider et al. 1978, 24).

But what about the other stocks—particularly 3309C—that are still enormously important in France? Interestingly enough, although the slopes of the curves on 3309C are roughly the same as—and in two cases better than—AxR1, absolute measures are always less. This supports the claim that AxR1 is generally more productive than the Couderc stock. Yet, in trial six, the dry, poor phylloxerated site, the final observations show 3309C closing rapidly on AxR1. It would have been extremely interesting had this site been examined just once more, say, in the mid-1980s.

Lider, Ferrari, and Bowers's 1978 paper did nothing but increase confidence in AxR1. Wolpert, noting Lider's 1957 remarks about AxR1's sensitivity to phylloxera and previous failures around the world, sums up the situation clearly and succinctly:

> Nevertheless, over the 25 years which followed [the 1957 paper], AxR1 performed consistently well in California vineyards (presumably under significant phylloxera pressure) and grew to be preferred by nurserymen and appreciated by viticulturalists and winemakers. Thus AxR1 became the predominant rootstock in North Coast vineyards. Any of the original concern about phylloxera resistance expressed in Lider's [1957] article was forgotten or ignored. . . . In 1974 Lider and co-workers returned to six of the trials included in the 1957 report and found that AxR1 continued to perform adequately. (Wolpert 1992, 52)

Thus, as the planting boom continued, it continued to be a boom for AxR1 as well. Unfortunately, disaster loomed.

THE BUG AS PHOENIX

Phylloxera's return to California has already become the stuff of legend, recounted in books and journals and the subject of a very thorough and fair analysis in a feature article in the Sunday New York Times Magazine (Lubow 1993). There is not much to be added to the chronicle of events, but the analysis of the events and of the chronicle itself remain unfinished.

The disaster began slowly. Four of John Baritelle's vines showed symptoms of stunted growth during the 1980 season. The following season the Napa grower found sixteen sick vines, but no amount of investigation indicated the cause. Not until a year later, when Davis researcher Austin Goheen pulled up one of Baritelle's vines and saw the dreaded yellow bugs, was the sickness understood: phylloxera was back (Lubow 1993).

Baritelle's vineyard was only fifteen years old and it was planted on AxR1, so how could it possibly be attacked by phylloxera (Lewin 1993, 31)? Examination of Baritelle's records, however, indicated that he couldn't fully account for the provenance of all the rootstocks in every vineyard. Relief—the phylloxera might not be the result of using AxR1 after all—was immediate among the concerned parties: "University of California researchers initially believed that the infested vineyard was not actually on AxR1 rootstock but on a related rootstock that was known to be susceptible to phylloxera" (Smith and Weber 1999). Lider, who had become involved with the new phylloxera incident, was particularly relieved, for obvious reasons (Lubow 1993, 5). But the relief was to be short-lived: by the following summer (1983), "'oil spot' die out symptoms, typical of grape phylloxera damage, began being noted in the Napa Valley" (Fergusson-Kolmes and Dennehy 1991). That fall, Davis entomologist Jeffrey Granett took time off from his work on mites to collect phylloxera samples from Baritelle's vineyard for investigation in his lab.[30] "They're very easy to raise," says Granett, "You just put them on the rootstock and they're fine" (Lubow 1993, 25).

GRANETT

Unfortunately, they were far too fine. Initially, Granett planned to raise a colony of the bugs on Cabernet roots, but by accident, some of the

FIGURE 21. Jeffrey Granett in his lab. Courtesy Jeffrey Granett.

bugs were left alone with AxR1. The results were fearsome: Baritelle's bug "was feasting as happily on the AXR as it was on the cabernet" (Lubow 1993, 25). Granett's inference was immediate: this wasn't the familiar old phylloxera that had left Lider's AxR1s in peace all those years. Thus, "Dr. Jeffrey Granett and his colleagues at U.C. Davis soon described a new phylloxera, Type B, in the Napa Valley" (Fergusson-Kolmes and Dennehy 1991). According to the experiments Granett and his colleagues carried out, there indeed was a newly emerged strain— which Granett called a *biotype*—that flourished on AxR1, while the familiar bug did not (Granett et al. 1985).[31]

Granett's claim that a new biotype had emerged was controversial. Over the next few years different researchers came down either for or against the claim, providing sometimes quite divergent reasons for their positions. Fergusson-Kolmes and Dennehy, for example, argue that "there is good reason for expecting biotypes to occur in grape phylloxera populations," namely, phylloxera resembles aphids in significant ways, and aphids have biotypes (Fergusson-Kolmes and Dennehy 1991). Walker, Granett's colleague at Davis, argued, on the other hand, that "biotypes are irrelevant to the collapse of AXR. . . . [I]t is not a resistant rootstock." In other words, biotype or no biotype, AXR1 was destined to fail (Lubow 1993, 25). Galet, the world-renowned ampelographer from Montpellier, said of the Davis scientists "they even resorted to the old—and false— theory of the German Borner (1910) that two types of Phylloxera existed, one of which was more aggressive than the other" (quoted in Perdue 1999, 54). Galet went even further, adding a purported motive to the Davis position: "It is my opinion that they have adopted this in order to cover up their incompetence" (quoted in Perdue 1999, 54).

Galet was not alone in having suspicions. Some suspected that the university itself had a vested interest in the new biotype. After all, if biotype B were an evolutionary outcome, an unpredictable virulent surprise, who could blame Lider and the other Davis scientists for having recommended AxR1, a recipe for disaster (Lubow 1993, 59)? But regardless of whether a new biotype had evolved or not, AxR1 was failing. Early on, in 1985, Granett had believed that scientists and growers had about five years to solve the problem: "While he feels there is no emergency now, there will be 'if we don't figure it out,' Dr. Granett said" (Bishop 1985). Others felt the same way: according to "Keith W. Bowers, the farm advisor in Napa, the Type B Phylloxera is a concern, he said, 'but it's going to be a couple of years before we know a lot more about it, it's not something that's going to eliminate Napa Valley grape growing overnight'" (Bishop 1985). With what now looks like incredible slowness, the university dithered over the next several years. There was no suspension of the AxR1 recommendation, no urgent rootstock field trials, no sense of impending disaster. According to some, behind closed doors the university researchers were in a desperate state, even while presenting a calm front to industry and the public: "In its official bulletins the university spoke with oracular unity, but within its walls there was polyphony, if not cacophony" (Lubow 1993, 60). No consensus opinion could be reached. In 1988 a phylloxera task force was constituted at Davis, but the group could come to no swift consensus, leaving the rootstock situation muddy and dangerous.

The stakes were enormous. Rhonda Smith and Ed Weber, UC Extension agents for Sonoma and Napa counties, respectively, estimated that at the time the infestation began to produce observable damage, Napa had twenty-two thousand acres on AxR1, while Sonoma had twenty thousand, roughly two-thirds of the total acreage in both counties (Smith and Weber 1999). Reconstitution of this many acres of grapes would be a massive job, expensive in sweat, tears, and cash. Unfortunately, there was no alternative to reconstitution: "After the first sign of infestation is found, no one can afford to take a 'wait and see' approach and watch their neighbors try one of the alternatives to a total replant" (Smith 1994).

The suspense finally ended in December 1989 when the Davis phylloxera task force issued an emphatic press release warning that "AXR#1 IS NOT RESISTANT to this NEW PHYLLOXERA BIOTYPE . . . DISCONTINUE PLANTING AXR#1 ROOTSTOCK" (quoted in Lubow 1993, 60; emphasis in original). From that point onward, planted AxR1 grafts were considered doomed by the great majority of the California wine industry.[32] They would have to go. As an editorial in the industry journal *Wines & Vines* noted, "Farm advisers believe that the biotype B phylloxera will spread anywhere in California where vines are planted on AXR rootstocks. It's just a matter of time" ("Could Phylloxera" 1991).

Several options existed. The simplest method, and the only procedure that was recommended, was a total replant. But some growers tried other, untested methods that offered the hope of a more drawn-out reconstitution, preventing the single-season double shock of a complete loss of productivity plus the cost of replanting and waiting for the new vines to mature. One method attempted was interplanting new grafted vines between the AxR1 grafted vines. Although theoretically this method had promise, there were physical problems related to doing the actual planting, as well as biological problems of poor growth of the new vines (Smith 1994). Another technique was the approach graft. A resistant rootstock[33] was established in close proximity to the AxR1 grafted vine and grafted slightly above the graft union of the original rootstock. The hope was that the new, resistant rootstock would take over the job enough so that when either phylloxera damaged the AxR1 or the AxR1 stock was severed, the scion vine could thrive (Smith 1994).

Most growers chose the total replant. Napa and Sonoma soon resembled the Midi of 1872. In a scene frighteningly like that reported by Joseph de Pesquidoux,[34] a vignoble came to resemble something from Dante: "All over California wine country, hulking, gnarled mounds of grapevines are being bulldozed into funeral pyres as big as two-story houses; they stand in the middle of bare-dirt fields, smoldering the ashes of lost dreams into the sky" (Perdue 1999, 46). Thus, "The skies over

Napa and Sonoma are commonly filled with the smoke from the burning phylloxera-infested vines that have been uprooted" (Reiss 1991). By 1997 16,500 acres of AxR1 had been removed from Napa, leaving only a residual 15 percent of that county's total acreage on the doomed rootstock. Over in Sonoma, both infestation and reconstitution were progressing more slowly: ten thousand acres had been removed, leaving 20–25 percent of the county's acreage left to remove (Smith and Weber 1999). According to one expert analysis the job of reconstitution was well begun and the future was clear:

> Both Napa and Sonoma counties are well on their way towards completing the replanting of AXR#1 vineyards. In Napa County, most of the remaining AXR#1 acreage will likely be replanted in the next five to seven years. Replanting in Sonoma County will probably continue for at least 10 years. Eventually, AXR#1 will be eliminated in both counties and declining vineyards due to phylloxera will be a thing of the past. (Smith and Weber 1999)[35]

Estimates of the cost vary, with $1 billion at the low end and about $6 billion on the high end ("Could Phylloxera" 1991). No matter how much it cost, phylloxera redux was a disaster of the first magnitude. Now the issues are how and why did it happen?

WHAT WENT WRONG BIOLOGICALLY?

Two distinct but related questions arise from the phylloxera redux affair. First is the question about the bug itself: what happened in the bug population that precipitated this disaster? Second is the question about the people involved: what went wrong in the scientific institution—both research and applied—that allowed this disaster?

Although the original explanation for the failure of AxR1 was given in terms of the emergence of a new biotype, recent work has suggested strongly that the issue is much more subtle and complicated. During the time of the phylloxera redux (1985–95), not much was known about the basic biology of the bug or of the mechanisms of rootstock resistance. Thus, as Fergusson-Kolmes and Dennehy noted in 1991, "There are critical deficiencies in our understanding of the biology of the most important arthropod pest of grapes worldwide, grape phylloxera" (Fergusson-Kolmes and Dennehy 1991). Davis's Andrew Walker stated in 1992 that "phylloxera resistance remains a black box, that is, we know it exists, but our understanding of how and why is not complete"

(Walker 1992, 61). The situation has improved markedly in the years since, especially in our knowledge of the genetics of the bug. While we still do not know much about the mechanisms of resistance,[36] studies from around the world—for example South Africa (Downie 2005), Australia (Corrie and Hoffmann 2004), Europe (Vorwerk and Forneck 2007), and California (Lin at al. 2006)—have begun to unravel some of the enigmas of phylloxera evolution, thereby illuminating what went wrong in California's misadventure with AxR1.

Several puzzles need solving. First, what is the structure and dynamic—the demographics, as it were—of California phylloxera populations? Second, how could Lider's trials of AxR1 not have shown the weakness of the stock? Finally, what are the diffusion mechanisms that could have allowed the nearly simultaneous appearance of phylloxerated AxR1 in widely separated locations in north coastal vineyards?

From the start, California researchers were confident that the winged form of phylloxera did not occur in the state. This fact was also associated with the belief that reproduction of the bug in the state was parthenogenic—asexual or clonal—which implied that there would be no genetic recombination, a major source of heritable variation to be acted upon by selective forces.[37] This suspicion looks as if it is correct. Recent work has shown little or no evidence that sexual reproduction occurs with any frequency in cultivated vineyards worldwide (Granett et al. 2001; Corrie and Hoffmann 2004; Vorwerk and Forneck 2006). Indeed, data suggest that phylloxera populations in South Africa have no more genetic variability than they had when they first infested the country.

Earlier views about clonal diversity and reproduction implied that the forces of selection would have little variation to work with. That is, it was believed that members of phylloxera populations in California and elsewhere were as alike as maternal twins or quadruplets—except in this case we would be talking about maternal trillionuplets. Recent work has seriously amended this view.

Using various molecular markers researchers have shown that any given vineyard might contain anywhere from three to twenty different (although closely related) genotypes, clones, subpopulations, of the bug (Corrie and Hoffmann 2004).[38] This amount of variation, needless to say, was unexpected[39] and indicates that the ordinary processes of evolutionary selection are well and truly at work. Population adaptations, especially local ones, are inevitable. This mechanism and its consequences, albeit widespread and well known, were certainly not explicitly fore-seen in phylloxera by grape scientists anywhere.

In addition to these typical processes of selection and adaptation, other factors influence the evolutionary trajectory of our bug. First is host diversity (Fergusson-Kolmes and Dennehy 1991) There are some fifty species of *Vitis*. They grow in wildly diverse terrains, climates, and terroirs. The native range of phylloxera stretches from the high North-east United States to southern Mexico, the Rocky Mountains to the Atlantic. Within this range are a large number of quite distinct grape species, many of which are sources of quite distinct populations of phylloxera.[40] Investigation of invasive populations in vineyards far re-moved from the native range has shown clear and undeniable links to American populations (Downie 2005; Corrie and Hoffmann 2004). Multiple introductions into a vignoble provide ample genetic diversity for selection to work upon (Corrie et al. 2002; Downie 2005).[41]

Moreover, different clones of phylloxera show differential fitness on host plants, and recent work in Australia revealed that there are strong associations between a given vine type and a given clone in a vineyard (Corrie et al. 2003). Added to this is the possibility that various clones are "specialists" in certain niches, whereas others are more "general pur-pose" (Corrie and Hoffmann 2004).

Another mechanism at work is the bottleneck effect, which occurs when a population suffers a drastic reduction in genetically distinct in-dividuals for whatever reason (Vorwerk and Forneck 2006). Donnie's work suggests that grape phylloxera invasive introductions themselves provided bottlenecks (Downie 2002). A related but distinct mechanism is the founder effect, which occurs when a subset of a given population, differing in gene distribution from the parent population, separates from the original and starts its own colony. Both effects can work, sepa-rately or together, to bring about population/genetic shifts. In Australia, phylloxera overwinter in the root zone of vines. Very few members of the population survive the winter, and the small number of survivors constitute founders, with their associated effect (Corrie and Hoffmann 2004).[42]

So, while introduced phylloxera rarely (most likely never) engage in sex, other mechanisms are constantly at work to provide the raw mate-rial, the genetic diversity among populations, upon which evolutionary selection can work its wonders. As Vorwerk and Forneck remark, "There is significant potential for adaptation behind the high genetic diversity monitored [in phylloxera populations], allowing genotypes to quickly adapt to new environmental conditions or habitats when and where possible (Vorwerk and Forneck 2006, 686).

Several specific adaptations of individual phylloxera genotypes aid their viticultural devastations. Interestingly enough, there is strong evidence that some populations are specialists in leaf living, while others specialize in root dwelling (Corrie and Hoffmann 2004). This lifestyle difference is important since it again offers a source of genetic variability: just insofar as leaf and root genotypes stay separate, they can provide variations independently. A predictable outcome is that eventually a specialist will evolve into a type that can make a living either as an aerial or as a terrestrial being. Sure enough, there are some clones that are happy living either lifestyle (Corrie and Hoffmann 2004).

More immediately threatening are the evolved clones adapted to specific hosts. It has become quite clear that any given vineyard population comprises subpopulations—a neighborhood of closely related but distinct clones—that exhibit "preferences" for one varietal or another (Downie 2002). Thus, lurking in a Sonoma vineyard might be a collection of clones, some capable of making a living on only one or two different hosts—Cabernet, *riparia,* AxR1—and others of a "general-purpose" nature, some adapted to aerial living, others adapted to life among the roots. When given the chance via the planting of a preferred host type, the phylloxera clone, the subpopulation in the vineyard adapted to life on that particular host, will reproduce much more successfully than the other clones: "Some of these genotypes probably existed at low frequency until increased planting of a host they could do well on, while others couldn't. Classic natural selection."[43] Within a very short period of time, the successful clone will dominate the population and will soon extend its dominance to other vineyards containing that same host rootstock.

This is the most plausible mechanism to explain what happened in the case of AxR1. Out there, lurking in a vineyard—or maybe several vineyards—were clones that were ready, willing, and able to successfully colonize AxR1. The stock was never resistant, or so several researchers think (Lubow 1993). It was simply a question of getting enough AxR1 stocks into an area where there were clones that could make a living on AxR1. Not a question of biotypes, rather it is a question of population dynamics and evolutionary processes.[44] Once there were enough AxR1 stocks to support a selected clone, evolutionary dynamics produced a burgeoning subpopulation of adapted bugs ready to be exported to other vineyards.[45]

Why didn't Lider's trials of AxR1 show the stock's weakness? It is entirely plausible that Lider got lucky (or unlucky, as far as California viticulture is concerned). Since AxR1 was not widely planted—indeed it was a rare, "experimental" stock until the boom of the late 1960s—it is quite easy to suspect that the various trial sites, although they were close to, and perhaps even within, phylloxerated areas, simply did not contain the relevant clone, the subpopulation adapted to life on this particular rootstock. Moreover, at each site the number of stocks of a given variety was quite small. Thus, the trial sites did not present a big enough "target" for the widely dispersed and small AxR1-adapted clone.[46] Only when the target became large enough—the thousands of acres of AxR1 planted during the boom years—did the selective forces of evolution grow the clone to critical proportion. Once critical mass had been achieved, the well-known typical human-assisted dispersal vectors seeded vineyards in widely scattered areas of the north coastal vignoble.

But why did it take fifteen or so years for the effects to show up? First, it takes time for selection to grow the virulent clone. Second, more time is taken for the clone to locally disperse: phylloxera move quite slowly—by crawling—in California. Finally, once several vines are infected, it will take some years, especially in a fertile, well-watered site, for the bug to sicken the vine. Fifteen years is not long for such a cascade of events to occur.

With the basic biology of grape phylloxera finally revealed, it becomes clearer how and why the AxR1 disaster occurred *biologically*. What remains unclear is how *institutional* processes failed and contributed to the disaster. We now turn to this issue.

WHAT WENT WRONG INSTITUTIONALLY?

Lider was nearly alone; indeed, Lider was only the latest of the lonely rootstock researchers in California. From the time that Bioletti engaged Flossfeder to begin trials, and, later, realizing the scope of the job, persuaded USDA's Husmann to initiate trials as well, there have never been more than a handful of rootstock researchers in California at any given time. Rootstock research is not singular in this regard: "Funding for enology and viticulture research lags because the University of California system will not underwrite it and the industry won't open its own checkbook" (Perdue 1999, 55). To make matters worse, enology and viticulture research, and rootstock research in particular, slips through the funding cracks at all levels, not just at the university. Fergusson-Kolmes and Dennehy describe the situation with painful accuracy:

> Unfortunately, research groups like ours, that desire to work on practical problems relating to grape pest biology, find it extremely difficult to keep financially afloat even limited research programs on grape pest biology. Experience has shown us that there is a lack of funding sources from federal, state or industry groups for sustaining the sort of long-term undertakings that are needed with grape phylloxera biology. Research on pest biology is too "applied" to be attractive to the principal scientific grant programs while at the same time it is perceived by grower groups as being too "basic" to receive funding priority. (Fergusson-Kolmes and Dennehy 1991)

The National Science Foundation and USDA are looking for research that is "basic" by their standards: for them, phylloxera research is "applied." But for industry groups, phylloxera research is too "basic" to be funded: "Funding for this kind of fundamental research depends on the wine producers' willingness to support it. . . . Not surprisingly, they would prefer to spend their money on new rootstock trials than on uncovering what is—to them—the arcane genetics of the tiny yellow pest" (Lewin 1993, 30). In 1992, at the height of the disaster, Davis researcher James Wolpert issued an ominous warning about California's lack of knowledge about phylloxera: "In the interim we are utilizing knowledge from Europe, Australia, and South Africa. . . . [H]owever, this places California in the position of being a 'net importer' of viticultural information, an uncomfortable position at best and a dangerous one at worst" (Wolpert 1992, 53).

In the end, the knowledge produced by the hard but lonely work of Lider and his predecessors simply was not enough. California did not know enough about phylloxera under its conditions, and it especially did not know enough about rootstocks and their interaction with the bug in California.

Yet underfunded and nonexistent research is not nearly a sufficient explanation—nor, perhaps, even a necessary condition of such an explanation—in the peculiar case of AxR1. This very rootstock had failed in France, Spain, Sicily, and South Africa, and Lider (and some of his predecessors) knew about the failures—indeed, they referred to them in their reports—well before the 1957 recommendation.[47] Moreover, shortly after the AxR1 recommendation, many international colleagues, especially two eminent French researchers, explicitly warned their Davis colleagues to stay away from AxR1. In 1963, Denis Boubals of the École Nationale Supérieure d'Agronomie de Montpellier warned Davis not to plant AxR1 ("Denis Boubals" 2007). As ENSAM's director, Jean-Paul Legros, recently put it, Boubals "was not listened to until the moment when the phylloxera, having passed the barrier of the Rockies, began enjoying all the tender roots offered to its appetites" (Legros 2005, 182). A few years later, ENSAM's Pierre Galet noted that "during my first trip to California, I tried very assertively to bring the dangers of [AxR1] to the attention of Napa Valley viticulturalists and UC Davis professors, but without success" (quoted in Perdue 1999, 53).

From our vantage point today, knowing what we know now, we can only lament that the historical evidence from Europe and South Africa and the explicit warnings of eminent international scientists went unheeded. One major reason for this seems unambiguous: California exceptionalism.[48] If America believes itself an exception to the normal rules of the game, then, of all the places within America, California believes itself the most exceptional. California's perceived exceptionalism is manifest in many arenas, from climatic superiority to lifestyle leadership, from beautiful terrain to Beautiful People. This exceptionalist perception extends right into viticulture. Wolpert points out one significant aspect of this, an "enduring belief" that "growing conditions in California were different enough from those in Europe to permit AXR#1 to escape damage" (Wolpert 2002, 7). Thus, according to Lider, "both the climate and the soils of California are natural agencies which tend to reduce the dangers of phylloxera" (Lider 1957, 58). Exceptionalism was frequently linked to soil type, since "from the early viticultural reports through the works of Bioletti, Husmann and Lider runs the thread of a belief, presumably based on a combination of experiment and experience, that phylloxera's ability to overcome resistant rootstocks was to a great degree dependent on soil type" (Wolpert 2002, 7). And California's soil types, or so went the belief, were exceptional.

Given the well-known European experience, this belief is unfounded. The only soils exempt from phylloxera were pure sands; every other imaginable soil type, everywhere in Europe, eventually succumbed to the invading bug. And with it went all the vines as well.[49] Since this fact was widely known, California scientists somehow must have thought that their vignobles would again be the exception to the rule.

Clearly, a weaker sort of exceptionalism has always been evident in the phylloxera tales, right from the start. Vignerons in Burgundy argued that it couldn't happen there because their climate and terrain were so different from the Languedoc. And Champagne, well, it couldn't happen *there* either, since Champagne's fastidious vignerons were morally superior to those careless vignerons of the Languedoc, whose slapdash southern vine tending allowed the malady to strike their vines. Everyplace, everyone, in turn, thought that they would be the exception. Even in the 1860s, during the first phylloxera disaster, California growers denied the impending peril. As Pinney describes it, "For years growers in the afflicted regions had pretended that the threat was not serious, or that it was under control, or that it did not exist, or that it would go away by itself" (Pinney 1989, 344). But a hundred and twenty years later, during phylloxera redux, California's exceptionalism had grown to be exceptional in its strength. Surely this is the reason underlying Jeff Granett's admission that "the earlier experience in South Africa and Europe was overlooked or forgotten, and, to be honest, tests for phylloxera resistance in the California trials weren't pushed as hard as they might have been" (quoted in Lubow 1993, 31). Indeed, perhaps what is operating here is, as some have indicated, "a Davis mindset" (Lubow 1993, 30).

Perdue's indictment is harsh, but it has a significant ring of truth to it: "Some observers believe there was a California arrogance at work, that somehow it was felt that what had failed all over the world would work here. 'I think it's an American phenomenon that despite the evidence, we still think we're different,' Napa Valley vineyard consultant Rich Nagaoka told *Wine Spectator* magazine" (Perdue 1999, 50). Nagaoka refers to American exceptionalism, justly enough; but, of course, California is no exception, and it did happen here . . . twice.

AxR1 had many advantages. As the field trials of Lider and others clearly showed, vines on this stock produced more crop than any other stock, in more locations, and with a wider variety of scion varietals. AxR1 really did look like a universal rootstock. Moreover, it had distinct advantages for the nurseries whose job it was to provide grafted vines to growers. This is no small point. Simply as a grapevine, AxR1 was a delight in the nursery. Mother vines were extremely productive, reliably turning out huge amounts of stock cuttings. The cuttings rooted and grafted well. For these reasons, as Wolpert notes, AxR1 "grew to be preferred by nurserymen" (Wolpert 1992, 52).

Good performance in the nursery is a necessary factor in any rootstock's success. If the nurseries cannot work easily and successfully—which is to say, profitably—with a stock, its viticultural performance is moot. And here is where a relatively unknown feature of AxR1 plays a major role. Until recently, varietal scions in California were infected with various viruses, many of which introduced incompatibility between stock and scion.[50] Only two stocks, St. George and AxR1, were compatible with the wide array of commercially important scion varietals. Thus even if, for example, 3309C or SO4 had shown better productivity than AxR1 in field trials, neither of these stocks could match the latter in graft compatibility. It would have taken special skills and techniques to use either of these two stocks successfully in the nursery. At the very least, profitability would have suffered. Only with the development of techniques for the production of virus-free scions during the 1960s did widespread use of stocks other than St. George and AxR1 become economically feasible (Duran-Vila et al. 1988).[51]

In addition, there were problems with the field trials, both in conception and in execution. From the start of the trials, phylloxera was not a major consideration: "The questions of resistance to Phylloxera and of adaptation to various climatic and soil conditions did not come within the scope of these investigations for reasons already given.[52] The principal points studied were the quantity and quality of the crop and the perfection and permanence of the unions" (Bioletti et al. 1921, 97). Bioletti and Fossfelder originated the whole rootstock trial series. And, from the start, they weren't interested in testing phylloxera resistance: resistance simply was not a goal. Lider follows in Bioletti's footsteps. He describes the goal of his trials as "to evaluate the performance of a few chosen experimental rootstocks with a number of fruiting varieties over as wide a range of environmental conditions as possible in the north coastal vineyards of California" (Lider 1957, 58). Phylloxera resistance is not mentioned as a specific criterion for evaluation. Ultimately, Lider bases his recommendation of AxR1 on several factors, most importantly "its great vigor" and the fact that "the yields of varieties grafted on it have been consistently the highest of any of the stocks used" (Lider 1957, 65). Indeed, in Lider's recommendation discussion, the word "phylloxera" does not appear.

From their very conception, then, the California rootstock trials never included evaluating phylloxera resistance as a goal. Whether evaluation of resistance was achieved was secondary to the main goal of the trials: finding the most vigorous, most productive, and best-grafting rootstock.

In addition to their conceptual shortcomings, the trials also had shortcomings in their execution. Bioletti and Husmann tested many different stocks at a small number of sites. Jacob's trials, which were completed by Lider, tested a much smaller subset selected from Bioletti's and Husmann's trials, at a much wider range of sites, using plots donated by growers across the state. But the Jacob and Lider trials had several methodological problems. First, they were not replicated. This is odd, since even the very first rootstock trials, at Montpellier in 1873, were replicated. Second, on sites where phylloxera were assumed to be present, no own-rooted controls were used (Walker 2007). Thus, as noted earlier, it is quite possible that a significant number of sites did not, in fact, produce phylloxera exposure. Finally, it is hard to see how the sample size involved in any (indeed, even all) trials would be high enough to provide statistical significance, let alone confidence of a useful sort. This is another way of looking at the small size of the "target" provided by Jacob and Lider's vines.

Another weakness of the methodology involves its inconsistent use of French (and other European) results. Bioletti begins his work resolving "to make as much use as possible of the previous work of European investigators by commencing where they had left off" (Bioletti et al. 1921, 84). He specifically notes his reliance upon the work of Prosper Gervais, "because of his acknowledged competence" and "because most of his work was done in Southern France, where the conditions are in many ways similar to those of California" (Bioletti et al. 1921, 84). Yet, by the time of Bioletti's work, the French had already disavowed most *vinifera*-American hybrids, AxR1 specifically, as being deficient in resistance, and AxR1 had failed in 1908 in Sicily and in 1915 in Spain (Galet 1979, 200; Granett et al. 2001, 399). But these failures are not noted in Bioletti's report. Thus, it is explicitly claimed that French and European experience will be relied upon in judging phylloxera resistance, yet it is not relied upon in the specific case of the *vinifera*-American stocks, AxR1 included. Bioletti, after mentioning possible problems with the resistance of these vines, accepts them in California "since they are sufficiently resistant where the soil and other conditions are favorable" (Bioletti et al. 1921, 85). Forty-six years later, Lider begins his report by describing the French rootstock research and acknowledging his debt to and use of this work. He then, however, denies the French results with AxR1 and recommends this very vine as the best choice for most California vineyards (Lider 1957).

Evidently, the workers involved did not see the inconsistency between accepting the French and European results and, at the same time, rejecting them as well. Taken together, the institutional failures provide a striking litany of errors, ranging from underfunded and nonexistent research, to psychological denial, to conceptual blindness, to methodological weakness. Human flaws aided and abetted the bug that led to the phylloxera redux disaster in California.

Conclusion

A few sickened vines in an obscure vineyard in the Rhône Valley rapidly became a worldwide disaster for grapes, wine, and the people whose life they were. In the end, everything changed. And things are still changing, because the tiny yellow bug will always be with us. California now thoroughly understands this reality. We will never be able to relax: Darwinian principles tell us that there is always a moving offense and a moving defense, each eternally trying to win the unwinnable contest. Saving the vine from phylloxera is a never-ending battle (Gale 2003).

It is not clear, and probably never will be, how the contest with the bug is to be scored. To the people in the Midi in the 1870s, the score went totally against them: they lost their source of wine, and thereby a central part of their lives. Many of them uprooted, moving to North Africa or to cities. The pain and suffering were widespread and intense. As the bug moved north, east, south, and throughout the world, it caused deep pain and suffering, dislocation, and, in several cases, the intercontinental migration of peoples.

Yet, at the end of the day, winegrowers—at least those who survived—ventured into a completely newborn viticulture, one in which vines were healthier, happier, and certainly more productive of higher-quality grapes than before the disaster. Even today, after the disaster of phylloxera redux, California winegrowers talk about the disaster's "silver lining," the sudden freedom that they have to reconstitute their vineyards according to new methods of growing, training, and cropping their vines, all in aid of producing better wine (Smith 1994). Perhaps this is another instance of Nietzsche's "Whatever does not kill me strengthens me."

Most likely the phylloxera disasters tell us as much about ourselves as they do about science or the bug itself. As different as the bug's attacks were in various places around the world, the initial reaction of the human beings was always and everywhere the same: denial. "We're the exception; it can't happen here." Much of the time the denial was rooted in a perceived moral superiority—region X was suffering the plague because it was morally inferior to region Y (or so thought the residents of Y) in its vine care, its lifestyle, its culture, its wine type. Thus X received God's deserved punishment, or so many perceived it. In this the phylloxera plague was no different than any other plague, old or new, from the ancient tribulations of the Egyptians, captors of the Jews, to the modern tribulations of the gays, sinners all. Plagues as punishment, a common theme underlying denial.

In the end, however, phylloxera kills the vines of the pious and the sinner alike. The Champenois' vines ended up just as dead as the Languedocian's, no matter their morality. Prayers proved ineffective as well, and for the same reason (Garrier 1989). The phylloxera disaster was not a punishment, divine or otherwise. It was, just as Planchon, Bazille, and Sahut had originally proposed, a purely natural affair, and it would take purely natural—not supernatural—means to deal with it. What was somewhat curious about the purely natural means were their organization.

Although the effective responses to the invader were technical—scientific, in their way—they were not discovered by scientists. All of the effective responses except sulfiding—submersion, sand, American direct producers, and grafting—were discovered by practical vine people such as Faucon, Laliman, and Bazille. What is more, these practical men made their discoveries within the first several years of the crisis, and fifty years later Bioletti found himself discussing precisely the same responses for California.

This is not to say that scientists such as Planchon and Riley, Millardet and Viala, did not have crucial roles to play. Legros hits the mark exactly:

> One of the major characteristics of the crisis was the absence of a research structure strong enough to manage the whole ensemble of problems. Thus everyone carried out acclimation or grafting trials on their own hook. This is not to say that researchers didn't intervene, far from it. But a large part of their interventions consisted in pulling what was of value out of the jumble of trials made by *viticulteurs* on the ground. (Legros 2005, 178)

Thus we find the vigneron Faucon observing his flood-saved parcel and asking Planchon, the scientist, "How long can these things swim?" Planchon, intervening via a careful experiment, answers "Twenty-two days." With this information in hand, the submersion treatment is well on its way toward scientific standardization. Until the late 1870s, when Montpellier, Bordeaux, and the other large centers came to be fully organized and operational, this was the way the research was done: practical people experimented, observed, and communicated with the scientists and professional researchers, who then sorted out and made standard what was valuable in the trials and errors of the practical men's work.

Of course, one significant brake on the scientists' early entry into applied work was the seven-year-long theoretical battle between the phylloxera-effect scientific establishment, and the phylloxera-cause rebels in Montpellier. As long as the theoretical battle was alive, the professionals kept each other busy, leaving the field to the practical men.

One of the very first elements of research structure to come on line was the departmental phylloxera committee. Modeled on the original Société centrale d'agriculture de l'Hérault, these committees organized the tasks of research and communications early on in the fray. Again, Legros:

> The departmental phylloxera commissions have especially served to realize such syntheses. The same goes for the viticultural congresses of the time. It is notable that these meetings had a format, at least at the beginning, which is now lost. One would pose a question to the assembly, for example, "Does Clinton resist drought?" and each notable would give his advice in turn. There were no individual communications. (Legros 2005, 178)

By the end of the 1870s, all the departmental committees were solidly established, and viticultural congresses were a regular and frequent occurrence. Moreover, by this time the École had set up regular workshops covering a vast range of crucial subjects, such as varietal selection, training techniques, and grafting methods. But this was just one of Montpellier's roles.

Everywhere one looks in the phylloxera disaster, Montpellier and the Montpellierians are there, researching, teaching, persuading, politicking—they are always busy battling the enemy in any way possible. When Spain needed help it brought Planchon down for a conference; Romania sent its vine researchers to work and study at the École; and Bulgaria brought Viala out for an extended consultation. Even California looked to Montpellier's Prosper Gervais for guidance in their initial scientific responses to the bug. (And, more recently, they certainly should have listened to Montpellier's Boubals and Galet for guidance about AxR1!) Even though Montpellier's vignoble was the first decimated by the bug, there was no necessity that it become the world leader in the successful response to the invader. That achievement is entirely to be laid at the feet of the resourceful men and women of that region.

Of course they had help—lots of it—particularly from America, where the curse originated. Notable was Missourian C.V. Riley's various interventions, first by mail and then in person, when he came to Montpellier to identify the bug himself (Carton et al. 2007). He was there as well when Planchon came to make his tour of America, and when Viala came later seeking a vine for the white soils. After a long visit with Jæger in Neosho, Missouri, Viala went to Texas, where Munson befriended the younger man, leading him to the *berlandieri* growing in the Bell County chalk, effectively firing the last salvo of the first battle with the bug. Central to the story as well were the grape people around St. Louis: Engelmann, Bush and Sons, Meissner, and Husmann in Hermann, all of whom welcomed Planchon for a long stay and then kept him and the Montpellierians supplied with vines, cuttings, and seeds for the next few decades.

From that point on, good solid establishments with good solid people came on line all around the globe. Phylloxera had become a global problem requiring a global solution. And so it remains. As Jeff Granett has recently noted only too well, "Finally, we wish to emphasize that phylloxera is a global problem, and worldwide cooperation of researchers would be a powerful force for understanding and managing it" (Granett et al. 2001, 405).

As we have seen in the recent work on phylloxera genetics, by 2008 workers in California, Australia, South Africa, and the European Union had clearly begun to cooperate, integrating each others' work and pushing the battle against the tiny but powerful pest to a new level. Only through such a unified front can we hope to keep up with the ever-resourceful phylloxera, leading enemy of our closest plant friend, the grape.

Life Cycle of Phylloxera

Phylloxera has an extremely complicated life cycle. First, the bug has two life-styles: *aerial,* in which it lives and reproduces on the leaves; and *terrestrial,* in which it lives and reproduces on the roots. Second, the bug has two forms of reproduction: *asexual,* in which each new generation is produced by partheno-genic females; and *sexual,* in which a new generation is produced by a female descendant from a fertilized egg. Both lifestyles can, under the right circum-stances, produce both forms. Most often, however, the asexual aerial form lives on the leaves of American species vines, while the asexual terrestrial form lives on the roots of *Vitis vinifera.* These two forms are called *crawlers,* in reference to their mode of locomotion. Aerial crawlers can move from leaf to leaf, and from leaf to earth, and then to another vine. Terrestrial crawlers can move from root to root, and to the surface, where they can move to another vine. This mode of diffusion is slow. However, crawlers may frequently be dispersed by the wind over short distances (tens of meters). More frequently, crawlers hitch-hike on equipment, vineyard materials, soil, and persons. Irrigation can also spread the crawlers.

Control of crawler hitchhiking is difficult, but, as Australia's phylloxera quarantine system has shown, it can be effective. Hitchhiking was a major fac-tor in the initial spread of phylloxera in Napa and Sonoma counties during the first infestation of the mid- to late nineteenth century. In Argentina, the bug was widely dispersed by the irrigation systems used in the desertlike conditions.

Under some conditions asexual terrestrial bugs will produce winged forms, alates, which emerge from the ground, and, when fertile, spread the infestation widely and rapidly. European infestations, especially in France, spread this way. South African conditions were especially hospitable to the fertile winged form, which accounts for the rapidity of that country's total infestation. While the winged form occurs in California, it is not fertile. In Australia, the winged form is extremely rare, which is significant for the success of their protection system.

American Wild Grape Species

North America has one of the richest troves of wild grape species in the world, rivaled only by the valleys of central China, west of Shanghai. According to legend, Leif Ericson, the first European to land on North American shores (ca. 1000 A.D.) was so impressed by the profusion of wild grapes that he called this place "Vineland." Because of the wildly varying geography and climate they face, the American species are localized, with no single species covering more than about one-third of the United States. It is generally accurate to say that there is a grape native to every region of the United States, southern Canada, northern Mexico, and the Caribbean. However, despite the diversity of terrain and climate, the European grape, *Vitis vinifera,* thrives in only a small slice of Pacific coastal North America, stretching from the tip of Baja California to mid–British Columbia. Until only very recently, *vinifera* grape plantings in other parts of the continent have perished time and time again from the harsh climate and disease attacks.

This meant that North Americans who wanted to make wine had to rely upon the native grapes they found in the meadows, woods, and prairies of the country-side. As time went on, when they came to find interesting new "native" grapes, such as Concord, Catawba, and Norton, they were able to make better wine than before, unaware that these improved grapes were accidental crosses between native grapes and imported *vinifera,* which had lived just long enough to fertilize a wild grape with pollen. Thus, the Old Americans and direct producers.

Most important for our story, of course, is the fact that many of the American species thrived in phylloxera-infested terrain. Because of this, they were an invaluable source of resistance against phylloxera, either as producers of wine grapes—direct or hybrid—or as rootstocks for *vinifera* scion vines. Not all of the twenty or so American species were equally important for the reconstitution of French and other vignobles. Among them there are six of major significance:

Vitis aestivalis (including *V. aestivalis lincecumi*), *V. berlandieri*, *V. labrusca*, *V. riparia*, and *V. rupestris*. These species are widely distributed geographically.

It is useful to discuss each of these species in turn, describing some of their characteristics, and pointing out how the species was used in the battle against phylloxera.

VITIS AESTIVALIS (AND V. AESTIVALIS LINCECUMII)

Aestivalis and its subspecies *V. aestivalis Lincecumii* are vines from upland country, frequently associated with oak trees. They are real climbers: I have seen native *aestivalis* vines seventy-five feet up an oak tree. Not especially hardy, this species ranges from mid-Michigan (in the form of the subspecies *V. aestivalis argentifolia*) to roughly mid-Georgia. While the species is quite resistant to phylloxera, it has one fault preventing it from functioning as a rootstock: it does not root at all well from cuttings. The main service provided by *aestivalis* in the war against phylloxera was the provision of numerous high-quality hybrid wine grapes. Of all the American species, *aestivalis* most closely approaches the wine quality of traditional *vinifera* grapes by a large margin. Norton, Jacquez, and Herbemont, among the Old Americans most used as direct producers, are *aestivalis* hybrids. Additionally, a surprisingly large number of the hybrid direct producers are descendants of the original J70 *aestivalis lincecumii* × *rupestris* hybrid vine created by Herman Jaeger in Neosho, Missouri, and sent to Eugène Contassot in France. Because of these four vines, *aestivalis* "blood" is now present in grapes in nearly every wine-growing region in the world.

VITIS BERLANDIERI

T.V. Munson led Pierre Viala to this vine deep in the heart of Texas, and the crisis of planting in France's white soils suddenly seemed headed toward a solution. Of all the American vines, *V. berlandieri* can best withstand alkaline, chalky soils. Unfortunately, as the long Berlandieri Wars illustrated, the solution was neither simple nor easy to reach. While pure *berlandieri* resists phylloxera well and thrives in chalky soils, it does not root easily, and thus it can be propagated only with great difficulty. After trying all sorts of methods to induce rooting in pure species plants, everyone gave up. Millardet's next proposal was to plant *berlandieri* seeds and graft to the first-year plant. This solution had been tried before, with *riparia*, and had failed due to the ordinary genetic variation among offspring. It failed with *berlandieri* for the same reason: a population of pure *berlandieri* raised from seed showed considerable variation in resistance, size, and vigor. It was finally realized that the only solution was to hybridize *berlandieri* with some other species and graft onto these hybrids. Millardet's offering was 41B, a cross between *berlandieri* and *vinifera*. Opposition was initially raised to this vine because of fears that its *vinifera* component would weaken its resistance to the phylloxera. But, even though the roots showed swellings due to the insects' attacks, the vines never succumbed to the bug. Hybrids of *berlandieri* with other species have also been successful. Paulsen in Italy made some *berlandieri* crosses and then selected for local conditions, especially in Sicily.

Berlandieri is most heavily used as an element in graft stocks. However, a few complex hybrid direct producers, such as the renowned Villard blanc, contain small amounts of the species, although it in no way contributes any of its predominant features to the phenotype of this variety.

VITIS LABRUSCA

V. labrusca's taste, smell, and mouthfeel are the first things that North Americans think about when they are prodded by the command "Think grape!" This organoleptic complex, exemplified in the skins of the Concord among reds, and the Niagara among whites, is the overpowering sign of the species.

Labrusca grapes generally find their home in the northeastern United States and southeast Canada. They evolved in the cool, wet summer and cold, snowy winter climate there and do not do well in hot, dry conditions, as the French vignerons of the Midi discovered. Moreover, although the species is fairly resistant to many of the fungal diseases, it is not particularly resistant to phylloxera, another unpleasant reality that contributed to the ongoing horror of the southern French vignerons. Unlike most native American species, *labrusca* varieties frequently have large berries, sometimes coupled with large clusters; such fruitfulness is highly appreciated by grape growers. But the species does not generally achieve high sugar levels, which results in a low level of alcohol in the wine. Although *labrusca* grapes are perceived as being quite sweet, this is due not to high sugar levels, but rather to a very low acid background.

For these and other reasons—such as their sensitivity to rot when they are ripe—*labrusca* varieties quite simply are not the best candidates for widespread adoption as grapes for wine. Yet, against all odds, during and after the worldwide phylloxera crisis, certain varieties descended from this species—the Old Americans—were planted nearly everywhere in the world, and millions upon millions of gallons of wine have been produced from them since. Thus, although *labrusca* varieties do not figure in the post-phylloxera reconstitution as a source of rootstock, they must be acknowledged as a permanent component of worldwide winemaking.

VITIS RIPARIA

Riparia is the most widely adapted of all North American species. When Canadians or Midwesterners say "wild grapes," they are typically referring to *riparia* grapes. *Riparian* means "by the river," and that's the favored habitat of this species. But it thrives everywhere, and it is typically visible all along fence lines as one drives the back roads. With its small grapes in small bunches, *riparia* is not a great place to begin from a grape grower's point of view. However, what is important is *riparia*'s all-purpose status: the species is acclimated everywhere from deepest Canada to Tennessee and North Carolina, from the Atlantic coast to the Missouri River valley. It can withstand terrible winters and hot, humid summers, all the while mightily resisting the depredations of the phylloxera bug. Its leaves are quite well protected against fungal diseases, and, nicely enough, it roots very well. As far as the winemaker is concerned, it has a very specific taste, not overpowering like the flavor of *labrusca,* but distinct, and sometimes wedded to extremely high sugar and acid levels. Luckily, the distinct taste is quite easy to breed out in hybrid offspring, and the high sugar is a virtue, although reducing the acidity can be a long-term project for the hybridizer.

Given all its pluses, *riparia* has functioned very effectively both as an element in rootstocks and rootstock breeding, and as a progenitor of many very successful hybrid direct producers, especially Baco noir, which has been planted widely in Europe and North America.

VITIS RUPESTRIS

Rupestris is a vine of meadows and prairies in the south-central United States. It grows low and bushy and is now somewhat endangered because of grazing cattle in its native habitat. Its small berries, low sugar, high acid, astringent tannin, and distinct but not bad taste combine to make the species not of compelling interest to winemakers in terms of its sensory qualities. However, it has good resistance to phylloxera and other diseases and it roots easily. Moreover, in certain cool, well-drained soils it is quite fertile and provides an excellent rootstock, which is extremely easy to graft onto for most *vinifera* varieties. There is one very widely planted pure rootstock, Rupestris du Lot (known in California as St. George), and a number of popular rootstocks hybridized with other species, including *riparia* and *berlandieri.* Its most successful hybrid direct producer varieties were created by Eugene Kuhlmann, at the Alsace Viticultural Institute in Colmar. These include Marechal Foch and Léon Millot, which are standards of high-quality cold-climate viticulture throughout Europe and North America.

Rupestris is another example of the old winemaking truism: a grape's best features come out when it is in a blend, not when it stands alone. The wine world is just lucky that some useful varieties were found before grazing cattle ate up all the vines.

This review has left out many native species—*cordifolia, cinerea, monticola,* and *mustangensis,* to name just a few—each of which has its own interesting features. But for the most part these species have not yet played significant roles in the phylloxera war. Yet who knows what the future might bring? Climate change, new diseases, or future mutations in phylloxera may draw these native species into a new battle, perhaps providing exactly what we need as a weapon against the new foe.

Old American Varieties

These grapes are all natural hybrids among American species and, frequently, *vinifera*. All were tried in Europe post-phylloxera. Most didn't succeed, but some were (and remain) widely successful.

DELAWARE Dark pink grape, fruity white wine; natural hybrid of *vinifera, labrusca,* and *aestivalis,* much appreciated for use in American sparkling wines; failed in Europe.

CATAWBA Dark pink grape, fruity pink wine; genetic background is controversial, but *labrusca* dominates; failed in Europe and parts of the United States; today widely grown in Missouri and Arkansas.

CLINTON Black grape, strong grapey flavors; hybrid of *labrusca×riparia;* once widely planted in Europe, now limited to small plantings in France and several small areas in Italy and Eastern Europe.

CONCORD The most widely recognized (some would say notorious) American grape, it has a flavor that defines *grape* in North America; typically used to make a sweet red wine; hybrid of *labrusca×vinifera;* widely planted in North America, it also exists in scattered locations in South America

ISABELLA Dark grape that produces a highly flavored wine strongly recalling strawberries; this hybrid of *labrusca×vinifera* is planted throughout the world, especially in tropical regions of South America (Brazil) and Asia (India); wines with Isabella's special flavor are undergoing a renaissance in Austria and Italy.

JACQUEZ, HERBEMONT, AND CUNNINGHAM Black grapes producing a dark red wine of good, ordinary table wine quality, with no off flavors; genetic background of these grapes is extremely controversial, although *aestivalis* dominates; still planted in France and Eastern Europe and along the Gulf Coast and Texas, where they resist Pierce's disease.

NOAH Strongly flavored greenish-white grape that makes a very grapey white wine and a faintly strawberryish eau-de-vie; once very widely planted in France and Central Europe; after being outlawed in France plantings decreased significantly, but it is still alive and well, especially in the Vendée, where it is widely used to produce a clandestine eau-de-vie.

NORTON/CYNTHIANA Another genetic mystery, although *aestivalis* clearly dominates; highest-quality native American, prized in France, especially the Midi, for its wine, but ultimately a failure because of its lack of adaptability to French terroir and poor propagation; increasingly popular from Missouri to Virginia for its sturdy performance in the vineyard and excellence in the bottle.

SOLONIS An American grape mysteriously found under this name in the Berlin Botanical Garden and brought back to France to be experimented with as a rootstock type; Munson and other American ampelographers argued that it represented a good species, and it is now universally accepted as *V. acerifolia* (= *V. longii*).

TAYLOR Greenish-yellow grape producing a grapey-flavored wine; hybrid of *labrusca* × *riparia;* not widely planted except as rootstock in France; most famous as parent of Noah.

Notes

1. First called *Phylloxera vastatrix*—"destroyer of vines"—it has been re-classified *Daktulosphaira vitifoliae*. Since the new name is not only unpro-nounceable but also has no historical cachet, I will continue to use the common name *phylloxera*.

2. The story is told from the entomological point of view by Ordish (1972) and from a journalistic perspective by Campbell (2004). This book, based upon new material found in the archives of the École Nationale Supérieure d'Agronomie de Montpellier, and the University of California, Davis, goes into far greater depth than either of these.

3. Although this idiom from economics began as a metaphor, it now is taken quite literally by ecologists.

4. See www.extension.umn.edu/yardandgarden/ygbriefs/p425dutchelm-resistant.html (accessed 12 September 2010).

5. The U.S. government now offers a website dedicated to the dangers of invasive species: www.invasivespeciesinfo.gov/ (accessed 30 November 2008).

6. In 1971 the Southern Corn Leaf Blight caused an epidemic in monoculture maize, resulting in the loss of 710 million bushels of corn (Tatum 1971, 1113).

7. Some might question my ranking. "Consider," they might propose, "the American chestnut blight, which eliminated a fine and economically valuable tree from North America, or the Irish Potato Famine, which caused enormous pain and suffering, economic loss, and migrations of people." But these invasive disasters, although horribly consequential, in the end do not compare to the vast changes that phylloxera caused worldwide.

8. Heretofore the only exception to this rule is the case of Australia. But see the discussion below, wherein it is predicted that Australia's quarantine system is now doomed by intensive and widespread new plantings of vines.

9. While Germany's take-no-prisoners, uncompromisingly militant (indeed, frequently military) eradication efforts succeeded until biological solutions could be effected, there is no question that its climate, generally unfavorable to the bug's life cycle, was an essential factor in the success of the eradication campaign.

10. The term *Big Science* is generally understood to refer to research developments during and after World War II, which President Eisenhower called "the military-industrial complex." I suggest here that an earlier "war"—the war on phylloxera—brought into being an original form of Big Science, justly so called.

11. The closest American analogue to these expositions would be the county fair, an ongoing organization charged with presenting the annual county agricultural show. It is not at all anachronistic to call organizations such as, say, SCAH and the agricultural expositions, respectively, NGOs and QUANGOs.

12. France's present biochemical-pharmaceutical industry finds its origins in this development.

13. A comparable structure did not exist in the United States until the Federal Co-operative Extension, a federal–state–county–research university entity, was authorized by the Smith-Lever Act. See www.csrees.usda.gov/qlinks/extension. html (accessed 25 March 2009).

14. It seems to me that to do an experiment is not solely a rhetorical exercise, although some rhetoricians of science would seem to think so (Bazerman 1997).

15. Burian argues that detailed analysis of arguments is essential to understanding theory change in particular cases (Burian 1990, 168).

16. Philosophers generally regard the historical work of Pickering and Pinch to be excellent. But, in a wonderful irony, most believe, as do Bogen and Woodward (1988), that "in most cases this empirical information can be readily separated from the relativist and constructivist conceptions of scientific knowledge which Pickering and Pinch defend" (314). Burian (1990) remarks that Pickering seems to have provided us with two books, one a historical account and the second the sociological account, "a project executed with far less care and success than the one just characterized [i.e., the historical account]" (164).

1. DISASTER STRIKES

1. Stevenson (1980, 47) claims 1862, but he gives no reference.

2. See the glossary for this and other foreign and technical terms. There is no simple English equivalent for *vignoble*: in effect, the French term refers to a region, perhaps a township at its smallest or a *département* at its largest, where wines are grown according to similar styles and practices. I will use the French term because of its technical precision.

3. Unless otherwise noted, all translations are the author's.

4. The "spot of oil" *(tache de l'huile)* metaphor rapidly became a technical descriptor for an essential behavior of the malady. When phylloxera returned to devastate the California vineyards in the 1980s, talk about "the spot" returned as well.

5. Roughly 2.5 acres.

6. Pouget 1990 is the recently published official summary history of the phylloxera crisis. It is sanctioned by both the French Institut National de la Recherche Agronomique and the Office International de la Vigne et du Vin. Unfortunately, although Pouget does a reasonable job telling the story for the general public, his history is far too Whiggish and, indeed, contains some telling errors.

7. Out of a grand total of ninety-five.

8. The SCAH is a local nongovernmental organization composed of growers, merchants, landowners, professionals (physicians, especially) and other interested individuals. It began life in 1800 (Société centrale d'agriculture de l'Hérault 1900) and soon assumed a very important role in the agricultural life of the Hérault. Not only did the organization serve as a venue for meetings, conferences, workshops, and other forums, it also published reports and, beginning with the oidium plague in the 1850s, acted as a quasi-official agency sponsoring investigations and research into agricultural problems. SCAH was not alone in its role, although it was certainly one of the very first regional societies and one of the most active and highly regarded. This explains why the neighboring society in the Vauclause called upon SCAH to appoint the investigatory commission. Similar local organizations played crucial roles in the battle against phylloxera. Unfortunately, this book must limit itself to only passing mention of their activities.

9. The term commonly means "fly" or "bug"; technically, it means "greenfly" or "louse," with the latter meaning becoming conventional for the phylloxera. The term is also used to mean "children" in the same way that "kid" is used by Americans. It's an evocative term, and I will frequently leave it untranslated where the context allows.

10. Although Hanson made the saying famous, he most certainly found it in Wittgenstein's *Philosophical Investigations,* part II, section II.

11. Marès was no lightweight. Fifteen years earlier, during the onslaught of the oidium, the leaf-destroying powdery mildew, the first of the deadly diseases imported from America, Marès' work with sulfur as a fungicide was original, pioneering, and, in the long run, highly efficacious. Sulfur is still used today in the role Marès discovered for it.

12. The term *pourri des racines* was incorrect, but in these early days it was frequently misapplied to the phylloxera infestation.

13. Pouget's pious declaration that "the polemic between the partisans of *Phylloxera-effect* and those of the *Phylloxera-cause* always remained within very reasonable limits, and did not in any manner retard research on a means to fight the war" is just that: pious. It is refuted, as will be seen below, by many observers of the controversy (Pouget 1990, 29).

14. Riley was a Darwinian (Riley 1871, 91; 1872c, 624), which turns out to be of some importance in the later solution of the phylloxera problem, circa 1880.

15. The variety was American (perhaps from California), hence the phylloxera were on the leaves, not the roots.

16. "Let Cornu, Planchon, and the rest of the scientists worry over whether the insect is the actual cause of the disease or not; *the time to act has come.* Some of the phylloxera discussions seem almost as absurd to me as the discussions in

Chicago as to whether a white cow or a black cow kicked over the lamp that caused the great fire. Let us rebuild our vineyards as the Chicagoans have rebuilt their city" (Guérin 1874, 186).

17. Faucon was not the first to try submersion. That honor goes to Chauzit and Trouchaud-Verdier, who tried it in 1868 and published a suggestive memoir. But the honor of making the method workable and providing a solid demonstration of its efficacy belongs to Faucon alone (Convert 1900, 452).

18. All the best vineyards are located on slopes, making flooding impractical. Although many schemes for dikes, pumps, and terraces were developed, few were workable (Borde de Tempest 1873). Water supply problems were eventually solved by the building of a huge system of waterways, ultimately linking the Gironde with the Rhône (Pouget 1990, 46).

19. Faucon ended up getting rather swellheaded about all this. In more than one venue, including a personal letter to Planchon, he argued that his method ought to be declared the only one that could possibly succeed in killing the phylloxera. Planchon's response is revealing: "I must say to you that I hold your work in high esteem . . . yours is a wonderful technique in submersible terrains. But I cannot follow you that vines located on ground that can't be flooded are simply doomed to be killed by phylloxera" (Planchon 1870–71, 2).

20. The Commission of 1869, while investigating the situation in Bordeaux, made the prescient observation that "some vines of the most susceptible variety [Grenache] seem to have survived on a band of very sandy soil" (Vialla 1869, 347). The observation was not forgotten.

21. Vines planted in sand typically do not result in characteristic wines. Indeed, such wines are usually quite bland and lacking in any distinction— Colares wines in Portugal providing the exception.

22. And they remain so until today. The old vineyards even advertise with roadside signs that read "Authentic Traditional Wines from Vines on Their Own Roots." Traditional though the wines might be, they still aren't very good. The reasons for the earlier hesitation to plant in sand are thereby made obvious.

23. We again remark Pouget's error in claiming that the controversy had no effect on the scientific work. If that were true, why would Planchon comment that "two years have been lost in vain discussion"? Lost years are distinctively costly.

24. The prize was never awarded.

25. Signoret was trapped in Paris during the siege, but he kept to his work, as Riley notes, with his usual style: "Nothing daunted by the siege, the former carried on his studies of this little louse, and wrote by balloon, that though he himself was reduced to cats, dogs and horse-flesh, the *Phylloxera*, which he had in boxes, kept well and in good health. No doubt our enthusiastic friend finds much solace in thus pursuing knowledge under difficulties" (Riley 1871, 86).

26. Riley's report had been instantly translated—"by the infatigable M. Lichtenstein," as Laliman put it—and published (Laliman 1872, 21).

27. This was the first of several visits by Riley. Moreover, when Planchon came to the United States for two months in 1873, it was Riley who hosted his visit. Ten years later, when Riley was the head of entomology for the USDA, he hosted Pierre Viala's extremely important six-month tour. Riley's setting up Viala with the Geologic Survey in order to find American regions with soils to

match France's huge areas of calcareous terrain was singularly responsible for the eventual success of American rootstocks in France.

28. "How, in effect, could it be imagined that the remedy could come from the same plants that were the cause of all the evil?" (Lachiver 1988, 428).

29. Official France was by no means Darwinian. Consequently, there were many scientific doubts about whether the American vines' resistance to the pest was a natural, inbred, heritable feature of the vine itself, or whether, alternatively, it was a feature of the interaction of the vine with the particular American environment. This question vexed nearly everyone, Planchon included, until many years of experimentation with American vines *and* acceptance of Darwin's theory of natural selection eased most minds.

30. The copy of this work held by the Missouri Botanical Garden is inscribed "To Dr. G. Engelmann, as an affectionate souvenir from the author, J.É. Planchon." Engelmann, one of America's great botanists, helped host Planchon in St. Louis, and his collection provided significant data for Planchon's mission. The two men continued to correspond thereafter. For further descriptions of Planchon's visit, see Morrow's excellent article (Morrow 1960).

31. The law got its first reading in 1873, at which point a commission was set up by the National Assembly to determine whether or not the law was justified. After the commission issued a strongly favorable report in early 1874, the law was soon adopted as "The Law of 22 June 1874" (Girard 1874, 92–116). Interestingly enough, the commission explicitly adopted the phylloxera-cause position as "today no longer possible to doubt" and chided "some men, whose opinion is of great influence, [who] have for too long persisted in wishing to find the cause in something other the presence of the Phylloxera" (Girard 1874, 105–6).

32. In one of the more awful ironies of the crisis, M. Fermaud took his rent money and bought American vines, which he planted right alongside the French vines being tested. At the conclusion of the field trials, four years later, the "luxuriance of his new vineyard" contrasted starkly with the "wretched aspect of the withered test vines" (Pouget 1990, 32).

2. LA DÉFENSE

1. The monsieur is identified only by his initials.

2. It is striking to note how often Mancey is cited, in one way or another, in the participant literature. It is not entirely clear why this is the case. The village is quite small and certainly in no way central to the Côte Chalonnaise, in which it is located.

3. Pouget sometimes seems puzzled that Dumas, Cornu, and others supported La Défense, even though they were not industrialists or vendors. He writes that the supporters "have not been exclusively recruited, as one might have thought, from the ranks of manufacturers and of merchants of carbon disulfide or insecticidal application material, but also from among the savants of the era, such as Cornu and Mouillefert, and above all from the heart of the administration of the Ministry of Agriculture and from the Commission supérieure du phylloxéra" (Pouget 1990, 88).

4. Except for in a very small region in Southwestern France, corn—maize—is not eaten, especially as bread, unless one is at the end of one's rope.

5. From 1861 to 1891 the population of Marseilles rose 49 percent, while in the surrounding Bouches-du-Rhône vignoble—one of the very first struck by the bug—it fell 11 percent. In Algeria, the population doubled between 1866 and 1901, but the population of persons of French origin more than tripled (Gachassin-Lafite 1882, 155).

6. Madder root was used as a red dye for rugs (see www.rugreview.com/13–3nest.htm) and as a shampoo and conditioner (see www.unmall.com/aus/montanadayspa/g4_lg.htm).

7. This was the favorite explanation of Faucon and his followers. Compare the arguments of M. Calvet, from Augers in the Charentais-Inférieure (Compte-rendu 1882, 353).

8. Espitalier planted at a density of 4,000 vines per hectare. Figuring conservatively at 80 liters per vine, that amounts to 320,000 liters of sand per hectare. Sand weighs 1.34 kilograms per liter. Hence, to cover one hectare, roughly 500,000 kilograms, or 250 metric tons, of sand had to be moved. Espitalier's system was not for the faint of heart—or weak of knee.

9. By the turn of the century approximately twenty thousand hectares were planted in sand; this constituted roughly 1.5 percent of the peak total vignoble surface (Garrier 1989, 73).

10. That honor belongs to Dr. Seigle, from Vaucluse, who experimented starting in July 1868. Seigle, however, attempted summer inundation, which was extremely difficult to carry out.

11. This amount is equal to fifteen acre-feet, or five million gallons per hectare.

12. One general rule was found to be applicable everywhere: the beneficial effects of submersion were not observed in vine behavior until the second season following the submersion. Correspondingly, the negative effects of suspending submersion were not observed in vine behavior until the second season following the suspension. Phylloxera counts on the roots, however, were affected immediately by either submerging or suspending submersion.

13. "Bridges and Roads," the national engineering department, was analogous to the American Corps of Engineers.

14. Most were puzzled by this. Consider the words of the official report from the 1881 congress: "From the start, a dark point always subsists in both of the treatments. At no moment is the vine completely free from the insects. . . . Summer treatments, be they carbon disulfide, be they sulfocarbonates, applied eight, then fourteen days before we come to inspect, have left subsisting such a large quantity of insects that one is forced to ask what has been the profit of the operation" (Plumeau et al. 1882a, 63).

15. Morrow's account of the trials, tribulations, and successes of CS_2 is certainly the best extant. His criticisms of the French accounts hold even for the contemporary versions of Pouget and Lachiver. Descriptions of the role of CS_2 written by modern historians, however, give short shrift to CS_2 because they depend far too much upon the original narrative and history done by grape scientists. Hence successes of CS_2 are minimized, and successes of American

vines are treated as if they occurred in a vacuum (Morrow 1961, 13–43). My discussion here is strongly informed by his work.

16. Unlike in most of France, the laws of inheritance in Burgundy work against any consolidation of vineyards.

17. Lachiver argues that the process was more individual, and more sinister, than I describe here. When the Burgundian and other northern smallholders saw the success with American vines experienced by those who had been hit first by the bug, they smuggled the American vines into regions not yet infected. This, of course, served to bring the plague in as well. What the northerners were trying to avoid, according to Lachiver, were the enormous costs and losses of the years following the death of the French vines, and before recovery of productivity in the newly planted American vines (Lachiver 1988, 427).

18. In Burgundy, the great mass of vignerons grew the lowly Gamay, whose product brought a very small price in the marketplace. The grand crus grew the famous Pinot noir, which sold at a great price. As the Committee of Vigilance noted in March 1885, sulfur treatment "costs too much and is thus not applicable to the gamay vines" (Laurent 1958, 341). Moreover, and even worse, as the consul general noted, it was "notorious that some of the grand properties treat the same vines several times in a year . . . ; they thus absorb a large part of the subsidy to the detriment of the small properties, whose less remunerative products do not permit making a double treatment" (Laurent 1958, 343). For these sorts of reasons, the small vignerons of Burgundy looked to the vignerons of the Midi: "the gamay proprietors hoped to follow their example and thus cared less to save than to reconstitute" (Laurent 1958, 343).

3. LA RECONSTITUTION

1. "Direct producer" is a term of art referring to a specific set of American vines. The "direct" term means that the vines are planted directly in the ground on their own roots; "producer" means that the vines produce their own wine and are not used as a means, e.g., as rootstock, for some other vine's production of wine. For the most part the direct producers were members of two American species. Thus Jacquez, Herbemont, and Norton were members of the species *V. aestivalis,* while Isabella and Concord were *V. labrusca.* See Appendix C.

2. Initially rootstocks had comprised either pure American species such as *riparia* or *rupestris* or cultivated varieties such as Concord or Jacquez. The second-generation rootstocks were purpose-bred by hybridizing various American species or by crossing American species with European varieties.

3. Pouget is instructive. Speaking of Planchon's initial proposal (see Planchon 1875) that the American vines might be suitable in two roles—namely, as graft stock and as direct producers—Pouget goes on to say, "This double objective, pursued since the debut of experimentation [with American vines], has continued throughout the decades since. It marks an incontestable change in viticulture. After the initial epoch, the objective has enticed a growing divergence, accentuated and maintained until the decade of the 1960s: on the one side, viticulturalists faithful to the traditional varieties of *Vitis vinifera* producing wines of high quality and great renown; on the other, the viticulturalists

adept with the hybrid direct producers for the production of ordinary wines for current consumption" (Pouget 1990, 56). Except for the fact that the divergence still exists, Pouget's words are certainly apropos. Part 1 of Harry Paul's *Science, Vine, and Wine in Modern France* (1996) provides the best overview of the debate for the first half of the twentieth century. Pierre Galet (1988), particularly chapter 5, tells the story from the Montpellier point of view.

4. Respectively, *Vitis riparia, V. rupestris,* and *V. aestivalis.*

5. North America's wild *Vitis* diversity derives from its equally wild climatic diversity.

6. Almost all of the Old American direct producers—Concord, Noah, etc.—were accidental hybrids of this sort.

7. There are several notable exceptions to this. Although their origins are still very controversial, the American grapes called Jacquez and Herbemont were held by many botanists to be purely American (Munson 1905). Galet, the leading contemporary ampelographer, argues that they have at least one European parent (Galet 1956–64). Galet also claims that Norton, a grape becoming ever more important today in the United States, is of mixed parentage. He is, however, more or less alone in this belief, as nearly every other ampelographer since 1850 has held that this variety is pure American *aestivalis.* Among the successful old grapes, only Clinton, apparently a *labrusca × riparia* cross, is uncontroversially pure American.

8. Once the American origin of the bug had been established, Laliman was accused—probably correctly—of having been one of the carriers. He always denied it, and he even got into several court battles to establish his innocence and reestablish his good name.

9. The genus is in the same class as *Vitis,* the genus for grapes. Familiar members of the genus include Boston ivy (from whence "Ivy League") and Virginia creeper (Galet 1956–64). I have found no evidence that Bazille's idea to graft *Vitis* upon *Parthenocissus* was ever tried.

10. Several theories of inheritance were on offer, for example, a "mosaic" view that different parental characteristics would be mixed and matched in the offspring, and Lamarck's notion that acquired characteristics would be inherited. Mendel's theory did not become known until after 1905.

11. Harry Paul's (1996, 81–87) discussion on this point is excellent.

12. According to Morrow, Planchon didn't think much of American architecture, nor "American Sundays." Planchon was treated extremely well by absolutely everyone he visited, and, as he makes clear, he learned a vast amount about America, American vines, and American wines during his trip.

13. Engelmann was an energetic and valuable scientist. Five years before Planchon's visit, Henry Shaw, under Engelmann's tutelage, had set up the Missouri Botanical Garden—a.k.a. Shaw's Garden to residents of St. Louis—which was already well started toward its world prominence by the time of Planchon's visit. In 1863 Engelmann was one of the founding charter members of the National Academy of Sciences. But his work wasn't limited to grapes: he later became the new world's expert on agave classification.

14. Planchon notes that of the California wines, "all of which are made from the European varieties, I can say that those I tasted in St. Louis are inferior to the

analogous wines made from indigenous wines" (Planchon 1875, 77). In addition to Engelmann, who was an expert in the vineyard, St. Louis and Missouri, which excelled with indigenous wine, had other experts on hand. Missouri-born and -bred Dr. George Husmann—who, later, after moving to California, became in great part responsible for connecting the University of California and that state's wine industry—was at that time still in residence at the University of Missouri, where he provided great service to the state's winegrowers and winemakers.

15. The Missouri Botanical Garden holds Engelmann's copy, dedicated with "affections of the author" and signed by Planchon. Although Engelmann's marginal comments are frequent, they are usually limited to an underline or a vertical line. But sometimes there are disagreements, as when Planchon says of the muscadines *"vrilles continues"* (every node has a tendril), and Engelmann underlines the *"continues"* and writes in the margin "no!"

16. Planchon and his colleagues were quite clear on this point. Only the *rotundifolia* actually repelled the phylloxera, which will not feed on the roots. All other varieties more or less supported the phylloxera eating away at their roots. Resistance thus was a measure of how well the two, vine and bug, could get along. Ultimately, the issue was how robust was the resistance: if the vine was stressed by soil or climate or other predations, could it retain its resistance to the bug during this stress? Although many attempts to produce a reliable means of predicting resistance were made, in the end it took a huge experimental trial—using all of La Belle France herself—to make the determination.

17. The actual numbers are: 1873–74, more than 500,000, mostly Clinton; 1874–75, 5.7 million, chiefly Concord; 1875, 14 million, including Cunningham, Jacquez, Herbemont, *aestivalis,* Taylor.

18. This is not the Darwinian sense of "adaptation." Rather, it is something closer to what we today call epigenetic variation, or phenotypic expressional differences under climatic influence.

19. I suspect that this is a veiled reference to Planchon and Riley, although one could never be certain that it was.

20. It was not clear to either the Americans or the French that these two varieties were distinct from one another. Whether they are continues to be controversial.

21. In a footnote on page 6, Planchon remarks that he felt the reservations of his colleagues were excessive. As far as he was concerned, he "retained no doubt about the fact of the general resistance of *aestivales, cordifolia, rotundifolia,* and even the majority of the *labrusca*. My reservations concern exclusively the *degree* of this resistance in such and such particular variety" (Vialla and Planchon 1877, 6–7).

22. The nature of the soil is also an essential issue. For example, until Viala's trip to America in 1887, there were no American vines that could survive in the extremely chalky soils of the Cognac country. Moreover, certain graft stocks have less affinity for certain scions and vice versa. Again, it would be a long-term experiment to find which graft stocks would be successful where, and with which European scions.

23. Planchon and Laliman were right about the resistance; what they did not expect was the inability of *cordifolia* to stand the least amount of chalk in the

soil. It fails above 3 percent chalk content; hardly a soil in France is this free of chalk (Galet 1988, 112).

24. Over the next few years Millardet modified his view. As he noted in 1886, some first-generation hybrids are precisely as resistant as their parents (Millardet 1888, 29).

25. Again we see a problem resulting from the lack of an appropriate theory of inheritance. Millardet believed that pure wild types would breed true to type. But when the seed propagation was carried out, variations appeared, indicating that the true-to-type concept was not correctly applicable to wild populations.

26. It is difficult to know where Engelmann got the seeds. Ordinarily, seeds are gathered from the residue thrown out after winemaking. But, since the Missouri winemakers were using cultivated varieties, this would not have been a source of wild seeds. Perhaps Engelmann had others gather bunches of wild berries, which he then fermented.

27. It is worth noting that the French edition was published jointly by the two leading medical-botanical presses in the country, Coulet in Montpellier and Delahaye in Paris. Distribution was worldwide. UC Davis's copy of the volume had been earlier deaccessioned from the UC Berkeley library, where it had arrived immediately upon publication.

28. A typical example: on the occasion of a conference on the American vines at Châteauneuf-sur-Charentais—located in the midst of that difficult chalky terrain northeast of Bordeaux—local physician Achille Aubert made the review presentation. Concluding his speech optimistically, Aubert remarked that surely, given the three hundred varieties in the catalogue, among them were some "that offer the highest guarantee . . . and are appropriate for our terrains" (Aubert 1880, 16).

29. The first issue of this journal, which was dedicated to all aspects of viticulture using American vines, appeared in early 1877. It was started up by the famous Beaujolais viticulturalist and proprietor Victor Pulliat. Pulliat's journal was a direct challenge to the *sulfuristes* then controlling Beaujolais with the iron hand of official bureaucracy. Counted among Pulliat's editorial board were such "illustrious collaborators as Gaston Bazille, Laliman, Ganzin, Foëx, Viala, Planchon, Lichtenstein, Bouschet de Bernard—all the grand names of the antiphylloxera war" (Garrier 1989, 110).

30. This was likely from chlorosis, a deficiency caused in part by the inability of some American vines to acquire sufficient nutrients from a chalky soil. Chlorosis became the leading issue in research from the early 1880s on, as will be discussed below.

31. The journal's position here is extremely conservative: surely by 1880 the immunity of the American vines to "the pricks of the insect" could be taken as demonstrated? *La Vigne française*'s reluctance to accept the evidence even at this late date suggests that other attitudes—a straightforward anti-Americanism, for instance—should be inferred.

32. Bazille's foregrounding of the length of time of his—and the other Montpellierians'—experience is significant. Many of his listeners were only newly attacked; others had suffered for somewhat longer. But in the end, none had the seven to eight years' experience of the Midi growers.

33. Perhaps at one time there was something to the accusation. Galet reports that "during the time of the crisis, French growers were willing to pay high prices (1 to 2 gold francs) for cuttings of this direct producer [Othello, a black hybrid created in 1859 by Charles Arnold by crossing Clinton × Black Hamburg]" (Galet 1979, 161). But this would have been several years earlier than Bazille's speech, when demand for Othello was at its peak.

34. As noted earlier, Dumas was only converted to this view after his hopes for La Défense were dashed.

35. Bazille's guess was not far off. Despite repeated condemnations, legislation, and active attempts to search it out and destroy it, Jacquez stubbornly hung on in the Midi and other regions in France. At the time of an official survey (1955) there were still 546 hectares of the vine planted (Galet 1988, 346). Although most of this area may be presumed uprooted during the subsequent campaign, Jacquez persists in the Cevennes region, especially in the region's national park, where it has become an element of the official patrimony and can be legally made and sold. And, of course, Eastern Europe down to Greece has innumerable plantings of Jacquez, which still produce its sturdy, rustic wine.

36. Balbiani's theory denied the importance of parthenogenesis and argued that the virginal generations must be interspersed with sexual generations. The latter occurred in the winter and resulted in the laying of an egg in the bark of a vine. "Hence Balbiani was sure, and convinced others—particularly leaders in the French government—for many years that destruction of the winter egg would mean the ultimate extirpation of phylloxera" (Morrow 1960, 28). But Boiteau bred virginal bugs to the twenty-fifth generation without any diminution in the population's vigor. Unfortunately, it was not until almost the end of the 1880s that efforts inspired by Balbiani's theory—heavily treating the vines with insecticide during the winter, scraping the bark with special chain mail gloves, and so on—finally ceased for lack of results. "But so great was Balbiani's authority, as the discoverer of the life cycle of phylloxera, that according to Grassi the failure of his plan was never published in as explicit terms as it should have been" (Grassi 1915, 1271). It is not clear when or even whether de Lafitte gave up Balbiani's theory.

37. This claim, if true—as it very well might be—is odd. Balbiani was a well-respected scientist, especially following his brilliant work in deciphering the arcane life cycle of the bug. One can only think that solutions other than his were more direct, and hence more attractive. Had Balbiani's theory been true, it would have taken years of painstaking labor to eliminate the bug. Only those who were heroically committed to an absolutely pure French viticulture would be prepared to endure it.

38. It is worth remarking that the *Revue* was Planchon's venue in 1874 and 1877 when he wanted to take his views about the origin and cause of the phylloxera crisis to the most important audience.

39. De Lafitte's argument doesn't just present a slippery slope; it also explicitly covers up what had been admitted for years, at least since Planchon's and Riley's publications six years earlier, namely, that resistance varies among American species, and *labrusca* especially is not resistant. Fitz-James calls de Lafitte on this rhetorical ploy in her response published one month later.

40. Anatomical studies were carried out chiefly by Foëx at Montpellier and Millardet at Bordeaux. Each criticized the other's theory, although in the end Foëx confirmed that Millardet's scheme for characterizing a variety's resistance (although not the reason or reasons for it) was operationally sound (Pouget 1990, 65–66).

41. Not only was de Lafitte's criticism trenchant at the time, it continues to hold. We still don't know the basis of the resistance.

42. That is, the local committee has reported a phylloxera infestation.

43. Later in the year, at the Bordeaux Congress, this distribution according to geography was affirmed by the Committee on American Vines: "The commission has been struck to see that Jacquez was remarkably more vigorous and more fruitful in the meridional region than in the southwest.... Herbemont and Cynthiana/Norton seem, on the contrary to succeed better in the southwest than in the Midi" (Gachassin-Lafite 1882, 124). Just as would be expected, Jacquez does better in the hotter Midi, and Herbemont and Cynthiana/Norton do better in the cooler Gironde.

44. Here she writes in English.

45. De Lafitte's account of inheritance is extremely cursory, limited to the observational or phenotypical level. In part, his lack of theory may be explained by the fact that he needs only the most cursory facts about inheritance to make his argument plausible. However, there did exist more developed—non-Mendelian, of course—theories of inheritance in grapes, most particularly those of Millardet and others at Bordeaux. It must be assumed that de Lafitte knew about them and chose not to deploy any of their features. And he certainly would have stayed away from the notion then coming to the fore in Millardet's thought, namely, that among first-generation hybrids, a small number return to parental type completely in some organ systems.

46. But de Lafitte is kind enough to note that "Millardet takes the pain to alert the reader to his 1877 work" (de Lafitte 1883a, 539).

47. Galet puts its western limit at Indiana (Galet 1979, 49).

4. THE UNDERGROUND BATTLE

1. Paul's comment is directly on target: "Given our wisdom from historical hindsight, we are surprised that the École d'agriculture de Montpellier did not play much more of a role in hybridization and in viticultural genetics than it did" (Paul 1996, 86). Reasons given for the École's absence from this area vary considerably. In my view several issues were involved. First, the intense criticism the École received around 1870–71 for concentrating on science rather than practice brought about some research reticence. Second, its initial lack of success with higher-quality direct producers (especially the failure of Norton to grow locally), even while Bazille's success with grafting from 1872 on was so impressive, pointed interest toward grafting. Finally, the "blending" theory of inheritance held at Montpellier strictly predicted that any hybridization would produce decreased resistance to phylloxera.

2. "After 1879, viticultural meetings, organized by the school and the Société centrale d'agriculture, attracted people from all over the Hérault as well as

winegrowers from Italy and Spain. The advice of the school's professors was sought everywhere in France. Viala and Ravaz were consulted in most of the viticultural world. The works of the professors on viticultural topics were best-sellers" (Paul 1996, 88).

3. Garrier looks at costs from the Var, Alpes-et-Haute Provence, the Midi, Bordeaux, and Beaujolais. The cost in each department averaged nearly three thousand francs per hectare, so the variation among regions was not great.

4. This is one of the few points Ordish gets flat wrong. He complains that the notion of adaptation "suggests a Lysenkoist belief that any given cultivar through successive vegetative propagation gradually 'adapted' itself to local conditions and became more suitable, which, as we know today, is nonsense" (Ordish 1972, 129). The French knew this notion to be nonsense as well; if they were not in-clined toward Darwinian thinking already, the unceasing pressure in that direc-tion by Riley and other Americans gave them no other choice. Careful reading of either a straight practitioner such as Sahut or a theoretician such as Millardet would have shown Ordish the error of his interpretation. Probably he just read too much into the term itself.

5. Foëx's argument is unfortunate. At that point neither he nor any other of the viticultural scientists knew enough about the interaction between soil ions. This topic was only just coming into prominence in agricultural chemistry in general. It would take several more years before the viticulturalists would learn that the action of the bicarbonate ion (HCO_3) tied up the available iron in the white chalky soils, rendering it unavailable for the plant.

6. Gloire is a varietal selection from among wild *riparia* made by the faculty at Montpellier in light of its excellent performance as a graft stock.

7. Just over a year later Sahut reprised this pamphlet during a presentation at a congress in Toulouse. Although his recommendations there were no differ-ent from those seen here, his statement of the problem was particularly acute and filled with sympathy for the poor vigneron: "How many times have I not viewed with sadness in my excursions, be it round the environs of Montpellier, be it in several other regions, plantings of American vines made in terrain that was absolutely improper for them, where I see them subsisting miserably during some years and finally then disappearing! I take pity upon the poor cultivators who, having been nearly ruined a first time by the phylloxera, employed all their finances to reconstitute their small vignoble, only to watch with sadness all their hopes being annihilated" (Sahut 1888, 8).

8. It should be noted that true hybridization had been carried out in North America since early in the century. Men such as Charles Arnold and Edward Rogers had produced many different varieties by crossing among both Ameri-can varieties and European varieties. Arnold's Othello, produced in 1859, in fact was widely expanded in France during the 1870s and 1880s as a direct producer. But it was Millardet who understood enough about inheritance to predict the eventuality of the hybrid direct producers (HPDs), which combined American resistance to phylloxera and cryptogamic diseases with European fruit quality.

9. This vine was "discovered" by Engelmann, circa 1868, in the Botanical Garden of Berlin, its provenance unknown. Argument immediately erupted

over what it was and where it came from. It soon began to play a significant role as a graft stock in the Midi. American ampelographers, most significantly Munson and Bailey, maintained that it was a species to be called *V. longii,* while European ampelographers, namely, Viala, Ravaz, and Millardet, maintained equally forcefully that it was a complex American hybrid. It was not until 1985, when the world's leading ampelographer, Pierre Galet, visited Munson's old stomping grounds in northern Texas and southern Oklahoma, that the issue was settled once and for all. Galet observed "numerous samples" of Solonis "completely isolated without any other species anywhere around" (Galet 1988, 62). Galet's observations were sufficient to raise Solonis to the rank of a species, *V. longii.*

10. Apparently, the idea that pollinating could be done by insects was controversial at the time. In a footnote describing insects he had seen on blooming grapevines in sufficient number to do the job, Millardet posed the following rhetorical question: "In any case, if it is the wind that must do the pollen transporting, why is it that all vines have so penetrating an odor at the moment of blooming?" (Millardet 1882, 470). Why, indeed, if not to attract insects? And the odor really is penetrating—it can be smelled in a closed bedroom thirty yards from the vine, as I have verified myself many times! In the next decade, credit would be given to Millardet for settling the issue of how insects pollinated so many plants.

11. Millardet takes time to refer to the looming possibility of a *riparia* disaster equal in scale to that of the *labrusca* disaster, all because "the *riparia* cannot flourish in either wet clay, or dry, thin chalky soils" (Millardet 1882, 478).

12. AxR1's failure to resist phylloxera is responsible for the California disaster of the 1980s, about which more in the following pages.

13. Millardet teamed up with Marquis Charles de Grasset of Pézenas (in the Hérault), and from 1880 to 1892 the two made and tested thousands of hybrids, from which three vines of permanent importance—41B, 101–14Mgt, and 420A—were selected. 41B is France's most important rootstock today.

14. Mendel's work would not be of any significance in France until much later (Burian et al. 1988). Interestingly enough, Mendel was already being cited in the nearly contemporaneous work of Munson (see Munson 1905).

15. Paul calls Kuhlmann a "technician" (Paul 1996, 84). In fact, Eugene Kuhlmann was director of the Institute of Viticulture in Colmar, Alsace, a position in which he succeeded his father-in-law Christian Oberlin. Although Kuhlmann's impact upon Alsatian viticulture was not inconsiderable, it is in America that his work has come to full fruition. His hybrids Marechal Foch and Léon Millot (named for a friend in the senate) are planted widely in the United States and Canada, where they produce "the best red wine among the French hybrids" (Galet 1988, 441). Interestingly enough, Kuhlmann's American parent of these two sibling crosses with Pinot Noir Précoce is a seedling of 101–14Mgt, a Millardet–de Grasset hybrid.

16. All three are Franco-American hybrids; Oberlin and Baco are F1s—first-generation crosses.

17. I suspect that Millardet here is thinking of Darwinian variation, in the sense of the basic material for evolution. Millardet, like many others at the time

(including Mendel himself), apparently held that stable natural hybrids could found new species.

18. Galet claims otherwise (Galet 1988, 287). According to Galet, Millardet made the cross in 1882 with the goal of finding a hybrid rootstock suitable for the white soils, but this claim is not credible. First, Millardet was looking for direct producers when he first made the cross (Millardet 1881); moreover, even as late as 1888 he was still looking for directs (Millardet 1888, 26). Second, Millardet was looking for phylloxera resistance. Third, *berlandieri*'s extreme tolerance for chalk was not, in 1882, its highlighted property; rather, it was prized for being highly resistant. Finally, the general and overwhelming need for a rootstock adapted to white soil had not yet come to occupy the foremost place in consciousness, even among the scientists. All this would be changed by Viala's mission.

19. Of course, no matter what he says here, he brought back scads of specimens, including cuttings, seeds, and rooted plants. How could he have failed to do so?

20. One wonders why, exactly, he visited these states. None was predicted by the Geological Survey to have chalky soils; moreover, these were almost precisely the same states—with the exception of Missouri, whence Viala was bound next—that Planchon had visited fourteen years earlier. Planchon had observed no chalky terrain, so why did Viala think it worthwhile to revisit them?

21. This is the parentage of the justly renowned Jaeger #70, which founded so many lines of extremely successful HPDs.

22. Very recently *V. candicans* has been renamed *V. mustangensis* (Lamboy et al. 1998). Although the species most likely got its name from the fact that wild horses—mustangs—enjoyed eating it, locals claim that it equally comes from the fact that the species has a very strong and unpleasant taste, one with a real kick to it.

23. This is doubtful prima facie. Most likely it is Montpellier speaking again its view on hybrids.

24. Munson was certainly America's greatest contributor to world viticulture. Unfortunately, there is no completely satisfactory biography of him. Jacobs 1975 is typical: neither very complete nor very accurate, it fails to appreciate the main point, Munson's significance in the reconstitution of French viticulture. The one book-length study (Renfro and McLeRoy 2008) is thoroughly researched but contains many errors. Greyson County College in Munson's hometown of Denison, Texas, has inaugurated a Munson Memorial Vineyard in conjunction with a program in viticulture and enology. And in view of Munson's essential aid to Viala's saving of the Charentais vignoble, Denison and Cognac have become sister cities.

25. Munson's own creations were not particularly successful in France (Galet 1988, 385). This is not unexpected since Munson did not select for wine quality; however, according to Ravaz, at least one of his creations, Hopkins, produced wines as good as Norton, which is a very high quality indeed. Unfortunately, it was not adapted to most French terrains (Ravaz 1902). One of Munson's table grapes, Bailey, has been widely successful around the world in such varied spots as Brazil, India, and Korea.

26. Berget's comparison of this debate to that of the *sulfureurs* versus the Americanistes is apropos (Berget 1896, 41).

27. Galet's view is more nuanced but nonetheless reveals the opposition of the École to hybrids, especially in regions where quality wines were grown (Galet 1988, 369).

28. Foëx—then director of the École—strictly followed the Montpellier line that French viticulture could be saved by grafting all the old varieties upon only a very few selected American rootstocks. Moreover, these rootstocks must either be pure American species or Americo-American hybrids. No Franco-American crosses were even to be considered. For this reason, the hybrids of Couderc and others never entered into Montpellier's thinking (Paul 1996, 70).

29. Paul Gouy, founder of *Revue des hybrides américains,* derided Foëx's officially settling for a paltry half-dozen Americo-American crosses as the rootstock weapons against phylloxera (Paul 1996, 70).

30. This was a daring cross. It is not clear why Millardet even tried this pairing in the first place. Given the view on inheritance he held at the time, there was a clear risk that phylloxera resistance—his goal at that moment—would be superseded by the *vinifera* heritable factors. However, since he was certainly aware of both the elevated resistance of *berlandieri* and the difficulty of propagating it, it is plausible to speculate that he hoped to keep the resistance even while achieving some improvement in the propagation problem. Nicely enough, this was exactly the outcome.

31. "Work and science" came to be a mantra as the disaster spread out from the south. The phrase's origin is obscure, but it certainly came from someone at Montpellier.

32. The hybrid 41B is a cross between a widely grown *vinifera* wine grape, Chasselas, and a *berlandieri* seedling from Texas. Millardet made five series of crossings between these grapes; 41B comes from the second series, made in 1882. It is the best choice for chalky soils even today. Although the major reason that pure *berlandieri* were not used is their poor propagation from cuttings, 41B can be tricky as well. Galet reports that "the take is often mediocre, from 15 percent to 40 percent, depending upon the care taken by the nursery and the rainfall." At Galet's experimental vineyard near the École his take was 53 percent, but the rainfall that spring was high (Galet 1956–64, 265). The 41B hybrid reached its greatest planting expanse in 1973, but by 1984 it had dropped to third place, which is roughly where it remains today (Galet 1988, 210).

33. Today the park is called Place Planchon. It is at the most important crossroads of modern Montpellier, directly across the street from the train and intercity bus stations on one side, and, on the other, the city bus headquarters and the major downtown bus stop and transfer point. The park itself is calm and restful, full of people sitting on benches or strolling the paths under the old, very large trees. Planchon's statue dominates the main entry to the park, where it can even be seen from the street. Just past the monument is a large shield labeled "Place Planchon" that lists the rules for use of the park. Unfortunately, it carries no information about Planchon himself. Most likely the monument gives all the information needed for those interested.

34. In a wonderfully revealing interpretation, the sketch of the statue in the agricultural students' alumni yearbook has Planchon gazing down fondly into the eyes of the viticulturalist (*Annuaire* 1895, 55).

35. Twenty years later there was a quiet, relatively muted dispute. From the start, Planchon had gotten all the credit for the discovery of the phylloxera. He did nothing to deny the attribution; indeed, in his written reports, especially the articles in the *Revue des deux mondes* (Planchon 1874), he frequently refers to himself alone. Sahut finally became unhappy about his exclusion from the discovery and decided to set the story straight. Donnadieu, Planchon's specimen preparer on the fateful occasion in question, eventually took Sahut's side, arguing that it was Sahut who first pulled up an apparently healthy vine in order to look at its roots and pointed out the yellow points to Planchon, whereupon Planchon looked at the points, declared they were insects, looked again with his loupe, and declared that they were *pucerons* and, more importantly, the cause of the disease (Donnadieu 1887). So who, then, *really* discovered the phylloxera? By 1887 the question probably wasn't worth arguing; it certainly isn't worth arguing today. In his obituary for Planchon, Foëx said that "after eighteen years . . . the truth of his being the discoverer has been contested," but "for those who have known his disinterested love of, and scrupulous respect for, the truth, his own word will suffice." Foëx then reprints verbatim the passage from the *Revue des deux mondes*. (Foëx 1890, 4). Dean Sabatier, in his elegy, is very careful, very fair: "Arriving on the property, the members of the commission pulled up some vines attacked by the malady in order to examine the roots. M. Sahut was the first to perceive and to announce the yellow points on a root that he passed to Planchon. After a rapid examination, Planchon immediately cried out, 'These are scale insects,' and an instant later, 'These are *pucerons*,' and, at that first moment, he designated them as the cause of the malady, as the enemy" (*Inauguration* 1895, 46). Sabatier solidifies Planchon's claim a page later with the remark, "Not only was Planchon the first to say that the yellow points perceived initially by M. Sahut were an insect, not only did he denounce them as the cause of the malady, but he also made with his brother-in-law J. Lichtenstein . . . a persevering study of this insect and its diverse forms."

5. PHYLLOXERA MAKES THE EUROPEAN GRAND TOUR

1. There is controversy over the date. Although the official date given is 1879, Paulsen has provided strong evidence that in fact the region north of Milan was infected by 1870 (Paulsen 1933, 153–57).

2. It must be noted at the outset that Gervais' review limited itself to considering "the method which everywhere . . . has been recognized as the most sure and the best," namely, "the grafting of indigenous vines upon American rootstocks" (Gervais 1903, 11). Gervais explicitly rejects direct producers—Old Americans or new hybrids—as a plausible method of reconstituting phylloxerated vineyards: the Old Americans are "quickly fallen from memory" and the new hybrids are "in the future, which, because it is the future, is uncertain" (Gervais 1903, 90–91). This narrow perspective misses some very significant

aspects of the reconstitution, not just in France but all over Europe, as will be made clear below.

3. Most of the rootstock research conducted in California took Gervais' work as its jumping-off point (Bioletti et al. 1921, 84).

4. Four years later Gervais would be called to investigate the revolt of vignerons in the Languedoc. The riots occurred because of the disastrous slump in wine prices due to overproduction, which would become a chronic problem in the southern vignoble, now rationalized, well managed, and efficient—indeed, everywhere improved in the wake of changes made while responding to the crisis. See Clavel 2007.

5. Ironically enough, Clinton, which failed as a rootstock, was enormously successful as a direct producer. Indeed, large plantings of Clinton continue in Italy today, as will be discussed below.

6. Selection is an excruciatingly localized phenomenon. For example, suppose that Millardet crossed American vine A with European vine V and planted out one hundred seedlings in his test plot in Languedoc. Several years experience might prove that seedlings number 13, 17, and 19 were well adapted to his local conditions. But Paulsen, in his environment in Palermo, using the same A×V cross, might find that seedlings number 71, 83, and 91 were best adapted to his conditions. The ultimate conclusion is that there is no universal rootstock.

7. As we now know, at this time most European vines were infected with viruses. Such infections produce severe compatibility problems with most rootstocks, in particular the *berlandieri* stocks required for chalk soils, which predominate in France. *Riparia* and *rupestris,* however, are quite tolerant of virus-infected scions.

8. Interestingly enough, olives were a frequent—although not the majority—choice of alternate crops. This choice was distinct enough that the government survey called it out individually, while other alternative choices—beets, cereals, and forage—were all grouped together.

9. Here again, Gervais' limited perspective leaves out a significant part of the story, namely, the wide expansion of surface planted to Old American vines, vines that are still there today as producers of traditional Uhudler—wine with an old-fashioned strawberry flavor—stubbornly held on to by the vignerons against the efforts of the EU to rip them out, as will be seen below.

10. Eke, head of the Hungarian department of plant protection, disagrees, setting the date as 1876 (Eke 2004). However, since de Horváth was there and Eke is speaking about the *official* first date, the discrepancy is negligible.

11. According to Eke, the Hungarian Phylloxera Station is "considered as the first nucleus of the Hungarian plant protection organisation" (Eke 2004).

12. Speaking of seemingly ridiculous processes, it has recently been found that applications of powdered shrimp and other crustacean shells are effective against nematodes in fruit crops (Brown et al. 1979).

13. Wild *riparia,* Taylor, Clinton, Clinton Vialla, Solonis, Elvira, Jacquez, Herbemont, Cunningham, Rulander, Norton's Virginia, Cynthiana, Concord, Ives Seedling, and York's Madeira.

14. Interestingly enough, Viala also recommended that certain of the indigenous varieties, which he thought of high quality, be emphasized in the reconstitution and that certain less valuable local varieties be phased out (Bulgarian Wine 2008).

6. THE BUG GOES SOUTH

1. Chile, for reasons that are still unclear, was never invaded by the phylloxera, although stories are currently circulating that this holiday has ended, since many replantings are now done with grafted vines. Chileans explain that this is because stocks provide protection against nematodes and allow better control of crop loads (Gilby 2001).

2. France did not precede the Australian colony in Victoria by much, a decade at most. And for many reasons, mainly political ones, the Victorians didn't avail themselves of French knowledge until very late into their own disaster, as we will see shortly.

3. The comments of John Whiting were very helpful in writing this section.

4. Known phylloxerated areas are classified as Phylloxera Infested Zones (PIZ); areas where the phylloxera status is unknown (but which are presumed to be free from the bug) are Phylloxera Risk Zones (PRZ); and areas known to be free of phylloxera (historically or by inspection) are Phylloxera Exclusion Zones (PEZ).

5. The first outbreak since the early 1900s occurred in the Goulburn Valley in 1987, and then there was a subsequent outbreak in 1991 in the King Valley, where eradication of a portion of the vineyard was attempted but the phylloxera was not contained. Various other infestations have occurred in the Ovens Valley (1998), Strathbogie Ranges (2000), Buckland Valley (2003), Yarra Valley (2006), Murchison (2007), and Macedon Ranges (2008) (Whiting, private communication, 28 April 2008; Dunstan 1994).

6. Although the winged form is observed in the most common biotypes, it is, at most, in a 1:10 ratio with the crawler ("wanderer") form. Moreover, the winged form does not reach sexual maturity under Australian conditions (Whiting, private communication, 28 April 2008).

7. The point here is that since both stations are in the phylloxera-excluding sands, the only source of infection is beyond the exclusionary zone, namely, the nearest upwind vineyard.

8. The ungrafted vines were pulled out eight years later, dead or dying from phylloxera.

9. This notion of "one stock, one variety" is widespread among the researchers worldwide. It is particularly evident in the California research programs at this time, as will be pointed out later. It is the loss of this notion that is in some important ways responsible for the tragic phylloxera redux in California in the 1980s–90s.

10. The mechanisms underlying such precision are only now coming to notice and examination via topics such as developmental plasticity and epigenetics. See, for example, West-Eberhard 2003.

11. A Couderc cross of *V. vinifera* × *V. rupestris*.

12. Yet, as Morton so devastatingly points out, even given the research results, growers—at least many California growers—don't necessarily pay attention (Morton 1994).

13. The correct date is 1877.

14. This failure should have served as a caution to the Californians, since it was widely reported in the literature, and, as Southey remarks, at that time it was suspected that "differences in the susceptibility of certain rootstocks to phylloxera in different regions in South Africa suggested the possibility of the occurrence of biologically different races of phylloxera" (Southey 1992, 28). Perold discussed this possibility in the 1920s: "I am inclined to believe that a *new biological race of phylloxera* has evolved on Aramon [AxR1] roots in Helderberg, where Aramon has been the almost exclusive stock" (Perold 1927, 54; emphasis in original). Somehow the Californians missed these events and their publicity.

15. This was also one of the major factors pushing the use of AxR1 in California (Walker, private communication, 17 October 2007).

16. The exact nature and provenance of this grape is highly controversial: Robinson notes that it is, in fact, the famous Dolcetto, as well as other claims identifying it as California's Charbono (Robinson 1996, 32). In any case, it makes a solid, flavorful wine under Argentinean conditions.

17. This quartet of defenses—insecticides, submersion, sand, and grafting—deserves to be called "the four remedies." They were discovered, tested, and perfected by the French between 1870 and 1890. Every invaded country, including California (Bioletti 1901, 9), subsequently tested all four, found them effective under their conditions, and used them. But lack of resources, whether water, sand, or the money necessary to support CS_2, ultimately led most to rely upon resistant vines, whether as direct producers, as rootstock, or as both.

7. THE OLD AMERICANS

1. Since these vines had been grown commercially for some time in America, and were either full-blooded natives or hybrids containing large percentages of original native American species, the name "Old American" soon stuck. Galet, in his careful, thorough, and thoroughly negative recounting of the plight of these vines in France, consistently calls them *"cépages prohibés,"* the prohibited varieties (Galet 1988, 377).

2. Although there are many tales about why the taste of *labrusca* grapes is called "foxy" and the grapes themselves are called "fox grapes," the real origins of the vulpine reference is a mystery.

3. This origin story applies to such famous successes as Concord, Delaware, and Catawba, among others.

4. There are countless names for this vine, which is usually known as "Lenoir" or "Black Spanish" in the United States and as "Jacquez" by the French. The latter name will be used in what follows.

5. T.V. Munson, America's foremost grape man, coined a new species name for these two vines, *bourquiniana,* after the owner of the estate where they were found (Munson 1905, 143). Today many think that the two are *aestivalis*-based

complex hybrids, like Norton, but this opinion is not universally held among grape taxonomists.

6. The one exception to this was Norton/Cynthiana, which, although prized by the French, was a failure because of its difficulty of propagation and its very limited soil adaptation.

7. An obvious exception is sherry, which uses oxygenation as a part of its manufacture.

8. There is no white Isabella in North America. The origins of the Italian white variety are unknown. While there is the possibility that it is a bud sport of the original black Isabella, the physical characteristics of the white cultivar are so different that they suggest a mistaken name acquired long ago.

9. The meaning of *Uhudler* is the subject of many tall tales. In the end, no one knows for sure what it means.

10. An exception was at the German wine and vine research establishments. Thus, for example, Geisenheim and Geilweilerhof kept their hybrid vines and continued their research, which is today producing rich payoffs, with mildew resistant, fine-flavored varieties such as Regent and Solaris.

11. A later decree was quite explicit and direct: the vines must be uprooted because "they expose you to sanctions, they give you bad wine, and they are no longer à la mode: they are relics of the past" (Galet 1988, 378).

12. Galet notes that this method of reporting completely eliminated any possibility that the government would have a handle on how many hectares of these vines were planted and where they were.

13. Needless to say, this is a huge area by American standards. It comprises a number of vines greater than that grown in all but the top three American wine-growing states.

14. This is most likely Oberlin Noir, a second-generation—Franco-American—hybrid from Alsace. While it is not, strictly speaking, prohibited from being planted in France, as are the Old Americans, it is nevertheless not authorized for commercial traffic.

15. And, obviously, regardless of their objective qualities.

16. In this context, the term *l'Antan* is best rendered in English as "yesteryear"; hence, the name of the website is "The Vines of Yesteryear."

17. He had. Not only that, the following June an informal tasting of wine from the Cévennes was held in an important restaurant on the Right Bank. Agreement among the assembled critics, chefs, and wine industry members was universal: the wine was well balanced, properly fruity, and actually of quite good quality. There was absolutely nothing in the wine to rank it as anything other than a good-quality French *vin de pays*.

18. Couderc is a descendant of the famous French hybridizer Georges Couderc, who produced some of the most important rootstocks and HPDs.

19. While the methanol extract is indeed higher in the press wine from a few Old American cultivars (e.g., Concord and Ives), even here it remains within permissible limits and is comparable to the press wine of some traditional European red varieties (Lee et al. 1975, 184).

20. Just recently a special bottle and label for Uhudler were approved. Sales are increasing, along with membership in the association.

8. PHYLLOXERA BREAKS OUT (TWICE) IN CALIFORNIA

1. Mission grapes, ubiquitous throughout Spain's American colonies under various names, were of mysterious origin until recently. In 2007 a Spanish research team conclusively linked Mission to the little-known Listan Prieto, which has almost disappeared in Spain due to phylloxera depredation. However, it is still a dominant grape in the Canary Islands—last stop for galleons bound for South America—where the phylloxera has never appeared (Tapia et al. 2007, 242).

2. Today some of Wisconsin's finest wines are produced on the Hungarian's old estate, where he inaugurated grape growing.

3. Most likely these were Mission grapes, probably along with some Eastern grapes such as Catawba.

4. Haraszthy's son later claimed that California owed its most important red grape, Zinfandel, to this expedition; he was believed by one and all. Sullivan has recently dismissed this story as a myth, especially in the face of evidence that Zinfandel existed on Long Island before it was in California (Sullivan 2003, 51).

5. Bioletti, in 1901, indicated that the surface of Old American vines was insignificant. One wonders what happened to all those vines in the intervening twenty-six years? Perhaps, as he suggests, at least some of them, *labrusca*-based vines such as Concord and Catawba, proved insufficiently resistant in California soil (Bioletti 1901, 12). But even then, others such as Isabella and Clinton would surely have been resistant, even under California conditions.

6. Contemporary DNA evidence suggests that Davidson was right (Downie 2002).

7. Davidson and Nougaret later call it the "migrant" form (Davidson and Nougaret 1921, 8).

8. Although Wolpert comments here on the AxR1 experience (to be discussed later), his point can be applied as well to the similar doublethink of growers facing phylloxera for the first time.

9. And apparently sterile (Davis IPM 2009).

10. This is evidently a reference to phylloxera appearing for the first time in the southern Central Valley.

11. Although, as Bioletti notes, it is precisely this strict exclusion that in Algeria has kept the Algiers district safe from infection from neighboring Constantine Province (Bioletti 1901, 7).

12. It should be noted, however, that these numbers include a sizable portion of hybrid direct producers, the successors to the Old Americans. For example, in the mid-1930s the very successful hybrids of Seyve-Villard, such as Seyval and Villard blanc, began to appear.

13. The problem most likely was not Wetmore's advice, but rather was linked to the fact that most of the easily found so-called *californica* was in reality hybridized with *vinifera*. Pure *californica* is as resistant as standard rootstocks (Granett et al. 1992, 249).

14. According to Bioletti, Lenoir has insufficient resistance in all but the richest soils (Bioletti 1901, 14). This is an interesting contrast with South Af-

rica, where Lenoir/Jacquez remained the stock of choice late into the twentieth century, until it was replaced by newer hybrids such as 99R.

15. Their work covered only eight years. Thus, as Lider notes, "the very important aspect of longevity of the combinations was never determined" (Lider 1957, 59). Only with Lider's good fortune in finding a few still-existing trials twenty years later was the longevity problem even partially addressed (Lider et al. 1978).

16. This claim is disputed by Blunno (1914, 16).

17. The basis for Lider's claim here about phylloxera resistance is not at all clear from Husmann et al.'s report.

18. Not all of Jacob's trial sites survived: "During the period of study a few trials were uprooted, others became intermixed with replanted and missing vines and were abandoned" (Lider 1957, 61).

19. Rupestris St. George is known as Rupestris du Lot in Europe.

20. Since none of the other stocks are recommended, it is odd that Lider and his colleagues later claim that "limited recommendations were also made [in the 1958 paper] for some of the others on the list of experimental stocks" (Lider et al. 1995, 14). The stock 99R is "doubtful"; 3306C and 3309C are "not recommended"; 420A is "not promising"; 5A is "an erratic performer" and "must be investigated further"; and, of Dogridge and 1613C, "neither can be recommended" (Lider 1957, 64–66). These are the only stocks other than St. George, 1202C, and AxR1 discussed by Lider. How these negative appraisals can be called "limited recommendations" is difficult to see.

21. The recommendation is limited to "heavy-producing" varieties because Lider felt that St. George's potential for reducing crop would too seriously affect shy-bearing varieties.

22. It must be noted that until very recently valley floors were the preferred place to plant for a number of reasons: they are more fertile and productive, and they are easier to plant, work, and pick. Only with the drive for the highest-quality grape production was it finally realized that the finest wines came from grapes planted on the difficult hillsides.

23. The plural "these" is used here because Lider's discussion begins with AxR1 and 1202C together; it ends focused solely upon AxR1.

24. "Experiments" is an extremely odd term to use for the widespread and costly failure of thousands of square kilometers of AxR1 grafted vines worldwide.

25. One was in Mendocino, three in Sonoma, and two in Napa (Lider et al. 1978, 20).

26. The six varieties were Sauvignon Vert, Cabernet Sauvignon, Barbera, Sylvaner, French Colombard, and Zinfandel. The stocks were AxR1, 1202C, 3309C, St. George, and, on three sites, 420A.

27. Just as in the 1957 paper: "Replete with tables and data, Lider's paper on the virtues of AXR was a classic specimen of the Davis mind-set. The experiments had been run and the results were in, with AXR the clear winner" (Lubow 1993, 28).

28. Lubow claims that the original sites weren't phylloxerated: "The field trials that led to the 1958 recommendation for AXR in California had been

conducted in fields in which grapes had not been grown (and grape phylloxera may not have thrived) since Prohibition" (Lubow 1993, 59). But the 1974 paper revisits six of the original sites, and only one of these was not phylloxerated. Hence, both the 1958 and the 1974 recommendations were based, at least in major part, on phylloxerated sites.

29. "A survey of the data collected has shown that the most popular phylloxera rootstock used commercially, Rupestris St. George, is not the most suited under the conditions to which it was tested, whereas, the experimental stock, A×R1, is generally the most vigorous and productive of the array of stocks under consideration" (Lider 1957, 58).

30. "Phylloxera was a hobby," Granett said in 1993 (Lubow 1993, 28). I doubt he would say the same today.

31. "Because the term biotype has been so loosely applied in the literature, its descriptive power has been greatly diluted. We therefore urge that more precise terminology be employed to reflect the mechanisms underlying differentiation" (Diehl and Bush 1984, 472). Diehl and Bush identify five different sorts of populations to which the term *biotype* might be ascribed.

32. Yet AxR1 plantings continued for several more years in the North Coast region. Evidently some growers relied on the fact that phylloxera spread more slowly in California than in Europe, and that the distance of their vineyards from infected sites gave them enough protection to gamble on planting (Lubow 1993, 60).

33. In an unfortunate irony, the newly recommended resistant rootstocks are, for the most part, the original "usual suspects" from 1880s France. Indeed, leading the list—which includes, for example, Millardet's 420A and Richter's 110, among others—is none other than Georges Couderc's tried and true 3309C, now California's "most widely used rootstock for ultra premium programs" (Duarte Nursery 2010).

34. "One downsizes equipment and material, lets people go, reduces expenses. One retreats into oneself as in a depression. The beast wins everywhere. In its wake solitude invades all the land. And the horizon takes on an unfamiliar aspect, made up of empty and desolate space. As a palpable sign of the plague, one sees all along the roads huge carts overladen with dead vines, leading one to the funeral pyre" (quoted in Garrier 1989, 50).

35. Since no grower surveys have been taken since that reported here in 1999, the present status of AxR1 acreage is unknown (Rhonda Smith, personal communication, 14 January 2010).

36. "It should be pointed out that the evolution of resistance of native North American grapevines to grape phylloxera has received absolutely no research attention" (Downie 2006). At the start, it was Riley's hypothesis of coevolution of grape phylloxera and American vines that shored up the belief that the resistance was both real and reliable.

37. Genetic recombination occurs when the genetic complements of both parents mix during sexual reproduction. In the beginning, the fact that California phylloxera were parthenogenic yet could evolve was extremely puzzling: "The source of this extensive genetic variation is unknown (and puzzling for an organism that spends much of its time reproducing asexually)" (Lewin 1993).

38. These molecular markers include FARKS, AFLP fingerprinting, and DNA and mtDNA sequencing, among others.

39. Native populations, "putatively intraspecific," exhibit "considerable structure," and even though introduced populations have reduced structure, there is still plenty for evolution to work on (Downie 2002).

40. And some that are not so distinct. Arguments persist, and probably always will, about how many good species there are in this vast territory. But settling this argument is not necessary for our point here: all we need to note is that there are lots of species of *Vitis*, each adapted to a different ecological niche.

41. South Africa, California, and Europe evidently each had more than one introduction (Downie 2002; Vorwerk and Forneck 2006).

42. It is hypothesized that phylloxera, like other aphids, have an ecological advantage in cold climates when they adopt sexual reproduction, since the so-called "winter egg" is capable of surviving temperatures that kill the overwintering parthenogenetic kin. This hypothesis is controversial (D..A. Downie, personal communication, 29 July 2008).

43. D.A. Downie, personal communication, 29 July 2008.

44. It is here where phenotypic description collides with genotypic description. Biotypes in phylloxera are defined behaviorally: type A cannot make a living on AxR1, but type B can. However, molecular investigation shows as much genetic difference *among* A and B populations as *between* A and B populations. Thus, the phenotypic grouping into biotypes is not presently supported by associated genotypic grouping (Downie 2002). In the end, epigenetics rather than genetics might be involved, although it is far too early to do anything but speculate about this possibility.

45. If the belief is that all phylloxera are the same, then there is no value in maintaining controls or quarantines on material transfers between different sites.

46. D.A. Downie, personal communication, 29 July 2008.

47. For example, in 1943 Lider's immediate predecessor, Jacob, "acknowledged that AxR1 was less resistant to phylloxera than St. George and even failed in dry sites 'where phylloxera attacks are likely to be most severe' " (Wolpert 1992, 52).

48. Obviously, here I play on the well-known historical theme of American exceptionalism, which commonly refers to the deeply held American belief that the nation is unique, blessed, superior in such a way that the ordinary socioeconomic and political standards that apply to other nations simply don't apply to America. De Tocqueville was among the first to call attention to this facet of the American character.

49. To be fair, many of the researchers, from Bioletti on, mention the fact that only root-type phylloxera exist in California, another difference that leads toward California exceptionalism. However, this exception speaks to diffusion, not to host infestation. Hence it is not relevant to the issue being discussed here.

50. M.A. Walker, private communication, 17 October 2007.

51. Davis's Foundation Plant Material Services, which was responsible for producing and distributing virus-free material, was founded in 1958.

52. Namely, Bioletti is relying upon the French and European researchers to have provided reliably resistant stocks.

Glossary

AMÉRICANISTE One who held that winning the battle against the phylloxera in France would necessarily involve using American vine material, either as direct producers, rootstocks, or hybrid parents.

AXR1 A hybrid rootstock produced by Victor Ganzin by crossing the *vinifera* variety Aramon with a selection of the American species *rupestris*. This was effectively the first hybrid rootstock; it later became notorious for its role in the 1980 phylloxera disaster in California.

CÉPAGE Roughly, a variety of grape, typically one used in winemaking; the term as used in France includes all clones and closely related varieties, e.g., pinot blanc and pinot gris are part of the same *cépage*.

LES CHEMISTES, LES SULFURISTES Participants in La Défense who were proponents of use of the insecticide CS_2 against the bug; led by industrialists in rail and chemical manufacturing, and abetted by physical scientists on the national antiphylloxera commission.

CORDON SANITAIRE A barrier (physical or administrative) to prevent the spread of disease.

LA DÉFENSE The first phase of the war against the bug, ca. 1875–81, characterized by the beliefs that the bug could first be defended against by planting in sand and vineyard submersion, and then by using insecticide to eradicate them, the goal being to protect and sustain traditional viticulture in its entirety. The attempt failed because there wasn't enough plantable sand area, submersion was costly and localized, and insecticides required money and skill to succeed. Moreover, Le Reconstitution was succeeding.

DÉPARTEMENT One of one hundred French administrative regions, functionally similar to an English district or an American county, although its average size (2,300 square miles) is approximately 3.5 times that of the average U.S. county, and its average population (500,000) is approximately 20 times as large.

DIRECT PRODUCER A wine grapevine planted directly on its own roots. Old American varieties such as Concord, Catawba, Jacquez, Clinton, and the like, which were in the first wave of phylloxera-tolerant vines available in France, served as direct producers.

GRAFTED VINE A grapevine with parts from two vines: the *scion*, or leafed part, usually a traditional French/European variety; and the *stock*, or rooted part, usually an American variety.

HYBRID DIRECT PRODUCER A hybrid grapevine planted directly on its own roots, created by crossing a vine with a significant amount (up to 100 percent) of *vinifera* genes with a vine with one or more American species parents. These vines came into use following 1885.

PHYLLOXERA-CAUSE VS. PHYLLOXERA-EFFECT Two opposed theoretical positions held by, respectively, the Montpellierians and the entomological establishment; the former argued that the phylloxera bug was the direct cause of the disease, while the latter argued that the phylloxera was only a symptom, attracted to an already sickened plant

LA RECONSTITUTION The second phase of the war against phylloxera, commencing around 1881, during which the French vignoble was reconstructed from the ground up, using direct producers, grafted traditional varieties, and hybrid direct producers.

VIGNERON A person who works with vines; usually a grower, but can also be a manager, tenant, or smallholder who makes his or her own wine.

VIGNOBLE A wine-growing region with standardized styles and practices; of variable size, ranging from a township at the small end to a *département* at the large end.

VINIFERA The ancient and traditional European vine species; includes such well-known varieties as Cabernet and Chardonnay.

Bibliography

Ackerknecht, E.H. 1948. "Anticontagionism Between 1821 and 1867." *Bulletin of the History of Medicine* 22: 562–93.

———. 1949. "Recurrent Themes in Medical Thought." *Scientific Monthly* 69: 80–83.

———. 1957. "Medical Education in 19th Century France." *Journal of Medical Education* 32: 148–52.

Aîné, Boutin. 1877. "Rapport." *Extracts des mémoires présentés par divers savants à l'académie des sciences* 25(6): 2–3.

Ainsworth, G.C. 1981. *Introduction to the History of Plant Pathology.* Cambridge: Cambridge University Press.

Alley, L. 2007. "Researchers Uncover Identity of Historic California Grape." Wine Spectator Online, available at www.winespectator.com/Wine/Features/0,1197,3638,00.html (accessed 16 April 2008).

Alley, L., D. Golino, and A. Walker. 2000–2001. "Retrospective on California Grapevine Material." *Wines & Vines* November 2000 (part 1); January 2001 (part 2); and April 2001 (part 3).

Annuaire de l'Association Amicale des Anciens Élèves de l'École d'Agriculture de Montpellier. 1895. Montpellier: Imprimerie Serre et Roumégous.

Aubert, Achille. 1880. *Conférence sur les vignes Américains.* Pons: Imprimerie Noel Texier.

Azam, M. le Prof. 1882. "Pourquoi la vigne plantée dans le sable pur résiste au phylloxéra." *Compte-rendu 1882,* 497–501.

Baptista, Alfredo, and Edmundo Suspiro. 1955. *O Problema Filoxérico em Portugal.* Lisboa: Direcção-Geral dos Serviços Agrícolas.

Barral, J.–A. 1883. *La Lutte contre le phylloxéra.* Paris: Marpon.

———, ed. 1868–70. *Journal d'Agriculture* 49–51. Paris: Bureaux du Journal de l'Agriculture.

Barry, Brian. 1979. *Wine Talk, Australian Winemakers and Their Art: A Collection of Papers from a Symposium Held at the Australian National University Staff Centre, September 1979.* Canberra: Action Press.

Bazerman, Charles. 1997. "Reporting the Experiment: The Changing Account of Scientific Doings in the Philosophical Transactions of the Royal Society, 1665–1800." In *Landmark Essays on Rhetoric of Science: Case Studies,* ed. R.A. Harris. Mahwah, NJ: Hermagoras Press.

Bazille, Gaston. 1878. *Exposé de la question du phylloxéra.* Vienne: E.-J. Savigné.

Becker, Helmut. 1960. "Letter." Appendix IV, dated 10 June, in "Phylloxera and Its Relation to South Australian Viticulture," by B.G. Coombe. In *Technical Bulletin No. 31.* Adelaide: South Australia Department of Agriculture, 1963.

Berget, Adrien. 1896. *La Viticulture Nouvelle.* Paris: Félix Alcan.

Bioletti, F.T. 1901. *The Phylloxera of the Vine.* Sacramento: State Printer.

Bioletti, Frederic T., F.C.H. Flossfeder, and A.E. Way. 1921. *Phylloxera-Resistant Stocks.* Berkeley: University of California Press.

Bishop, Katherine. 1985. "Insect That Ruined Vineyards Reappears in California." *New York Times,* July 1. Available at http://query.nytimes.com/gst/fullpage.html?res=9F01E7DE1F39F932A35754C0A963948260&sec=&spon=&pagewanted=print (accessed 16 July 2008).

Blunno, Michele. 1914. *The Use of Phylloxera-Resistant Stock, Part 1.* Farmers' Bulletin No. 80, Department of Agriculture, New South Wales. Sydney: William Applegate Gullick, Government Printer.

Bogen, James, and James Woodward. 1988. "Saving the Phenomena." *Philosophical Review* 97: 303–52.

Bonner, Bill. 2007. "The Art of Homemade Booze." Available at http://dailyreckoning.com/bipolar-markets/ (accessed 13 August 2007).

Borde de Tempest, Earnest. 1873. *Phylloxéra: sa Destruction Certaine.* Montpellier: Imprimerie Centrale du Midi.

Brown, L.R., D.J. Blasingame, C.M. Ladner, C. Teichert, and S. Brown. 1979. "The Use of Chitinous Seafood Wastes for the Control of Plant Parasitic Nematodes." BMR Project No. GR-ST-78–003 and GR-ST-78–004, Bureau of Marine Resources, Mississippi Department of Wildlife Conservation, Long Beach, Mississippi, September.

Buchanan, Gregory A. 1992. "The Biology, Quarantine, and Control of Grape Phylloxera." *Study Tour Reports Series No. 37.* Melbourne: Victoria Department of Agriculture.

Bulgarian Wine. 2008. "Liberation." Available at www.bulgarianwines.com/index.php?action=history (accessed 28 January 2008).

Burian, Richard. 1990. "Review of *Constructing Quarks: A Sociological History of Particle Physics* by Andrew Pickering." *Synthese* 82: 163–74.

Burian, Richard, Jean Gayon, and Doris Zallen. 1988. "The Singular Fate of Genetics in the History of French Biology, 1900–1940." *Journal of the History of Biology* 21: 357–402.

Bush and Sons. 1876. *Les Vignes américaines,* trans. Louis Bazille. Montpellier: Coulet.

Campbell, Christy. 2004. *The Botanist and the Vintner: How Wine Was Saved for the World.* Chapel Hill, NC: Algonquin Books.

Carrière, F. 1884. *Rapport sur les vignes Américaines du Gard & de l'Hérault en 1883.* Royan: Imprimerie Victor Billaud.

Carton, Y., C. Sorensen, J. Smith, and E. Smith. 2007. "Une coopération exemplaire entre entomologistes français et américains pendant la crise du Phylloxera en France (1868–1895)." *Ann. soc. entomol. Fr. (n.s.)* 43: 103–25.

Cazalis, F. 1869a. *Le Messager Agricole* 9. Montpellier: Imprimerie Typographique de Gras.

———. 1869b. "De la maladie de la vigne causée par le Phylloxéra." *L'Insectologie agricole* 3: 29–33.

Cesar, A. 1948. "Letter." Appendix III in "Phylloxera and Its Relation to South Australian Viticulture," by B.G. Coombe. In *Technical Bulletin No. 31.* Adelaide: South Australia Department of Agriculture, 1963.

Chauzit, B. 1884. *Rapports sur les concours de greffage et de viticulture organisés par la Société d'agriculture du Gard en 1883.* Nîmes: Imprimerie Clavel & Chastanier.

Chavée-Leroy, M. 1885. *La maladie de la vigne, les microbes et la commission supérieure du phylloxéra.* Paris: Librairie Centrale des Sciences.

Clavel, Jean. 2007. http://1907larevoltevigneronne.midiblogs.com/ (accessed 24 September 2010).

Cohen, Henry. 1961. "The Evolution of the Concept of Disease." In *Concepts of Medicine,* ed. Brandon Lush. Oxford: Pergamon.

Comice Agricole d'Épernay. 1877. "Bulletin mensuel." March.

Comisión de Estudio del Problema Filoxerico. 1938. *Informe.* San Juan: Compañia Impresora Argentina, S.A. for el Gobierno de la Provincia de San Juan.

La Comisión Organizadora. 1880. *Congreso Internacional Filoxérico de Zaragoza.* Zaragoza: Imprenta del Hospicio Provincial.

Comité central d'étude et de vigilance contre le phylloxera, Dept. de Lot-et-Garonne. 1884. *Compte-rendu de la Séance du 12 mars.* Agen: Lentheric.

———. 1888. *Compte-rendu de la Séance du 13 oct.* Agen: Lentheric.

Commission Départmentale de l'Hérault pour l'étude de la maladie de la vigne. 1877. *Résults Pratiques de l'application des Divers Procédés.* Montpellier: Grollier.

Commission du Phylloxéra. 1877. *Avis sur les mesures à prendre pour s'opposerà l'extension des ravages du phylloxéra.* Paris: Académie des Sciences, Institut de France.

Compagnie des Chemins de fer de Paris à Lyon et à la Méditerranée. 1878. *Instructions pour le traitement des vignes par le sulfure de carbone.* Paris: Paul Dupont.

Compte-rendu général du Congrès international phylloxérique de Bordeaux du 9 au 16 octobre 1881. 1882. Bordeaux: Féret et fils.

Convert, F. 1900. "La Viticulture après 1870: La crise phylloxérique." *Revue de viticulture* 14: 337–39 (part 1); 449–52 (part 2); 512–17 (part 3).

Coombe, B.G. 1963. "Phylloxera and Its Relation to South Australian Viticulture." *Technical Bulletin No. 31.* Adelaide: South Australia Department of Agriculture.

Cornu, M., and J.B. Dumas. 1876. *Instruction pratique. Commission du Phylloxéra, sèance du 17 jan. 1876. Institut de France, Academie des Sciences.* Paris: Gauthier-Villars.

Corrie, A.M., R.H. Crozier, R. van Heeswijk, and A.A. Hoffmann. 2002. "Clonal Reproduction and Population Genetic Structure of Grape Phylloxera, *Daktulosphaira vitifoliae,* in Australia." *Heredity* 88: 203–11.

Corrie, A.M., and A.A. Hoffmann. 2004. "Fine-scale Genetic Structure of Grape Phylloxera from the Roots and Leaves of Vitis." *Heredity* 92: 118–27.

Corrie, A.M., R. van Heeswijk, and A..A. Hoffmann. 2003. "Evidence for Host-associated Clones of Grape Phylloxera *Daktulosphaira vitifoliae* Hemiptera: Phylloxeridae in Australia." *Bulletin Entomological Research* 93: 193–201.

Couderc, Freddy. 2000. *Les vins mythiques.* Pont-Saint-Esprit: Editions La Mirandole.

"Could Phylloxera Cost $70,000 per Acre?" 1991. *Wines & Vines,* February. Available at http://findarticles.com/p/articles/mi_m3488/is_n2_v72/ai_10641715 (accessed 24 September 2010).

"La crise phylloxerique." 1879. *La vigne française* 1–4: 57–58.

"Croatians." 2008. *Encyclopedia of Chicago,* available at www.encyclopedia. chicagohistory.org/pages/353.html (accessed 21 January 2008).

Cushing, James T. 1982. "Models, High-energy Theoretical Physics, and Realism." In *PSA: Proceedings of the Biennial Meeting of the Philosophy of Science Association,* vol. 2, *Symposia and Invited Papers,* 31–56.

Dalmasso, G. 1956. "General Report for Europe." *Bull. L'Off. Int'l. Vin* 29(308): 5–30. Appendix 1 in "Phylloxera and Its Relation to South Australian Viticulture," by B.G. Coombe. In *Technical Bulletin No. 31.* Adelaide: South Australia Department of Agriculture, 1963.

Darwin, Charles. 1986. "Journal of Researches." In *The Works of Charles Darwin,* vol. 1, ed. Paul H. Barrett and R.B. Freeman. New York: New York University Press.

———. 1987. *Notebooks, 1836–1844.* Edited by Paul H. Barrett, P.J. Gautrey, S. Herbert, D. Kohn, and S. Smith. Ithaca, NY: Cornell University Press.

Daurel, Joseph. 1886. *Quelques mots sur les vignes américaines, leur greffage, les producteurs directs dans la région du Sud-Ouest; Étude pratique sur cet important moyen de reconstitution des vignobles.* Bordeaux: Féret et fils.

Davidson, W.M., and R.L. Nougaret. 1921. *The Grape Phylloxera in California.* Washington, DC: U.S.D.A. Bulletin No. 903.

Davis IPM. 2009. "UC IPM: UC Management Guidelines for Grape Phylloxera on Grape." Available at www.ipm.ucdavis.edu/PMG/r302300811.html (accessed 9 January 2010).

de Castella, François. 1942. *The Grapes of South Australia: François de Castella's Reports to the Phylloxera Board.* Adelaide: Mail Newspaper.

de Ceris, A. 1870. "Chronique agricole." *Journal d'agriculture pratique* 34: 610–11.

———. 1873. "Chronique agricole." *Journal d'agriculture pratique* 37: 673–76.

de Gironde, Le Vicomte. 1886. *La vérité sur les cépages américains en 1886.* Montauban: Imprimerie Montalbanaise.

de Horváth, G. 1882. *Rapport annuel de la station phylloxérique hongroise.* Budapest: Société d'imprimerie par actions de Pest.

deKlerk, C.A. 2000. "Vineyard Terrorist Gets New Foot in the Door." Wynboer. Available at www.wynboer.co.za/recentarticles/0802terrorist.php3 (accessed 17 August 2008).

de Lafitte, Prosper. 1881. "La question du phylloxéra et le rôle des vignes *américaines.*" *Revue des deux mondes* 51: 196–208.

———. 1882. "Essai sur une bonne conduite des traitements au sulfure de carbone." In *Compte-rendu général du Congrès international phylloxérique de Bordeaux du 9 au 16 octobre 1881.* Bordeaux: Féret et fils.

———. 1883a. "Les vignes *américaines* obtenue de semis." In *Quatre ans de luttes pour nos vignes et nos vins de France,* by Prosper de LaFitte, 535–49. Paris: G. Masson.

———. 1883b. *Quatre ans de luttes pour nos vignes et nos vins de France.* Paris: G. Masson.

de la Loyère, Viscount. 1874. *Le Phylloxéra dan le centre de la France: Réponse de la Commission du Phylloxéra de la Société d'agriculture de Chalon-sur-Saône.* Chalon-sur-Saône: Imprimerie et Lithographie de J. Dejussieu.

de Leybardie, M. 1882. "Travaux de submersion." In *Compte-rendu 1882.* Bordeaux: Féret et fils.

de Lorenzo, A. Andreoli, R.P. Sorge, L. Iacopino, S. Montagna, L. Promenzio, and P. Serranó. 1999. "Modification of Dietary Habits (Mediterranean Diet) and Cancer Mortality in a Southern Italian Village from 1960 to 1996." *Annual of the New York Academy of Science* 889: 224–29.

Demole-Ador, J. 1877. *Le Congrès phylloxérique international de Lausanne.* Berne: Imprimerie Staempfli.

"Denis Boubals." 2007. Obituary, *London Times.* Available at www.timesonline.co.uk/tol/comment/obituaries/article3018372.ece (accessed 4 August 2008).

des Hours, L. 1894. "Souscription au monument Planchon." *Revue de viticulture* 2(40): 291–92.

Diehl, S.R., and G.L. Bush. 1984. "An Evolutionary and Applied Perspective on Insect Biotypes." *Annual Review of Entomology* 29: 471–504.

Donnadieu, A.-L. 1887. *Les véritables origines de la question phylloxérique.* Paris: Librairie J.-B. Baillière et fils.

Downie, D.A. 2002. "Locating the Sources of an Invasive Pest, Grape Phylloxera, Using a Mitochondrial DNA Gene Genealogy." *Molecular Ecology* 10: 2013–26.

———. 2003. "Effects of Short-term Spontaneous Mutation Accumulation for Life History Traits in Grape Phylloxera, *Daktulosphaira vitifoliae.*" *Genetica* 3: 237–51.

———. 2005. "Evidence for Multiple Origins of Grape Phylloxera *Daktulosphaira vitifoliae* Fitch Hemiptera: Phylloxeridae in South African Vineyards." *African Entomology* 13: 359–65.

———. 2006. "Comment: Grape Phylloxera and the Grapevine: Searching for the Science." *Journal of Experimental Botany.* Available at http://jxb.oxford journals.org/cgi/eletters/56/422/3029 (accessed 17 August 2008).

Duarte Nursery. 2010. "Grapevine Rootstocks." Available at http://duartenurs ery.com/grapevines/grapevine-rootstocks/ (accessed 14 January 2010).

Duclaux, Émile. 1872. "Etudes sur la nouvelle maladie de la vigne dans le sud-est de la France." *Memoires présentés par divers savants à l'Academie des Sciences de l'Institut National de France* 22(5).

Dumas, J.B. 1876. *Etudes sur le Phylloxéra et sur les sulfocarbonates.* Paris: Gauthier-Villars.

Dunstan, David. 1983. "Irvine, Hans William Henry, 1856–1922." *Australian Dictionary of Biography,* vol. 9, 437–38. Melbourne: Melbourne University Press.

———. 1993. "De Castella, François Robert, 1867—1953." *Australian Dictionary of Biography,* vol. 13, 604–5. Melbourne: Melbourne University Press.

———. 1994. *Better than Pommard—A History of Wine in Victoria.* Kew, Victoria: Australian Scholarly Publishing.

Duran-Vila, N., J. Juárez, and J.M. Arregui. 1988. "Production of Viroid-Free Grapevines by Shoot Tip Culture." *American Journal of Enology and Viticulture* 39: 217–20.

Eke, István, 2004. "Selected Passages from the History of the Hungarian Plant Protection Administration." The Jubilee Conference, Ministry of Agriculture and Rural Development. Available at www.fvm.hu/main.php?folderID=1564 (accessed 23 January 2008).

Espinouse, A. 1874. *Du phylloxéra, son traitement dans le Midi.* Bordeaux: Féret et fils.

Espitalier, Silvain. 1874. *Ensablement avec addition d'engrais: instructions pratiques pour l'emploi de ce procédé.* Montpellier: C. Coulet.

Falières, E. 1874. *Du phylloxéra et d'un nouveau mode d'emploi des insecticides.* Bordeaux: Imprimerie Nouvelle A. Bellier.

Faucon, Louis. 1870. "Nouvelle maladie de la vigne." *Journal d'agriculture pratique* 34: 512–18.

———. 1872. *Notes sur la maladie des vignes dites du phylloxéra.* Avignon: Amé du Chaillot.

———. 1874. *Instructions pratique sur le procédé de la submersion.* Montpellier: C. Coulet.

Fergusson-Kolmes, L., and T.J. Dennehy. 1991. "Anything New under the Sun? Not Phylloxera Biotypes." *Wines & Vines,* June. Available at www.winefiles. org/WineHistory/results.cgi?topic=%22Phylloxera%22&f=advanced&rec_per_page=10&start=31 (accessed 7 July 2008).

Fitz-James, Marguerite Duchesse de. 1881. "Les vignes américaines." *Revue des deux mondes* 51: 685–94.

Foëx, G. 1882. *Mémoire sur les causes de la chlorose chez l' Herbemont.* Montpellier: Coulet.

———. 1890. *J.-É. Planchon.* Montpellier: Charles Boehm.

———. 1900. "La crise phylloxérique en France." In *Congrès International de Viticulture,* 34–89. Paris: Compte rendu in-extenso. Société des viticulteurs de France et d'ampélographie.

Fong, G., M.A. Walker, and J. Granett. 1995. "RAPD Assessment of California Phylloxera Diversity." *Molecular Ecology* 4: 459–64.

Fontanari, Iris. 2004. "L'uva Isabella dal sapore di fragola." *Terra Trentino* 9: 45–46.

Franklin, Allan. 1986. "Experiment and the Development of the Theory of Weak Interactions: Fermi's Theory." *PSA: Proceedings of the Biennial Meeting of the Philosophy of Science Association,* vol. 2, *Symposia and Invited Papers,* 163–79.

———. 1990. "Do Mutants Have to Be Slain, or Do They Die of Natural Causes?" In *PSA: Proceedings of the Biennial Meeting of the Philosophy of Science Association,* vol. 2, *Symposia and Invited Papers,* 487–94.

Fuller, Thomas. 2004. "Winemakers Protect Outlawed Vines: The Grapes of Wrath." *International Herald-Tribune,* September 25.

Gachassin-Lafite, L. 1882. "Rapport de la commission des vignes américaines et des sables." *Compte-rendu 1881,* 100–143.

Gale, G. 1979. *Theory of Science.* New York: McGraw-Hill.

———. 2003. "Saving the Vine from Phylloxera: A Never-ending Battle." In *Wine: A Scientific Exploration,* ed. M. Sandler and R. Pinder, 70–91. London: Taylor & Francis.

Galet, P. 1956–64. *Cépages et vignobles de France,* vols. 1–4. Montpellier: Paysan du Midi.

———. 1979. *A Practical Ampelography,* trans. and adapted by Lucie T. Morton. Ithaca, NY: Cornell University Press.

———. 1988. *Cépages et vignobles de France, 2e édition, tome 1: Les vignes américaines.* Montpellier: Charles Déhan.

Gamulin, Stjepan. 1996. "Historical and Cultural Background of Health Care on Croatian Islands." *Croatian Medical Journal* 37(3), available at www.cmj.hr/1996/37/3/01.htm (accessed 21 January 2008).

Ganzin V. 1887. "Les Aramons-rupestris: Porte-greffes hybrides inédits." *La vigne américaine* 11: 359–65.

———. 1888. "Les Aramon-rupestris: Porte-Greffes hybrides inédits, Pt. 1, Pt. 2." *La vigne française* 1: 15–16; 2: 23–25.

Garrier, Gilbert. 1989. *Le Phylloxéra: Une guerre de trent ans, 1870–1900.* Paris: Albin Michel.

Gervais, Prosper. 1903–4. "La crise phylloxérique et la viticulture européenne." *Revue de Viticulture* 20(501) (July 23, 1903): 89–93; 20(507) (September 3, 1903): 259–63; 21(546) (2 June 1904): 606–9; 21(547) (9 June 1904): 633–38; 21(548) (16 June 1904): 657–64; 22(551) (7 July 1904): 10–15.

Gilby, Caroline. 2001. "Chilean Rootstocks." Available at www.cix.co.uk/~cgilby/chilerootstock1.pdf (accessed 6 January 2010).

Giliomee, Hermann. 1987. "Western Cape Farmers and the Beginnings of Afrikaner Nationalism, 1870–1915." *Journal of Southern African Studies,* 14: 38–63.

Girard, Maurice. 1874. *Le Phylloxéra de la vigne: son organisation, ses moeurs.* Paris: Hachette.

Gites with Pools. 2007. "Wines of the Vendee." Available at http://gites-with-pools.co.uk/ (accessed 16 May 2008).

Gouy, P. 1903. "L'Ecole de Montpellier et l'hybridation." *Revue des hybrides franco-américains* 5: 189–90.

Granett, J., P. Timper, and L.A. Lider. 1985. "Grape phylloxera *Daktulosphaira vitifoliae* Homoptera: Phylloxeridae Biotypes in California." *Journal of Economic Entomology.* 78: 1463–67.

Granett, J., J. De Benedictis, and J. Marston. 1992. "Host Suitability of *Vitis californica* Bentham to Grape Phylloxera, *Daktulosphaira vitifolia* Fitch." *American Journal of Enology and Viticulture* 43: 249–52.

Granett, J., M.A. Walker, L. Kocsis, and A.D. Omer. 2001. "Biology and Management of Grape Phylloxera." *Annual Review of Entomology* 46: 387–412.

Grassi, Battista. 1915. "The Present State of Our Knowledge of the Biology of the Vine Phylloxera." *Monthly Bulletin of Agricultural Intelligence and Plant Diseases* 6(10): 1269–90.

Guérin, Pierre. 1874. "Lettre d'avice." *Journal d'agriculture pratique* 38: 186–87.

Guy, Kathleen. 1997. "Chapter 3: Uprooted Vines and Contentious Communities: Phylloxera and Peasant Protest in Champagne, 1890–95." Ph.D. diss., University of Texas, Austin.

Hannaway, Caroline. 1993. "Environment and Miasmata." In *Companion Encyclopedia of the History of Medicine,* vol. 1, ed. W.F. Bynum and Roy Porter. London: Routledge.

Hanson, N.R. 1965. *Patterns of Discovery.* Cambridge: Cambridge University Press.

Harris, Randy Allen. 1997. "Introduction." In *Landmark Essays on Rhetoric of Science: Case Studies,* ed. Randy Allen Harris. Mahwah, NJ: Hermagoras Press.

Hilgard, E.W. 1880. "War on the Phylloxera." *New York Times,* 3 August.

Holmes, Oliver W. 1883. *Medical Essays.* New York: Houghton, Mifflin, and Co.

Husmann, G.C. 1910. "Grape Investigations in the Vinifera Regions of the United States with Reference to Resistant Stocks, Direct Producers and Viniferas." *U.S. Department of Agriculture Bulletin* 172: 1–86.

Husmann, G.C., E. Snyder, and F.L. Husmann. 1939. "Testing Vinifera Grape Varieties Grafted on Phylloxera-Resistant Rootstocks in California." *USDA Technical Bulletin* 697: 1–63.

Inauguration du monument élevé a la mémoire de J.-É. Planchon le 9 Décembre 1894. 1895. Montpellier: L. Grollier.

Isnard, H. 1951. *La Vigne en Algérie.* Gap, France: Ophrys.

Izard, Jean. 1885. *Plus de cépages américaines ils meurent!!! Reconstitution de la vigne par les cépages français.* Carcassonne: Imprimerie Nouvelle J. Parer.

Jacobs, Julius L. 1975. "The Saga of T.V. Munson, 19th Century Texas Grape Breeder of 300 Crosses." *Wines & Vines:* 51–53.

Johnston, W.B., and I. Crkvenčič. 1954. "Changing Peasant Agriculture in North-Western Hrvatsko Primorje, Yugoslavia." *Geographical Review* 44 (3 July 1954): 352–72.

Keller, Jack. 2009. "Winemaking Home Page." Available at http://winemaking .jackkeller.net/ (accessed 19 April 2009).

Kellerviertel Heiligenbrunn. 2008. "Uhudler." Available at www.kellerviertel-heiligenbrunn.at/uhudler.htm (accessed 16 May 2008).

Kessler, H.F. 1892. *Die Ausbreitung der Reblauskrankheit in Deutschland und deren Bekämpfung.* Berlin: R. Friedländer & Sohn.

Kuhn, Thomas. 1993. "Introduction." In *PSA 1992*, vol. 2, ed. David Hull, Mickey Forbes, and Kathleen Okruhlik, 3.

Lachiver, M. 1988. *Vins, vignes et vignerons: Histoire du vignoble français.* Paris: Fayard.

Laennec, R.T.H. 1819. *De l'auscultation médiate, ou, Traité du diagnostic des maladies des poumons et du coeur fondé principalement sur ce nouveau moyen d'exploration.* Paris: J.-A. Brossan.

Lafon, R., J.L. Vidal, and M. Nayrac. 1930. "Cinquantenaire de la reconstitution du vignoble charentais." *Revue de viticulture* 73: 41–45.

Laliman, L. 1872. *Étude sur les divers phylloxéra et leurs médications.* Paris: Librairie de la Maison Rustique.

Lamboy, Warren F., and Christopher G. Alpha. 1998. "Using Simple Sequence Repeats SSRs for DNA Fingerprinting Germplasm Accessions of Grape Vitis L. species." *Journal of the American Society for Horticultural Sciences* 123(2): 182–88.

Laurent, R. 1958. *Les vignerons de la "Cote d'Or" au XIXe siècle.* Paris: Société les Belles Lettres, Publications de d'Université de Dijon XV.

Lecouteux, M.E., ed. 1870–73. *Journal d'agriculture pratique,* vols. 34–37. Paris: Librairie agricole de la Maison Rustique.

Lee, C.Y., W.B. Robinson, J.P. van Buren, T.E. Acree, and G.S. Stoewsand. 1975. "Methanol in Wines in Relation to Processing and Variety." *American Journal of Enology and Viticulture* 26: 184–87.

Legros, Jean-Paul. 1997. "Le phylloxéra, une histoire sans fin." *Sciences* 97(1): 32–39.

———. 1998. *L'Odyssée des agronomes de Montpellier.* Paris: Editagro.

———. 2005. "Les Américanistes du Languedoc 1868–1893." *Étude et Gestion des Sols* 12: 165–86.

Legros, J-P., and Jean Argeles. 1994. "L'invasion du vignoble par le phylloxéra." *Bulletin de l'Academie des Sciences et Lettres de Montpellier* 24: 205–22.

Lewin, Roger. 1993. "California's Lousy Vintage." *New Scientist,* 17 April, 27–30.

Lichtenstein, Jules, and Jules-Émile Planchon. 1870. "On the Identity of the Two Forms of Phylloxera." *Journal d'agriculture pratique* 34: 181–82.

Lider, L.A. 1957. "Phylloxera-Resistant Rootstock Trials in the Coastal Valleys of California." *American Journal of Enology and Viticulture* 82: 58–67.

Lider, L.A., N.L. Ferrari, and K.W. Bowers. 1978. "A Study of Longevity of Graft Combinations in California Vineyards, with Special Interest in the

Vinifera × Rupestris Hybrids." *American Journal of Enology and Viticulture* 29: 18–24.

Lider, L.A., M.A. Walker, and J.A. Wolpert. 1995. "Grape Rootstocks in California Vineyards: The Changing Picture." *Acta Horticulturae* 388: 13–18.

Lin, H., M.A. Walker, R. Hu, and J. Granett. 2006. "New Simple Sequence Repeat Loci for the Study of Grape Phylloxera *(Daktulosphaira vitifoliae)* Genetics and Host Adaptation." *American Journal of Enology and Viticulture.* 57: 33–40.

Lubow, Arthur. 1993. "What's Killing the Grapevines of Napa?" *New York Times Magazine,* 17 October.

Lyell, Charles. 1990. *Principles of Geology,* vol. 2. Chicago: University of Chicago Press.

Mach, Edmund. 1873. "Die Phylloxera vastatrix in Frankreich." *Annalen der Oenologie* 4: 462–86.

MacNeil, K. 2001. *The Wine Bible.* New York: Workman Publishing.

Magen, Ad. 1885. "La destruction de l'œuf d'hiver." *La vigne française* 6: 261–78.

Marescalchi, A., and G. Dalmasso. 1931. *Storia della vite e del vino in Italia,* vol. 1. Milan: Arti Grafiche E. Gualdoni.

Marion, Antoine Fortuné. 1879. *Rapport sur les experiénces contre le phylloxéra et les résultats obtenus, campagne de 1878.* Paris: Paul Dupont.

Marza, V.D. and N. Cerchez. 1967. "Charles Naudin, a Pioneer of Contemporary Biology (1815–1899)." *Journal d'agriculture tropicale et de botanique appliquée* 14: 369–401.

Mayet, V. 1890. *Les insectes de la vigne.* Paris: Masson.

Millardet, Alexis. 1877. *La question des vignes américaines au point de vue théorique et pratique.* Bordeaux: Féret et fils.

———. 1879. "Études sur quelques espéces de vignes sauvages de l'Amérique du Nond." *Journal d'agriculture pratique,* 81: 28–35.

———. 1881. *Notes sur les vignes américaines et opuscules divers sur le même sujet.* Bordeaux: Féret et fils.

———. 1882. "De l'hybridation entre les diverses espèces de vignes américaines à l'état sauvage." *Journal d'agriculture pratique* 2: 470–78.

———. 1887. *Notes sur les vignes américaines. Série II.* Paris: G. Masson.

———. 1888. "Nouveaux cépages hybrides: Résistant au phylloxéraet au mildiou." *La vigne française* 9(2): 25–29.

———. 1891. *Essai sur l'hybridation de la vigne.* Paris: G. Masson.

Millot, M. 1876. *Le phylloxéra dans Saône-et-Loire; Rapport de la Commission Departementale.* Macon: Imprimerie Romand frères.

Ministère de l'Agriculture et du Commerce, Commission Supérieure du Phylloxéra. Various dates. *Compte rendu et pièces annexes; Lois, décrets et arrêtés au phylloxéra.* Paris: Imprimerie nationale.

Ministerio de Fomento. 1911. *La invasión filoxérica en España y estado en 1909 de la reconstitución del viñedo.* Madrid: Imprenta de los Hijos de M.G. Hernández.

Mori, Edoardo. 2008. "Il Fragolino." Available at www.earmi.it/ricette/fragolino.htm (accessed 16 May 2008).

Morrow, Dwight W., Jr. 1960. "The American Impressions of a French Botanist, 1873." *Agricultural History* 34: 71–76.

———. 1961. *The Phylloxera Story*, chapters 1–6. Unpublished manuscript, Shields Library, University of California, Davis. Used with permission.

———. 1973. "Phylloxera in Portugal." *Agricultural History* 47: 235–47.

Morton, Lucie. 1994. "The Myth of the Universal Rootstock." *Wines and Vines*, June. Available at www.encyclopedia.com/doc/1G1–16098122.html (accessed 14 April 2008).

Munson, Thomas V. 1905. *Foundations of American Grape Culture*. Dennison, TX: Public Library.

National Library. 2003. "Cultivating in Good Hope." In *Fruits of the Vine*. Pretoria: National Library of South Africa. Available at www.nlsa.ac.za/vine/index2.html (accessed 15 April 2008).

National Vine Health Steering Committee. 2003. *National Phylloxera Management Protocol: Overview and Contact Details*. Melbourne: Australian Government, Grape and Wine Research and Development Corporation.

"The New Vine-disease in the South-east of France." 1874. Part 1. *Nature*, 503–6.

Oberlin, Christian. 1914. *La reconstitution du vignoble sans greffage*. Paris: Librairie agricole de la Maison Rustique.

Olby, Robert C. 1993. "Constitutional and Hereditary Disorders." In *Companion Encyclopedia of the History of Medicine*, ed. W. F. Bynum and Roy Porter, 412–37. London: Routledge.

Ordish, G. 1972. *The Great Wine Blight*. New York: Charles Scribner's and Sons.

Pasteur, Louis. 1854. Lecture in Lille. Available at http://en.wikiquote.org/wiki/Louis_Pasteur (accessed 2 April 2009).

———. 1922. *Oeuvres*. Vol. 2, *Fermentations et Générations dit Spontanée*. Paris: Masson et Cie, Editeurs.

Paul, Harry. 1996. *Science, Vine, and Wine in Modern France*. Cambridge: Cambridge University Press.

Paulsen, Federico. 1933. "Storia della invasione Fillosserica e ricostituzione del Vigneti in Italia." In *Nuovi annali dell'agricoltura*. Rome: Ministerio dell'agricoltura e della foreste.

Pearce, M. Rosaria di Nucci, and David Pearce. 1989. "Technology vs. Science: The Cognitive Fallacy." *Synthese* 81: 405–19.

Pelling, Margaret. 1993. "Contagion / Germ Theory / Specificity." In *Companion Encyclopedia of the History of Medicine*, ed. W. F. Bynum and Roy Porter, 309–34. London: Routledge.

Perdue, Lewis. 1999. *The Wrath of Grapes*. New York: Avon Books.

Pereira, Sertorio. 1913. "Reconstitution of Portuguese Vineyards by Means of American Stocks." *Monthly Bulletin of Agricultural Intelligence and of Plant Diseases* 4: 1–4.

Peringuey, L. 1887. "A Note on the *Phylloxera Vastatrix* at the Cape." *Transactions of the South African Philosophical Society* 4: 58–62.

Perold, A. I. 1927. *A Treatise on Viticulture*. London: Macmillan and Co.

Phylloxera Commission. 1893a. *Preliminary Report of the Phylloxera Commission, 1893 : Quarantine and Import Regulations: with Minutes of Proceedings, Minutes of Evidence, and Appendix*. Cape Town: W. A. Richards and Sons, Government Printers.

———. 1893b. *Final Report of the Phylloxera Commission, 1893*. Cape Town: W.A. Richards and Sons, Government Printers.

Pickering, Andrew. 1984. *Constructing Quarks*. Chicago: University of Chicago Press.

Piernavieja, Herminie. 2005. "Le Jacquez, un cépage charge d'histoire: son adaptation au terroir cévenol et les enjeux de son maintien." Ph.D. diss., Université du Vin, Suze-la-Rousse, Université de Provence, Aix-Marseille 1, Université de Franche-Comté, Besançon.

Pinch, Trevor. 1986. *Confronting Nature: The Sociology of Solar Neutrino Detection*. Dordrecht: D. Reidel Publishing Company.

Pinney, Thomas. 1989. *A History of Wine in America: From the Beginnings to Prohibition*. Berkeley: University of California Press.

Planchon, J.-É. 1870–71. Unpublished correspondence. Copies in the papers of Dwight Morrow, Shields Library, University of California, Davis.

———. 1873. *Le Phylloxera et les vignes américaines*. Montpellier: C. Coulet.

———. 1874. "Le phylloxéra en Europe et en Amérique, I; II." *Revue des deux mondes* 44: 914-43.

———. 1875. *Les vignes américaines, leur culture, leur résistance au phylloxéra et leur avenir en Europe*. Montpellier: C. Coulet.

———. 1877. "La question du phylloxéra en 1876." *Revue des deux mondes* 47: 241–47.

Plovdiv 2008. "Day of the Plant Protector." Available at www.plovdivguide.com/newsfiles/news.php?id=2320&lang_id=1 (accessed 28 January 2008).

Plumeau, M., Régis Gayon, Richier Princeteau, and E. Falières. 1882a. "Rapport sur le sulfure de carbone et les sulfo-carbonates." In *Compte-rendu 1881*. Bordeaux: Féret et fils.

———. 1882b. "Rapport sur la submersion." In *Compte-rendu 1881*. Bordeaux: Féret et fils.

Pouget, R. 1990. *Histoire de la lutte contre le phylloxéra de la vigne en France*. Paris: Institut National de la Recherche Agronomique.

Prial, Frank. 1993. "Wine Talk." *New York Times*, 3 February. Available at http://query.nytimes.com/gst/fullpage.html?res=9F0CE1DE173FF930A357 51C0A965958260 (accessed 16 May 2008).

"Rapports des Comités d'etudes et de vigilance." 1880. *La vigne francaise* 1–7: 110.

Ravaz, Louis. 1902. *Les vignes amèricaines: Porte-Greffes et producteurs-directs, caractères-aptitudes*. Montpellier: Coulet et fils.

Reiss, Stephen. 1991. "Phylloxera. Super Bug, or Super Greed?" *Wine Education*. Available at www.wineeducation.com/phyl.html (accessed 18 July 2008).

Renfro, Roy, Jr., and Sherrie McLeRoy. 2008. *Grape Man of Texas: Thomas Volney Munson and the Origins of American Viticulture*. San Francisco: Wine Appreciation Guild.

"Réponse du editeurs." 1880. *La vigne française* 1–9: 135–37.

Riley, C.V. 1871. *Third Annual Report on the Noxious, Beneficial and Other Insects of the State of Missouri*. Jefferson City, MO: Public Printer.

———. 1872a. *Fourth Annual Report on the Noxious, Beneficial and Other Insects of the State of Missouri.* Jefferson City, MO: Public Printer.

———. 1872b. *Fifth Annual Report on the Noxious, Beneficial and Other Insects of the State of Missouri.* Jefferson City, MO: Public Printer.

———. 1872c. "On the Cause of Deterioration in Some of our Native Grapevines, and One of the Probable Reasons Why European Vines Have So Generally Failed with Us." *American Naturalist* 6: 622–31.

Robinson, J. 1996. *Guide to Wine Grapes.* Oxford: Oxford University Press.

———. 2006. *The Oxford Companion to Wine,* 3rd ed. Oxford: Oxford University Press.

Sahut, Félix. 1886. *La jaunisse ou chlorose des vignes.* Montpellier: Imprimerie Centrale du Midi.

———. 1888. *De l'adaptation des vignes américaines au sol et au climat.* Montpellier: Coulet.

Saintpierre, Camille. 1860. *De la fermentation et de la putréfaction.* Montpellier: Typographie de Boehm et fils.

———. 1877. *Recherches sur les vins des cépages américains recoltes en France en 1876.* Montpellier: Coulet.

Semmelweis, I.P. 1983. *The Etiology, Concept, and Prophylaxis of Childbed Fever,* translated and with an introduction by K. Codell Carter. Madison: University of Wisconsin Press.

Signoret, V. 1870. *Phylloxéra vastatrix cause prétendue de la maladie actuelle de la vigne.* Paris: Félix Malteste.

Slate, Frederick. 1919. "Biographical Memoir of Eugene Woldemar Hilgard, 1833–1916". *Biographical Memoirs,* vol. 9, 95–165. Washington, DC: National Academy of Sciences.

Smith, Rhonda. 1994. "The Phylloxera Picture in Sonoma and Napa Counties." *Wines & Vines,* August. Available at http://findarticles.com/p/articles/mi_m3488/is_n8_v75/ai_15728323 (accessed July 16, 2008).

Smith, Rhonda, and Ed Weber. 1999. "Removal of Phylloxera in Napa and Sonoma Counties." *Wines & Vines,* April. Available at http://findarticles.com/p/articles/mi_m3488/is_4_80/ai_54432346 (accessed July 14, 2008).

Sobrado, Teresa S. 1990. "La crisis del viñedo: La filoxera en España." Ph.D. diss., Universidad Complutense de Madrid.

Société Centrale d'agriculture de l'Herault. 1879. *Conférences pratiques sur le greffage des vignes américaines a l'amphithéatre de l'École du agriculture.* Montpellier: Grollier

———. 1900. *Livre d'Or, publié a l'occasion du centenaire de la société.* Montpellier: Progrès agricole et viticole.

Société d'agriculture, commerce, sciences et arts de la Marne. 1890. "Phylloxéra." Annals du Département de Marne 1J203. Cited in Kathleen Guy, "Uprooted Vines and Contentious Communities: Phylloxera and Peasant Protest in Champagne, 1890–95," Ph.D. diss., University of Texas, Austin, 1997; p. 5.

Société Linnéenne de Bordeaux. 1869a. *Rapport sur la maladie nouvelle de la vigne.* Bordeaux: Coderc, Degréteau et Poujol.

————. 1869b. *Deuxième rapport sur la maladie nouvelle de la vigne*. Bordeaux: Coderc, Degréteau et Poujol.

Song, G.C. and J. Granett. 1990. "Grape Phylloxera Homoptera: Phylloxeridae Biotypes in France." *Journal of Economic Entomology* 83: 489–93.

Southey, J.M. 1992. "Grapevine Rootstock Performance under Diverse Conditions in South Africa." In *Proceedings Rootstock Seminar: A Worldwide Perspective, 1992*, ed. J.A. Wolpert, M.A. Walker, and E. Weber, 27–35. Davis: American Society of Enology and Viticulture, 1992.

Stevenson, I. 1980. "The Diffusion of Disaster: The Phylloxera Outbreak in the *Département* of the Hérault, 1862–1880." *Journal of Historical Geography* 6: 47–63.

Sullivan, C.L. 2003. *Zinfandel: A History of a Grape and Its Wine*. Berkeley: University of California Press.

Tapia, A.M., J.A. Cabezas, F. Cabello, T. Lacombe, J.M. Martínez-Zapater, and M.T. Cervera. 2007. "Determining the Spanish Origin of Representative Ancient American Grapevines." *American Journal of Enology and Viticulture* 58(2): 242–51.

Tatum, L.A. 1971. "The Southern Corn Leaf Blight." *Science* 171(3976) (19 March): 1113–16.

Temkin, Owsei. 1977. *The Double Face of Janus and Other Essays in the History of Medicine*. Baltimore, MD: Johns Hopkins University Press.

Tibbitts, Paul. 1988. "Representation and the Realist-Constructivist Controversy." *Human Studies* 11: 117–32.

Tisserand, E. 1885. "Commission supérieure du phylloxéra, rapport sur les travaux de service du phylloxera en 1884." *La vigne française* 6: 117–24.

Trimoulet, A.-H. 1875a. *5eme Mémoire sur la maladie de la vigne: remèdes preconises par les phylloxeristes et les antiphylloxeristes*. Bordeaux: Imprimerie F. Degréteau.

————. 1875b. *Lettres sur le phylloxéra*. Bordeaux: Imprimerie nouvelle A. Bellier.

Trlin, A. 1979. *Now Respected, Once Despised: Yugoslavs in New Zealand*. Palmerston North, New Zealand: Dunmore Press.

Uhudler. 2008. "Geschichte." Available at www.uhudler.org/ (accessed 7 April 2009).

"University of California: In Memoriam, 1949." 2007. Available at http://content.cdlib.org/xtf/view?docId=hb167nbo8j&doc.view=frames&chunk.id=div00007&toc.depth=1&toc.id= (accessed 24 June 2008).

USDA. 2009. "GRIN Crop-specific Query." Available at www.ars-grin.gov/cgi-bin/npgs/html/desc_find.pl?crop=GRAPE&174001=Y (accessed 19 April 2009).

Van Zyl, D.J. 1984. "Phylloxera Vastatrix in die Kaapkolonie, 1886–1900." *Suid-Afrikaanse historiese joernaal* 16: 26–48.

Vega, J. 1956. "General Report from Latin America." *Bulletin de l'Office international du vin* 29(308): 31–42. Published in B.G. Coombe 1963 as "Appendix II."

Vène, Alexandre. 1882. "Communication sur le plus ou moins de résistance du terrain sableux." *Compte-rendu 1881*. Bordeaux: Féret et fils.

Viala, Pierre. 1888. *Mission viticole en Amérique, rapport au ministre de l'agriculture.* Montpellier: Coulet.

———. 1889. *Une mission viticole en Amérique.* Montpellier: Coulet.

Vialla, Louis. 1869. "Rapport sur le nouvelle maladie du vigne." *Journal d'Agriculture* 4: 341–60.

———. 1879. *Des vignes américaines et des terrains qui leur conviennent.* Montpellier: Grollier.

Vialla, Louis, and J.-É. Planchon. 1877. *Les cépages américaines dans le département de l'Hérault pendent l'année 1870.* Montpellier: Pierre Grollier.

La vigne. 1880. "Les vignes de Croatie." *La vigne française* 1(7): 112.

Vignes d'Antan. 2008. http://vigneantan.com/en/index.htm (accessed 16 May 2008).

"Les vignes de Croatie." 1880. *La vigne française* 1(7): 112.

Vimont, G. 1880. *Petit manuel & calendrier phylloxériques à l'usage des vignerons de Champagne.* Épernay: L. Doublat.

"Une visite au vignoble de Mancey." 1879. *La vigne francaise* 1-3: 43-44.

Vorwerk, S., and A. Forneck. 2006. "Reproductive Mode of Grape Phylloxera (Daktulosphaira vitifoliae, Homoptera: Phylloxeridae) in Europe: Molecular Evidence for Predominantly Asexual Populations and a Lack of Gene Flow between Them." *Genome* 49: 678–87.

———. 2007. "Analysis of Genetic Variation within Clonal Lineages of Grape Phylloxera (Daktulosphaira vitifoliae Fitch) using AFLP Fingerprinting and DNA Sequencing." *Genome* 50: 660–67.

Walker, M.A. 1992. "Future Directions for Rootstock Breeding." In *Proceedings Rootstock Seminar: A Worldwide Perspective, 1992,* ed. J.A. Wolpert, M.A. Walker, and E. Weber, 60–66. Davis: American Society of Enology and Viticulture, 1992.

———. 1993. "Alternative Rootstocks: How Will They Perform?" *Wines & Vines,* December. Available at http://findarticles.com/p/articles/mi_m3488/is _n12_v74/ai_14894921/print?tag=artBody;col1 (accessed July 16, 2008).

———. 2000. "UC Davis' Role in Improving California's Grape Planting Materials." *Proceedings of the ASEV 50th Anniversary Meeting, Seattle, Washington, June 19-23, 2000,* 210–15. Davis: American Society of Enology and Viticulture.

Walsh, Gerald. 1979. "The Wine Industry of Australia, 1788–1979." In *Wine Talk, Australian Winemakers and Their Art: A Collection of Papers from a Symposium Held at the Australian National University Staff Centre, September 1979,* ed. Brian Barry. Canberra: Action Press, 1979.

"War on the Phylloxera." 1880. *New York Times,* 3 April, p. 3.

West-Eberhard, M.J. 2003. *Developmental Plasticity and Evolution.* Oxford: Oxford University Press.

Whetzel, Herbert Hice. 1918. *An Outline of the History of Phytopathology.* Philadelphia: W.B. Saunders Company.

Whiting, J.R. and G.A. Buchanan. 1992. "Evaluation of Rootstocks for Phylloxera-Infested Vineyards in Australia." In *Proceedings Rootstock Seminar: A Worldwide Perspective, 1992,* ed. J.A. Wolpert, M.A. Walker, and E. Weber, 15–21. Davis: American Society of Enology and Viticulture, 1992.

Winslow, C.-E.A. 1943. *The Conquest of Epidemic Disease*. Princeton, NJ: Princeton University Press.

Wolpert, J.A. 1992. "Rootstock Use in California: History and Future Prospects." In *Proceedings Rootstock Seminar: A Worldwide Perspective, 1992*, ed. J.A. Wolpert, M.A. Walker, and E. Weber, 52–59. Davis: American Society of Enology and Viticulture, 1992.

Wolpert, J.A., M.A. Walker, and E. Weber, eds. 1992. *Proceedings Rootstock Seminar: A Worldwide Perspective, 1992*. Davis: American Society of Enology and Viticulture, 1992.

Wolpert, Jim. 2002. *Field Evaluation of Wine Grape Rootstocks*. Napa, CA: American Vineyard Foundation.

Zacharewicz, C.M. 1932. "Cinquantenaire de la reconstitution en Vaucluse." *Revue de viticulture* 77: 270–85.

Index

5A, 226, 281n20
41B, 139, 143, 155–56, 254, 272n13, 273n18, 274n32
99R, 224, 226, 280–81n14, 281n20
101–14Mgt, 174–75, 221, 223, 272nn13,15
110 (Richter 110), 282n33
161–49C, 223
420A Mgt, 272n13, 282n33; in California trials, 221, 223, 224, 226, 230, 281n20
1202C, 191, 193, 222, 223, 224, 228, 230, 278n11, 281n23
1613C, 223, 226, 281n20
1616C, 223
3306C, 223, 224, 226, 228–29, 281n20
3309C, 174–75, 221, 222, 224, 226, 228–29, 281n20, 282n33

Academy of Sciences, 41, 45–48. See also National Phylloxera Commission
Ackerknecht, E.H., 25, 26
adaptation (grape and rootstock varieties), 120–21, 128, 267n22; Australia, 190–91; French ideas about, 87–88, 112, 127, 271n4; labruscas and riparias, 87–88; Lafitte and Fitz-James positions, 106–7, 110–12; Old American varieties, 87–88, 128, 270n43; research proposals and methods, 107, 117–18; resistance and, 106–7, 112; rootstocks, in California, 221, 222, 223, 227–28, 244; rootstocks, in Italy, 169; rootstocks, in the southern hemisphere, 190–91, 193, 199; selecting/hybridizing for, 137, 138, 276n6; white soils problem, 127–34, 149–56. See also chalky soils
adaptation (phylloxera populations), 237–40
agricultural expositions, 6, 260n11
agricultural press: American vine controversy in, 96, 97–98, 103; cause controversy in, 15, 17, 18, 29, 30–31, 37, 42; on control methods, 40; on phylloxera's spread, 52; on promise of HPDs, 140. See also specific publications
agricultural societies, 100–101, 171. See also SCAH; other specific groups
agriculture: monoculture and invasive species, 4; polycultures, 77, 123, 166–67, 170
Aigues-Mortes, 58–59
Alameda County, 230
Algeria, 58, 204, 264n5, 280n11
Alicante Bouschet, 134–35
Alsace Viticultural Institute, 256, 272n15
Amador County, 213
America: Nedeczky's study trip, 181; as origin of French phylloxera infestation, 43–44, 81; phylloxera in, 35, 86; Planchon's 1873 study tour, 45–46, 84–86, 89, 262–63n27, 266n12, 273n20; Viala's 1886 visit, 139, 144–49, 250, 262–63n27, 267n22, 273nn19,20. See also specific states

American black rot fungus, 7
American chestnut blight, 259n7
American cooking, 86
American-European hybrids, 81, 266n6,
272n15; debate over uses of, 80,
265–66n3; varieties listed, 257–58. *See
also* hybrid direct producers; hybrid
rootstocks; Old Americans; *specific
varieties*
American exceptionalism, 242, 243, 283n48
American grapes, 45–46; Millardet's
taxonomic work, 46, 84, 91, 93, 141;
phylloxera on, 35, 86, 238, 251;
Planchon's study tour (1873), 45–46,
84–86, 89, 262–63n27, 266n12,
273n20; range and species diversity, 80,
238, 253, 266n5, 283n40; species
descriptions, 253–56; Viala's study tour
(1886), 139, 144–49, 254, 262–63n27,
267n22, 273nn19,20. *See also* American
vine controversy; American vines in
France; Old Americans; phylloxera
resistance; reconstitution; *individual
species and varieties*
American hybrids, 81, 93, 112, 142, 266n7;
adaptation to French conditions, 87–88,
106–7, 110, 270n43; Italian-bred
rootstocks, 169; Jaeger hybrids, 136,
137, 146, 254, 273n21; natural hybrids,
115–16, 136–37, 138, 147, 148, 271n6;
in Portugal, 174–75; varieties listed,
257–58. See also *individual species and
varieties*
américanistes, 8, 95–103, 268n29, 269n33.
See also American vine controversy
American rootstocks. *See* resistant
rootstocks
American vine controversy (France), 8,
54–58, 95–103; Bazille's arguments,
99–103; concerns about importing other
pests and diseases, 146; concerns about
phylloxera resistance, 97, 98, 102,
105–7, 268n31, 269n39; concerns about
preservation of traditional viticulture,
54, 119; concerns about taste, 45, 91,
122, 201, 202, 204, 205, 208;
conversions to the *américaniste* camp,
96–97, 100; Fitz-James's position, 19,
109–13, 269n39; key opponents, 103–4;
key proponents, 38–39, 82–83,
100–101, 268n29; Lafitte's position,
103–9, 113–19; official/government
position, 54–58, 70, 71, 98, 102–3;
Planchon's position, 56, 90–91, 93,

95–96, 127; polemic in, 100, 131–32; in
the press, 96, 97–98, 103; twentieth
century, 80, 265–66n3. *See also*
adaptation; American vines in France;
Lafitte, Prosper de
American vines in France, 44–45, 79–80,
81–82; Charentes, 129–30; chlorosis
problems, 128–29, 130–31, 133–34,
153, 154–55, 268n30; demand and
prices, 269n33; as direct producers,
265nn1,3, 275–76n2; disease resistance,
146–47; early plantings in the Midi, 45,
87–88, 95–96, 263n32; grower
syndicates' interest in, 76, 77, 265n17;
Lafitte's study proposal, 117–18;
Laliman's Bordeaux plantings, 9, 38–39,
43–44, 82; legal restrictions, 6, 56–58,
77, 283n45; phylloxera on, 35, 261n15;
pre-phylloxera imports, 81; twentieth
century, 269n35; viticultural practices
and, 10, 94–95, 268n28. *See also*
adaptation; American vine controversy;
hybrid direct producers; Old Americans;
phylloxera resistance; reconstitution;
resistant rootstocks
American wines and viticulture, 80–81,
110–11, 112, 116, 202, 272n15;
cooperation to develop American vine
usage in France, 10; French opinions of
American wines, 91, 201, 202; Missouri,
85, 89, 110, 266–67n17; Ohio, 85, 87,
89; Planchon and, 84, 85; *Vitis vinifera*
in America, 81, 89, 110–11, 202, 253.
See also California; Old Americans
Annaberg Gardens, 176
*Annual Reports on the Noxious, Beneficial
and Other Insects of the State of
Missouri* (Riley), 35, 42
anthracnose, 185
Appleton, Henry, 212–13
Aragon, 171–72
Aramon, 97, 100, 112
Aramon × *rupestris* #1. *See* AxR1
Ardèche, 80, 206, 207
arenious acid, 72
Argentina, 164, 196–99, 278n16
argument, in science, 7–8
Aristotle, 7
Armagnac, 129. *See also* Charentais
Armillaria root rot, 14
Arnold, Charles, 135–36, 271n8
arsenic sulfide, 72
Aubert, Achille, 129, 268n28
Aude, 59, 69

Australian viticulture, 184–85, 188; phylloxera, 4, 164, 185–91, 199, 238, 251, 259n8, 277nn2,5; quarantine regulations, 186, 187–88, 199, 259n8; rootstock research and plantings, 186, 188–89, 190–91

Austria, 176, 178, 204, 209, 276n9

AxR1, 137*fig.*; California acreage after reconstitution, 236, 282n35; California plantings, 1980s–1990s, 234–35, 282n32; and California's second phylloxera disaster, 227, 231–36, 239–40, 272n12, 278n14; development and parentage of, 135, 138, 139, 140; early California research programs, 222, 223, 224, 244–45; failures outside California, 191, 193, 228, 241, 245, 278n14; French disavowal of, 223, 245; good points of, 193, 226, 228, 243–44, 278n15; Lider's 1957–1958 recommendation, 226, 227–28, 229–30, 241; Lider's 1974 reevaluation, 229–32, 281n27, 282n29; origins and parentage, 135; phylloxera population dynamics and, 239–40; phylloxera resistance, 193, 222, 223, 224, 228, 230–31, 239, 244, 278n14. *See also* California rootstock research

AxR2, 193, 223

Azam, M. le Prof., 61–62

Baco, François, 219

Baco hybrids, 142, 256, 272n16

Bages submersion syndicate, 69

Bailey (grape), 273n25

Bailey, L.H., 271–72n9

Balbiani, E.G., 103–4, 118, 269nn36,37

Barbera, 281n26

Baritelle, John, 232–33

Barnes, Barry, 8

Barral, J., 29, 68

Baussan, M., 157

Bayle, M.C., 59

Bazille, Gaston, 101*fig.*, 268n29; on first phylloxera commission, 16, 19; grafting experiments, 82–83, 120, 266n9; on National Phylloxera Commission, 41, 55; and phylloxera cause controversy, 30; as proponent of American rootstocks, 82–83, 93; on sulfiding, 77

Bazille, Jean-Frédéric, 101*fig.*

Bazille, Louis, 94

Beaujolais, 51, 96–97, 123, 268n29

Beaune Viticultural Congress (1869), 38–39, 82, 91, 121

Becker, Helmut, 187, 190

Bédarieux, 67–68

bench grafting, 124, 125–26

Berget, Adrien, 154, 156

Bernard, M., 154

Big Science, 1–2, 5–7, 260n10

Bioletti, F.T., 213, 215, 217*fig.*, 220–21, 280nn5,14; recommendations for California, 216–19, 220–21; rootstock research program, 221, 222–23, 223–24, 225, 240, 244–45, 281n15

biotypes, 282n31; phylloxera Type B identification, 233–35

black rot, 42, 146

Black Spanish. *See* Jacquez

Blaxland, Gregory, 185

blending theories of inheritance, 93, 142, 270n1

Bloor, David, 8

Blunno, Michele, 186–87, 188–89, 281n16

Board of State Viticultural Commissioners (California), 221

Bogen, James, 260n16

Bonarda, 197, 278n16

Bonner, Bill, 206

Bordeaux, 54, 74, 123, 126; phylloxera in, 9, 14, 262n20; Thénard's carbon disulfide experiments, 39, 46, 72. *See also* Gironde; University of Bordeaux

Bordeaux International Phylloxera Congress (1881), 65, 69–70, 75, 138, 264n14; Committee on American Vines, 58–59, 60, 63, 117–18, 270n43

Bordeaux Phylloxera Congress (1886), 138, 139

Börner, Carl, 234

Boston, Planchon in, 86

Boston ivy, 120, 266n9

botany education, 24

bottleneck effect, 238

Boubals, Denis, 241–42, 250

Bouches-du-Rhône, 58–63, 264n5

Bouffet, Maurice, 69

Bouschet de Bernard, Henri, 94, 134–35, 268n29

Bouschet de Bernard, Louis, 134, 135

Bouze, 76

Bowers, K.W., 231–32, 234

Boyle, Robert, 27

Brahe, Tycho, 23

Bridges and Roads (Ponts et Chaussées), 69, 96, 264n13

Brito, Francisco d'Almeida e, 174
Buchanan, Gregory A., 185–86, 187, 188
Bulgaria, 182, 183, 210, 277n14
Burgundy: inheritance laws and vineyard consolidation, 265n16; Italian vineyard workers in, 54, 74, 123; phylloxera in, 52; phylloxera response in, 5, 41, 76, 77, 242–43
Burian, Richard, 260nn15,16
Busby, James, 185
Bush and Son and Meissner, 93–94; European visitors to, 146, 148, 181, 250; grape orders from France, 45, 82, 87
Bushberg catalogue, 93–95, 116, 268nn27,28
bush grape. See Vitis rupestris

Cabernet, 197, 281n26
Calaveras County, 213
calcareous soils. See chalky soils
California: exceptionalism in, 213–14, 242; Viala in, 146
California Farmer, 213
California phylloxera infestation (nineteenth century), 164, 212–32; acreage losses, 215; Bioletti's recommendations, 216–19; control efforts, 216–18, 278n17; identification and early response, 212–16; initial denial, 213–14, 243; origin of, 213; reconstitution on resistant rootstocks, 218, 219–22; spread of, 213–14, 215
California phylloxera infestation (1980s–1990s), 2, 4, 232–47, 260n4, 277n9; AxR1 and, 227, 231–36, 239–40, 272n12, 278n14; biological factors, 236–40; Davis's turnabout and reconstitution, 235–36; Granett's research, 232–35; grower responses, 5, 214, 242, 243, 278n12; institutional/research failures and, 240–46
California rootstock research, 221–32, 240–46, 276n3, 277n9; adaptation in, 221, 222, 223, 227–28, 244; early programs, 221–25, 240, 244–45, 281n15; failures of, 227, 228, 234, 240–46; Lider's 1957–1958 recommendations, 226, 227–28, 229–30, 241; Lider's 1974 follow-up, 229–32, 281–82n28; Lider's research goals and methods, 224–26, 229, 230, 240, 244,
245, 281–82n28; one stock fits all notion, 221, 226, 227, 277n9; performance in, 225–26, 227, 228, 229, 230, 243–44; phylloxera resistance in, 222, 223, 224, 225, 226–29, 244. See also AxR1
California Viticultural Commission, 214, 216
California viticulture, 211–12; nineteenth century, 85, 211–12, 215, 266–67n14; Old American plantings, 212, 213, 280nn3,5,6; Planchon on, 266–67n14; planting boom, 1960s–1970s, 228, 230, 232; use of AxR1, 222, 272n12, 278nn14,15. See also California phylloxera infestations
Camargue, 59, 62–63, 264n8
Cameron, William Gordon, 194
Canada, 272n15
Canary Islands, 280n1
carbolic acid, 72
carbon disulfide (CS_2), 71–74, 75–78, 97, 99, 264n14; extinction treatments, 57–58, 165–67, 177–78, 217–18; outside France, 165–68, 174, 177–78, 198, 217–18; problems with, 10, 71–74, 75–77, 278n17; railroad research and distribution program, 72–74, 99; Thénard's experiments, 39, 46, 72; ultimate failure of, 5, 77–78
Carcassonne, 131
Carignan, 100
Carrière, M., 129–30
Castan, M., 156
Castella, François de, 186, 189–90
Castets, M., 157, 158–59
Catawba, 257, 278n3; in California, 212, 213, 280n3; phylloxera resistance, 85, 86, 87, 89, 202
cause controversy, 1, 8, 18–32, 249; development and settlement of, 9, 32, 40, 42, 47–50; in first months after discovery, 18–21, 23–24, 28–32; impatience with, 37, 261–62n16; national commission's position, 41, 48–49, 54; polemic in, 33, 40, 261n13, 262n23; range of positions in, 32–33; Riley's position, 42. See also phylloxera-cause theory; phylloxera-effect theory
Cavazza, D., 169
Cazalis, F., 15, 17, 18, 28, 30–31, 156
Central Committees for Study and Vigilance of Phylloxera, 7. See also departmental phylloxera commissions

Central Europe, phylloxera in, 57, 164, 176–82, 260n9. See also *specific countries*

Central Valley (California), 216, 280n10

Les cépages américaines dans le départe-ment de l'Hérault pendent l'année 1870 (Vialla and Planchon), 15, 89–90, 267n21

cépages prohibés, 205–10, 278n1. See also Old Americans

Ceris, M.A. de, 40

Cévennes, Old Americans in, 206, 207–9, 269n35

chalky soils, 267–68n23; *berlandieri* hybrids for, 8, 152, 153–56, 254; Chauzit's varietal recommendations, 155*table;* chlorosis and, 131, 133–34, 153, 154–55, 268n30, 271n5; 41B for, 139, 143, 155–56, 254, 272n13, 273n18, 274n32; in Italy, 169; recognition of adaptation problem, 129–31, 133–34, 143–44, 267n22; sand planting on, 60–61; in Texas, 147; Viala's mission to find suitable American vines, 139, 144–49, 254, 262–63n27, 267n22, 273nn19,20

Champagne, 52–53, 54, 243, 248

Champin, Aimé, 98

Champin grape, 155*table*

Charbono, 278n16

Chardonnay, 197

Charentais: American vines in, 129, 144–45, 153–56; grafting techniques in, 124–25; hybrid grape trials, 138, 139, 152; phylloxera in, 51, 129, 144, 152–53. See also chalky soils

Charente phylloxera commission, 129–30

Chasselas, 274n32

Châteauneuf-sur-Charente conference, 268n28

Chauvin, M., 130

Chauzit, B., 124, 130, 154, 155*table*, 262n17

Chavée-Leroy, M., 132

chemical controls. See carbon disulfide; pesticides; sulfiding

chemical fertilizers, 60, 65, 74. See also fertilization

chemistes, 70–71, 95–103. See also American vine controversy; pesticides

Chenôve, 76

Chenu-Lafitte, M., 68

chestnut blight, 259n7

Cheysson, Émile, 96–97

Chile, 164, 196, 277n1

China, *Vitis* in, 253

Chiroubles, 96

chlorosis, 128–29, 130–31, 133–34, 153, 154–55, 268n30

cholera, 26

Cincinnati wine industry, 85, 89

clay soils, 64

climate: climatic adaptation of American vines, 87–88, 106–7, 110, 221, 270n43; overwintering capability of phylloxera, 238, 283n42; pesticide applications and, 99; phylloxera behavior and, 177, 187, 197–98, 260n9; phylloxera-effect view and, 18, 25–26, 29, 30–31, 47

climate change, 3

Clinton, 104, 202, 204, 269n33; foxiness in, 91; French growing prohibition, 205; French importation and plantings, 87, 88, 89, 128, 129, 130, 267n17; Hungarian trials, 181; in Italy, 168, 170, 203, 276n5; parentage, 93, 266n7; phylloxera resistance, 87, 88, 89; as rootstock, 90–91; soil adaptation, 128; varietal characteristics, 257

Clinton Vialla, 181, 276n13

Cognac, 74, 129, 152; American and hybrid rootstocks in, 152, 153–54, 155–56, 267n22. See also Charentais

Cohen, Henry, 24

Colares, 173, 262n21

Colmar Institute of Viticulture, 272n15

comices agricoles, 52

comités d'etude et de vigilance départemen-taux. See departmental phylloxera commissions; *specific departments*

Commission supérieure du phylloxéra. See National Phylloxera Commission

Committee of Vigilance, 265n18

Committee on Submersions, 65, 66

Compagnie des chemins de fer de Paris à Lyon et à la Méditerranée. See PLM

Compagnie des Salins du Midi, 59

Comte, August, 159

Concord, 265nn1,2, 266n6, 278n3; French importation and plantings, 87, 88, 130, 267n17; French planting failures and consequences, 88–89, 90, 121, 127, 128, 129; Hungarian trials, 181; in Italy, 168; parentage, 88, 93; phylloxera resistance, 87, 88–89, 202; Planchon's observations about, 85, 87, 88–89, 106; varietal characteristics, 257

Confronting Nature: The Sociology of Solar Neutrino Detection (Pinch), 11, 260n16
Congrès scientifique de France, 31
Constantia Estate, 191
Constructing Quarks (Pickering), 11, 260n16
Contassot, Eugène, 141, 254
control efforts and methods, 5, 6, 37, 51–78, 278n17; Argentina, 197–98; Balbiani's egg-destruction theory, 103–4, 109, 118, 269nn36,37; California, 216–18, 278n17; calls for suggestions, 46–47, 180; cause controversy and, 37–38, 39, 47–48; control vs. destruction, 102, 269n34; cordons sanitaires, 5, 6, 56–57, 178, 199, 283n45; expense of, 69, 74, 76–77, 79, 102, 104, 265n18; fertilization, 10, 60, 62, 65; Germany, 177–78, 260n9; government funding and facilitation, 6–7, 41, 46–47, 56, 75, 76; Hungary, 179–80; impatience for, 37–38, 261–62n16; Italy, 165–67; local syndicates for, 6–7, 63–64, 69, 75–76; Portugal, 173–74; South Africa, 192, 193–96; and tensions between large and small landholders, 69–70, 71, 74, 75. *See also* carbon disulfide; pesticides; sand planting; submersion
Convert, F., 45
Coombe, B. G., 187, 191
cordons sanitaires, 5, 6, 56–57, 178, 199, 283n45. *See also* quarantines
corn, 4, 58, 264n4
corn blight, 4, 259n6
Cornelius, M., 94
Cornu, M., 54, 70–71, 72, 74
Cortés, Hernán, 196
Côte d'Or, 41, 76
Couderc, Freddy, 208, 279n18
Couderc, Georges, 122, 135, 139, 140, 141, 279n18
Couderc hybrids, 142, 151–52, 191, 272n16, 278n11; 1202C, 191, 193, 222, 223, 224, 228, 230, 278n11, 281n23; 1613C, 223, 226, 281n20; 1616C, 223; 3306C, 228–29; 3309C, 174–75, 221, 222, 224, 226, 228–29, 231, 281n20, 282n33; 7120C, 142, 272n16
Coulet (publisher), 268n27
Cousin, M., 159
Craig, O.W., 212–13
Criolla Chica, 196
Croatia, 176, 178–79, 204
CS₂. *See* carbon disulfide

Cunningham, 87, 181, 257, 267n17
Cynthiana. *See* Norton

Daktulosphaira vitifoliae. *See* phylloxera
Dalmasso, G., 165, 166–67, 170–71, 187
Dalmatia, 176, 178–79, 204
Darwin, Charles, 25, 141
Darwinian theory, 102, 261n14, 263n29, 271n4, 272–73n17; resistance dynamics and, 83, 105, 263n29
Daurel, Joseph, 132
Davidson, W. M., 212, 213, 214
Débouchaud, M. le Dr., 129
La Défense, 54–58, 63, 95, 263n3; ultimate failure of, 78, 79. *See also* American vine controversy; control efforts and methods
degeneration, 25, 26
Delahaye (publisher), 268n27
Delaware grape, 87, 89, 257, 278n3
Delorme, M., 14
demographic impacts of phylloxera, 58, 179, 247, 264n5
denial, 5, 242–43, 248, 283n48; American exceptionalism, 242, 243, 283n48; California, 5, 214, 242, 243; France, 5, 15, 51–53, 242–43; Italy, 170
Dennehy, T. J., 233–34, 236, 241
departmental agricultural societies, 100–101. *See also* SCAH; *specific departments*
departmental phylloxera commissions, 7, 41–42, 118, 249; first commission investigation (Hérault 1868), 16–18, 19, 28, 43, 261n8, 275n35. *See also specific departments*
Descartes, René, 27, 102
"Des vignes américaines et des terrains qui leur conviennent" (Vialla), 128–29
Develle, Jules, 145
Dezeimeris, R., 55
diatheses, 25, 26, 29
direct producers, 265n1. *See also* hybrid direct producers; Old Americans
disease, theories of, 18–19, 24–26; fermentation analogies, 27; French medical training and, 26–27; ontological theories, 24, 26, 27; physiological theories, 18–19, 24–26, 27, 29; resistance dynamics and, 84. *See also* cause controversy; phylloxera-cause theory; phylloxera-effect theory
disease dynamics, 25, 26

disease resistance, 146–47, 203, 219
Dogridge, 223, 224, 226, 281n20
Dolcetto, 278n16
Donnadieu, M., 275n35
dos Santos vineyards (Portugal), 174
Douro, 163, 172
Downey, John, 212
Downie, D.A., 238, 239, 282n36
Dreissena polymorpha (zebra mussels), 3
Duclaux, Émile, 72
du Lot. *See* Rupestris du Lot
Dumas, J.B., 102, 269n34; cause contro-
versy and, 48–49, 54; on National
Phylloxera Commission, 41, 48–49, 54,
55; as pesticide proponent/researcher,
54, 70–71, 74, 263n3, 269n34
Dunal, M., 160, 162
*Du phylloxéra et d'un nouveau mode
d'emploi des insecticides* (Falières),
48–49
Durand, M., 46
Dutch elm disease, 3, 4

Eastern Europe: Bulgaria, 182, 183, 210,
277n14; Hungary, 176, 177, 179–81,
204, 210; Jacquez in, 269n35; Romania,
181–82, 183, 210, 249
École Nationale Supérieure d'Agronomie de
Montpellier (École Agronomique), 6, 7,
27–28, 51, 155; foreign students at, 169,
249; grower education programs, 96,
121–24, 270–71n2; pesticide trials,
46–47, 263n32; and phylloxera
responses outside France, 183, 241–42,
249–50; position on American-European
hybrids, 122, 142, 147, 151, 270n1,
274nn27,28; research focus on resistant
rootstocks, 122, 148, 270n1, 274n28;
research vineyard, 46, 154; rootstock
research, 80, 121, 154, 245, 249. *See
also* French expertise; Galet, Pierre;
Planchon, J.-É; Viala, Pierre; Vialla,
Louis
economic issues: costs of reconstitution,
124, 126–27, 127*table*, 236; economic
impacts of phylloxera in France, 15–16,
41, 54, 265n17; economic impacts of
phylloxera in South Africa, 192, 194;
expense of control methods, 69, 74,
76–77, 79, 102, 104, 265n18; French
wine glut, 1920s, 204–5; grafting and
workforce changes, 123, 124, 126; wine
production in the French economy,
15–16, 54

education. *See* viticultural education
Eisenhower, Dwight D., 260n10
Eke, István, 276n10
elm disease, 3, 4
Elvira, 181, 276n13
endogenous theories of disease. *See*
physiological theories of disease
Engelmann, George, 266n13; Bushberg
catalogue contributions, 94, 95; and
Planchon's U.S. visit, 85, 250, 263n30;
as provider of seed and cuttings, 93,
220, 268n26; on range of *V. labrusca*
and natural hybrids, 115–17; Solonis
discovery, 271–72n9
England, 43
English graft, 125*fig.*, 126
ENSAM. *See* École Nationale Supérieure
d'Agronomie de Montpellier
Entomological Society of France, 19, 33
entomologists, entomology training, 19, 21,
24
environmental factors, disease theory and,
25, 26, 29. *See also* climate; soils
Épernay, 52–53
epidemics, 26
eradication efforts, 5; South Africa,
193–94. *See also* extinction treatments;
pesticides
Ericson, Leif, 253
Espitalier, Silvain, 62–63, 264n8
Essai sur l'hybridation de la vigne
(Millardet), 142–43
EU (European Union) American vine
prohibition, 203, 207–8, 209, 210,
276n9
Europe: phylloxera outside France, 163–65,
183. *See also specific countries*
European-American hybrids. *See* American-
European hybrids
European grapes. *See Vitis vinifera; specific
varieties*
exceptionalism. *See* denial
exogenous theories of disease. *See*
ontological theories of disease
extinction treatments, 57–58, 165–67,
177–78, 217–18

F1 hybrids, 138–40, 142, 272n16
Fabre, M., 88, 92
Falières, E., 48–49
Faucon, Louis, 9, 37–38, 249, 264n7; and
submersion, 9, 38, 40, 63, 65–66,
262nn17,19
Federal Co-operative Extension, 260n13

Fergusson-Kolmes, L., 233–34, 236, 241
Fermaud, M., 46, 263n32
fermentation, theories of disease and, 27–28
Ferrari, N.L., 231–32
fertilization, 10, 60, 62, 65, 99; chemical fertilizers, 60, 65, 74
field grafting, 124–25, 125fig.
Fitch, Asa, 35, 43
Fitz-James, Marguerite Duchesse de, 19, 109–13, 109fig., 119, 149, 269n39
flooding.' See submersion
Flore des serres et des jardins d'Europe, 162
Flossfeder, Frederic, 221, 222, 240, 244, 281n15
Foëx, G., 102, 122, 145, 152, 155, 268n29, 270n40; berlandieri crosses, 151; on chlorosis problems, 130–31; on parentage of Jaeger hybrids, 136, 137; on Planchon and the National Phylloxera Commission, 55; Planchon obituary, 275n35; views on American-European hybrids, 142, 274n28; views on inheritance, 142
Forneck, A., 238–39
Foundation Plant Material Services (UC Davis), 283n51
Foundations of American Grape Culture (Munson), 112, 149
founder effect, 238
four remedies, 278n17. See also grafting; pesticides; sand planting; submersion
Fourth Annual Report on the Noxious, Beneficial and Other Insects of the State of Missouri (Riley), 43
fox grapes. See Vitis labrusca
foxiness, 81, 91, 202–3, 278n2
Fragolino (strawberry wine), 203–4
Franco-Prussian War, 6, 41
Franklin, Alan, 8–9
Free Syndicat of the Maritime Agricultural Colony of the Bay of Capitoul, 69
French Colombard, 281n26
French expertise abroad, 249–50; Australia, 186; Bulgaria, 182, 249, 277n14; California, 214, 222, 223, 241–42, 245–46, 249–50, 276n3; Italy, 168; Spain, 171–72, 249
French phylloxera infestation (1860s–1900): arrival and early reactions, 3, 4, 11, 13–16, 43–44; socioeconomic impacts, 15–16, 41, 54, 58, 247, 264n5, 265n17; spread and acreage impacts, 14–16, 41, 42, 51–53, 99, 129

French phylloxera responses: birth of Big Science and, 1–2, 5–7; as birth of new viticulture, 164–65, 170–71, 196, 247–48; denial and exceptionalism, 5, 15, 51–53, 242–43; initial grower responses, 14–15; legal restrictions, 6, 56–58, 77; lessons of, 7–12; local/regional responses, 6, 32–33, 41–42, 51–53; national responses, 6–7, 30–32, 40, 41, 45; pesticide development, 6–7, 10, 260n12; phases of, 4–5. See also American vines entries; cause controversy; French expertise; phylloxera commissions
Fuller, Thomas, 208, 279n17
fungal diseases, 182, 255–56; Australia, 185; oidium, 35, 138, 146, 261nn8,11
fungicide, sulfur as, 261n11

Gachassin-Lafite, L., 58–59, 60, 63, 117–18, 270n43
Galet, Pierre, 265–66n3, 269n33, 274n27; on 41B, 273n18, 274n32; on Old Americans, 206, 208, 210, 266n7, 271–72n9, 278n1; on Type B phylloxera identification, 234; warning about AxR1 in California, 242, 250
Gamay, 265n18
Gamulin, Stjepan, 178–79
Ganzin 1. See AxR1
Ganzin, Victor, 122, 140, 268n29; AxR1 development, 135, 138, 139, 140
Gard, 15, 54, 100, 124, 129–30, 131
Gardener's Monthly, 94
Garrier, Gilbert, 56, 57
Gautier, M., 16
gelatin, 72
genetic variation: among seedling graft stock, 115, 150–51, 254; in phylloxera populations, 237–40, 282nn36,37
Germany: exclusion of American direct producers, 204; phylloxera in, 57, 164, 176–78, 183, 260n9
germ theory, 27. See also ontological theories of disease; phylloxera-cause theory
Gers, 57–58
Gervais, Prosper, 8, 164–65, 275–76n2, 276n4; California researchers' reliance on, 245, 249–50, 276n3; on phylloxera responses outside France, 169, 175, 177, 181, 182; on Viala and V. berlandieri, 150

Girard, Maurice, 49
Gironde, 51, 62, 69, 70, 100–101
Gironde, M. le Vicomte de, 132–33
Gironde Horticultural Society, 132
glassy-winged sharpshooter, 3
global warming, 3
Gloire de Montpellier (Riparia Gloire), 131, 151, 221, 223, 271n6
Goheen, Austin, 232
Gouy, Paul, 274n29
government responses: Argentina, 197, 198–99; Australia, 186, 189; Bulgaria, 182, 183; California, nineteenth century, 214–15; France, 6–7, 30–32, 40, 41, 45; Germany, 176–78, 183; Hungary, 179, 181; importance of, 183; Italy, 165, 168; Portugal, 172–73; Romania, 181, 183; South Africa, 192–93, 194; Spain, 171, 175–76. *See also* legal restrictions
grafting, 120–27, 278n17; Bushberg catalogue discussion, 94; French experience with, 120; problems and challenges, 120–21, 137–38; scion-rootstock affinity, 108, 121, 221, 244, 267n22, 276n7; techniques and equipment, 10, 121, 124–27; virus-infected scions and, 244, 276n7; workforce problems and issues, 123, 124, 126
grafting education, 121–24, 190, 193
graft stock. *See* resistant rootstocks; *specific varieties*
Grande Champagne (Cognac), 153, 155–56
Granett, Jeffrey, 232–35, 233*fig.*, 243, 250, 282n30
Grasset, Charles de, 138, 139, 140, 152, 155, 272n13
Greece, 204
greffe anglaise, 125*fig.*, 126
Grenache, 9, 58, 262n20
Grimaldi, C., 169
grower responses, 14–15; California, 5, 214, 242, 243, 278n12; Italy, 166–67; resistance to government requirements and prohibitions, 6, 56–57, 76–77, 205, 206–9; resistance to joining syndicates, 69, 75; role of practical solutions and trials, 37, 248–49; tensions between large and small landholders, 69–70, 71, 74, 75, 76–77, 265n18. *See also* denial; *specific individuals*

grower syndicates. See *syndicats de défense*
Guérin-Méneville, F.-E., 21, 21*fig.*; and cause controversy, 9, 19, 23, 24, 27, 47–48, 261–62n16; Planchon's opinion of, 40

Handbook on Viticulture for Victoria (de Castella), 186
Hanson, N.R., 22–23
Haraszthy, Agoston, 211–13, 280n2
Haute-Garonne Agricultural Society, 132–33
Hérault, 15, 51, 54, 95; American vine plantings in, 87–88, 89–90, 99, 129–30, 131, 268n32. *See also* SCAH
Hérault phylloxera commission: 1868 investigation, 16–18, 19, 28, 43, 261n8, 275n35
Herbemont, 87, 89, 112, 265n1, 266n7; flavor and quality, 91, 202; French importation and plantings, 202, 267n17; French planting prohibition, 205; Hungarian trials, 181; parentage, 202, 254; soil adaptation, 128; varietal characteristics, 257
Hilgard, Eugene, 214–15, 219
hillside vineyards: California, 227, 230, 281n22; submersion and, 64, 167
Histoire de la lutte contre le phylloxéra de la vigne en France (Pouget), 14–15, 45, 55, 261nn6,13, 263n3
Hofmeyr, Wynand, 194
Hooker, Elisabeth, 162
Hooker, William Jackson, 161–62
Hopkins grape, 273n25
Hortolès, M., 31–32, 122
Horváth, G. de, 177, 179, 180–81
host diversity, 238
Houdaille, M., 154
Hours, L. des, 158
Howlong research station, 188–89
Human Genome Project, 6
Hungarian Phylloxera Station, 179, 180–81, 276n11
Hungary, 176, 177, 179–81, 204, 210
Hunter Valley, 185
Husmann, George C., 215, 216*fig.*, 250; rootstock trials, 221–22, 223–24, 224–25, 240, 242, 244–45
hybrid direct producers (HPDs), 79, 80, 81, 265–66n3, 271n8, 280n12; *berlandieri* hybrids, 255, 280n12; Jaeger parentage in, 254, 273n21; Millardet's work, 135,

hybrid direct producers *(continued)*
138–40, 143, 146–47, 271n8; outside
France, 183, 218, 219; *riparia*
parentage, 256; *rupestris* hybrids, 256.
See also Old Americans
hybridization and hybridization research,
134–35, 136; Germany, 279n10;
heritability of phylloxera resistance,
91–92, 114–15, 138, 141, 142,
270nn1,45; Millardet's work, 122,
135–43, 146–47, 151–56, 272n10,
273n18, 274n30; Montpellier's position
on, 122, 142, 147, 151, 270n1,
274nn27,28; natural hybrids, 115–16,
136–37, 138, 147, 148, 271n6;
pollination mechanisms, 136–37,
272n10; reversion to type, 136,
270n45; selection for local conditions,
169, 276n6. *See also* hybrid direct
producers; hybrid rootstocks;
inheritance theories; *individual species
and varieties*
hybrid rootstocks, 80, 81; *berlandieri*
hybrids, 142, 151–56, 169, 221,
254–55, 256, 274n30; in California,
222–24; French disavowal of, 223, 245;
in Italy, 169, 254–55; Millardet's work,
93, 135–40, 143, 151–52, 154, 155–56;
Montpellier's position on, 80, 122, 142,
274n28; Old Americans, 220, 221, 223,
265n2; phylloxera resistance, 222;
riparia hybrids, 138, 153, 174–75, 221,
228–29, 256, 272nn13,15; *rupestris*
hybrids, 153, 169, 174–75, 191, 221,
256, 272nn13,15, 278n11; selection for
local conditions, 169, 222, 223, 276n6;
in South Africa, 191. *See also* hybridization and hybridization research; *specific
varieties*

île de Ré, 62, 153
île d'Oléron, 153
immersion. *See* submersion
import restrictions, 171, 174, 192, 194. *See
also* quarantines; transport restrictions
inheritance laws, 265n16
inheritance theories, 140–42, 266n10;
Millardet's ideas, 93, 141–42, 268n25,
270n45; Montpellier research and, 142,
270n1; phylloxera resistance and,
83–84, 93, 114, 141–42, 263n29
insecticides. *See* pesticides
L'Insectologie agricole, 30–31
insect pollination, 137, 272n10

institutional failures, 248–49; California,
240–46
Institut National de la Recherche
Agronomique, 261n6
*Instructions pratique sur le procédé de la
submersion* (Faucon), 65–66
invasive species, 1, 2–5, 259n5
Irish potato famine, 259n7
iron, in soils, 130–31, 154, 271n5
irrigation, 198, 199
Irvine, Hans, 186, 189
Isabella, 202–3, 204, 265n1; in France,
81, 86, 87, 205, 208; in Italy, 170,
203–4, 209–10, 279n8; phylloxera
resistance, 86, 87; varietal characteris-
tics, 257
Isère, 32–33
Italian vineyard workers, 123
Italian viticulture, 166, 168–69; Old
Americans, 168, 170, 203–4, 209–10,
276n5; phylloxera and reconstitution,
163, 165–71, 223, 275n1
Ives Seedling, 181, 276n13
Izard, Jean, 131–32

Jacob, Harry, 224, 225, 245
Jacquez (Jacquet), 265n2; alternate names
for, 220, 278n4; in the American South,
89, 112; in California, 220, 221, 223,
280–81n14; chalk tolerance, 155*table;*
current French plantings, 207–9,
279n17; flavor and quality, 91, 202,
207, 279n17; French planting
prohibition, 205, 207–9; Hungarian
trials, 181; older French plantings, 102,
111, 130, 131, 202, 267n17, 269n35;
parentage, 202, 254, 265n1, 266n7;
phylloxera resistance, 87, 193;
Planchon's observations, 87, 89; as
rootstock, 193, 220, 223, 280–81n14;
seedling variability, 115; in South Africa,
191, 193, 280–81n14; varietal
characteristics, 257
"Le Jacquez, un cépage chargé d'histoire:
son adaptation au terroir cévenol and
les enjeux de son maintien" (Piernavieja), 208–9
Jaeger, Hermann, 116, 136, 146, 148–49,
250
Jaeger hybrids, 136, 137, 146; Jaeger #70,
136, 254, 273n21
Jardin des Plantes, 19
"La jaunisse ou chlorose des vignes"
(Sahut), 133–34, 271n7

Jeannemot, M., 46
JL (speaker about phylloxera in Épernay), 52–53
Journal d'agriculture, 29, 40, 68
Journal d'agriculture pratique, 37, 98

Kelleys Island, 85, 87, 89
Kepler, Johannes, 23
Kessler, H.F., 177
Kew, Planchon at, 161–62
KSCS₂ (potassium sulfocarbonate), 73, 74–75, 99
Kuhlmann, Eugene, 142, 256, 272n15
Kuhn, Thomas, 28

Lachiver, M., 264–65n15, 265n17
Lafitte, Prosper de, 103–19, 127; on American vines' phylloxera resistance, 105–7, 114, 270n45; Fitz-James's response to, 19, 109–13, 269n39; opposition to American vines, 104–9, 114–17, 119, 150; as pesticide proponent, 71–72, 103, 104, 109, 118; research plot proposal, 117–18; support for Balbiani's egg-destruction theory, 103–4, 109, 118, 269n36
Lake Erie viticulture, 85, 87, 89
Laliman, Leo, 42, 82–83, 131; American vine plantings, 9, 35, 38–39, 43–44, 266n8; as American vine proponent, 82, 93, 127, 268n29; in Portugal, 174; warnings about nonresistant American varieties, 86–87; on wine made from American grapes, 91, 201
Lamarckian inheritance theory, 266n10
Languedoc, 54, 126, 243, 276n4
Latour, M.L., 77
Laurent, R., 77, 265n18
Laval, M., 31
law: French inheritance laws, 265n16. *See also* legal restrictions and requirements
Law of 2 August 1879, 56–58, 75
Law of 15 July 1878, 56–58, 75, 99
Law of 22 June 1874, 46, 263n31
Leenhardt, M., 124
legal restrictions and requirements, 5, 6, 56–58, 263n31; American varietal prohibitions, 203, 204–10, 276n9; Argentina, 197; compulsory chemical treatments, 57–58, 75–77, 99, 165–67; EU, 203, 207–8, 209, 210, 276n9; grower impacts and opposition, 6,

57–58, 76–77; Italy, 165–67; South Africa, 194; Spain, 171. *See also* government responses; quarantines
Legros, Jean-Paul, 19, 242, 248, 249
Le Messager Agricole du Midi, 15, 17, 18
Lenoir. *See* Jacquez
Léon Millot, 256, 272n15
Le Play, P.-G.-F., 96
Lesseps, Ferdinand de, 70
Lewin, Roger, 241, 282n37
Lichtenstein, Jules, 44*fig.*, 268n29; and phylloxera identification, 33, 35, 37, 275n35; and Riley, 35, 42–43, 262n26
Lider, Lloyd, 223, 224–32, 241, 281n15; 1957–1958 AxR1 recommendation, 226, 227–28, 229–30, 241; 1974 followup, 229–32, 281n27, 282n29; research program goals and methods, 224–26, 229, 230, 240, 244, 245, 281–82n28
Liebig, Justus, 27
lime, in soils, 154–55. *See also* chalky soils
Linnaean Society of Bordeaux, 19; 33
Listan Prieto, 196, 280n1
Lister, Joseph, 27
Longuerue, M. de, 68
Longworth, Nicholas, 85
Lot-et-Garonne phylloxera committee, 103, 113, 131
Loubet, M., 100
Lubow, Arthur, 232, 235, 281nn27,28
Lyell, Charles, 25
Lyons, 57
Lythrum salicaria (purple loosestrife), 3

MacOwan, Peter, 194
madder root, 60, 264n6
Madeira, 171
maize. *See* corn
Malbec, 196–97
Mancey, 52, 53, 263n2
Manhattan Project, 6
Marais, J.S., 194
Mardeuil, 53
Marechal Foch, 256, 272n15
Marès, Henri, 28–29, 31, 41, 62–63, 261n11
Marescalchi, A., 165, 166–67, 170–71
Marion, Antoine, 61, 73
Marseilles, 73, 264n5
Martin, 221
Mas de las Sorres, 46, 47, 154, 263n32
Mas du Roy, 62–63, 264n8

medical models of disease, 18–19, 24–26, 27, 29
medicine in France (1800–1875), 26–28; medical training for scientists, 24, 26–27; physiological theories of disease and, 24–26
Médoc, 14. *See also* Bordeaux
Mémoire de la Vigne, 208, 209
Mendel, Gregor, 83, 140, 266n10, 272nn14,17
Mendocino, 230
Mendoza, 197, 198, 199
mercury sulfide, 72
Merlot, 197
Le Messager Agricole, 42
methanol, 209, 279n19
Meynot, M. de, 67
Micé, M. le docteur, 100–101
Michel, M., 69, 70
Midi: American vine controversy and, 100–102; American vine plantings, 87–88, 110, 127, 268n32; economic impacts of phylloxera, 110; Old Americans in, 87–88, 102, 202, 269n35; pesticide application difficulties, 99; phylloxera in, 4, 6, 177
military-industrial complex, 260n10
Millardet, Alexis, 8, 46, 107, 130, 155
Millardet hybrids: 41B, 139, 143, 155–56, 254, 272n13, 273n18, 274n32; 101–14Mgt, 174–75, 221, 223, 272nn13,15. *See also* 420A Mgt
Millardet research: hybridization work, 122, 135–43, 146–47, 151–56, 270n45, 272n10, 273n18, 274n30; on phylloxera resistance of American species, 91–93, 106, 141, 268nn24,25, 270n40; on rootstocks from wild seed, 92–93, 114, 220, 254; on taxonomy of American species, 46, 84, 91, 93, 141, 271–72n9
Milleferet, M., 72
Millot, M., 52
Ministry of Agriculture, 6, 41, 84, 123, 145, 263n3; minister at Planchon monument dedication, 157, 159–60. *See also* National Phylloxera Commission
Miquel-Paris, M., 69–70
Mission grape, 196, 211, 212, 280nn1,3
Mississippi Valley grape species, 115–17, 220
Missouri: Nedeczky's visit, 181; Planchon in, 84–85, 250, 266–67n14; Viala in, 146–47; wild seed and cuttings from,

115–17, 220–21; wines and viticulture, 85, 89, 116, 266–67n14
Missouri Botanical Garden, 263n30, 266n13, 267n15
monoculture, 4, 18, 25
Montecristo vine nursery, 168
Montpellier: as cradle of new science, 158–59; Planchon monument, 157–60, 158*fig.*, 274n33, 275n34. *See also* École Nationale Supérieure d'Agronomie de Montpellier; French expertise; University of Montpellier
Morrow, Dwight W., Jr., 76, 263n30, 264–65n15
Mortillet, M. de, 100
Morton, Lucie, 278n12
mosaic theories of inheritance, 93, 141
Mouillefert, F., 54, 70–71, 74
mulching, 60
Munson, Thomas Volney, 112, 113*fig.*, 258, 271–72n9, 272n14, 273n24, 278–79n15; as nurseryman/breeder, 149, 273n25; and Viala, 139, 147, 148–49, 250, 254
muscadines *(Vitis rotundifolia),* 86, 267nn15,16,21
mussels, 3

Nagaoka, Rich, 243
Nantais, 124–25
Napa, 215, 230, 232–36
Narbonne, 58
National Academy of Medicine (France), 26
National Phylloxera Commission, 6, 9, 41–42, 54–58; Planchon on, 6, 41, 55–56; position on American vines, 54–58, 71, 98; position on cause controversy, 41, 48–49, 54; support for control methods, 56–58, 70–71, 263n3
National Science Foundation, 241
natural selection: resistance dynamics and, 83, 263n29. *See also* Darwinian theory
Naudin, Charles, 19, 21, 48, 132, 141
Nedeczky, M.E., 181
New South Wales, 185, 188–89
newspapers. *See* agricultural press
New York: Planchon in, 86; Viala in, 146
New York Times Magazine: Lubow article, 232, 281nn27,28
New Zealand, 179
Niagara Falls, Planchon's visit to, 85–86
Noah, 202, 204, 205, 258, 266n6; *pineau de Noah,* 206–7

North Carolina: Planchon in, 84; Viala in, 146
Norton (Cynthiana, Norton's Virginia), 258, 267n20; in France, 128, 270nn1,43, 273n25, 279n6; Hungarian trials, 181; in Missouri, 89, 112; parentage, 89, 254, 258, 265n1, 266n7; Planchon on, 91
"Nouvelle maladie de la vigne" (Faucon), 37

Oberlin (Oberlin Noir), 206, 279n14
Oberlin 595, 142, 272n16
Oberlin, Christian, 141–42, 272n15
Office International de la Vigne et du Vin, 261n6
Ohio, 85, 87, 89, 146
oidium (powdery mildew), 35, 138, 146, 203, 261nn8,11
Old Americans, 79, 80, 183, 201–10, 248, 265n1, 278n1; in California, 212, 213, 218–19, 280n3,5,6; in Central Europe, 180–81, 204, 209, 276n9; Fitz-James's recommendations, 111table; flavor and wine quality, 45, 91, 122, 201, 202, 204, 205, 219; foxiness in, 81, 91, 202–3; in France, 79, 80, 122, 183, 202, 203, 205–9, 219, 269n35; French and EU prohibition and protection efforts, 203, 204–210, 276n9; French positions on, 91, 102, 122, 183, 275–76n2; in Greece, 204; in Italy, 168, 170, 203–4, 209–10, 276n5; methanol in, 209, 279n19; origins and parentage, 81, 201, 253, 254, 255, 266n7, 271n8, 278n1; phylloxera resistance, 203, 219; as rootstock, 93, 97, 122, 220, 221, 223, 265n2; in South Africa, 191, 193; variability in seedling offspring, 115; variety descriptions, 257–58; vinifera parentage in, 81, 202. See also American vine controversy; specific varieties
Oliveira, Manuel de, 173
olives, 87, 123, 166, 176, 276n8
Onderdonk, G., 116
"On the Identity of the Two Forms of Phylloxera" (Lichtenstein and Planchon), 35, 37
ontological theories of disease, 24, 26, 27. See also phylloxera-cause theory
Ophistoma ulmi (Dutch elm disease), 3, 4
Ordish, G., 271n4
Othello, 93, 155table, 204, 205, 271n7

Pagézy, M., 122, 124
Pais, 196
Pal injectors, 73–74, 73fig., 74fig.
Palomino Negro, 196
Parc National de Cévennes, 207–8, 209
Parthenocissus, 82, 120, 266n9
Pasteur, Louis, 9, 21–22, 27, 45–46, 132
Patterns of Discovery (Hanson), 23
Paul, Harry, 141, 142, 265–66n3, 266n11, 270nn1,2, 272n15
Paulsen, Federico, 167–68, 169–70, 254–55
Paulsen 1103, 169
Pearce, David, 9, 10
Pearce, M. Rosaria di Nucci, 9, 10
Pedro Ximénes, 138
Pelling, Margaret, 27
Pemphigus vitifolii, 35, 43
Perdue, Lewis, 240–41, 242, 243
Permanent International Viticultural Commission, 164
Perold, A.I., 278n14
Pesquidoux, Joseph de, 16, 235, 282n34
pesticides, 5, 70–78, 276n12, 278n17; calls for public suggestions, 46–47, 180; in the Charentais, 153; early experiments, 31–32, 39; government projects and funding (France), 6–7, 10, 46–47, 56, 75, 76; high costs of, 74, 76–77, 102, 265n18; outlandish treatments, 46, 99, 180. See also carbon disulfide; sulfiding
Petit Bouschet, 134
Petite Champagne (Cognac), 153
petroleum treatments, 178
Phillip, Arthur, 184
Philosophical Investigations (Wittgenstein), 261n10
phosphoric acid: fertilization with, 65
phylloxera (Daktulosphaira vitifoliae), 2, 14, 20fig., 36fig., 259n1; Balbiani's research, 103–4, 118, 269nn36,37; biotypes, 233–35, 283n44; discovery and identification, 16–18, 33–37, 43, 159, 162, 275n35; evolution and genetic variation, 237–40, 282nn36,37; genotypes, 236–39, 283n44; Granett's research and Type B identification, 232–35; leaf- vs. root-dwelling forms, 35, 239, 251; life cycle, 36fig., 251, 269n36; overwintering capability, 238, 283n42; reproduction, 237, 251, 269n36, 282n37, 283n42; winged forms, 20fig., 33, 36fig., 43, 57, 187, 195, 197, 213–14, 237, 251

"Le phylloxéra, une histoire sans fin"
(Legros), 19
phylloxera behavior, 251; Argentina, 197;
Australia, 187–88, 277n6; California,
237, 240, 282n32; climate and, 177,
187, 197–98, 260n9; irrigation and,
198, 199
Phylloxera Board (California), 214, 220
phylloxera-cause theory, 9–10, 18–19, 24,
27–28, 249; growth of support for, 32,
40, 42, 47–50; official/national
positions, 41, 48–50, 54, 263n31;
testability of, 29–30, 31–32. See also
cause controversy
phylloxera commissions, 7, 41–42, 249; and
American vine controversy, 98;
Argentina, 198–99; California, 214; first
commission's investigation, 16–18, 19,
28, 43, 261n8, 275n35; Hungary, 179;
Portugal, 173; South Africa, 193,
194–96. See also departmental
phylloxera commissions; National
Phylloxera Commission; specific
departments
phylloxera-effect theory, 9, 10, 249; early
expressions of, 28–29; Guérin-
Méneville's 1873 paper, 47–48; key
proponents, 18, 19, 21; multicausality
in, 25, 29, 48; as physiological theory,
24, 25–26, 29; Planchon's views on, 18,
40; sources of, 26–27. See also cause
controversy
"Le phylloxéra en Europe et en Amérique, I;
II" (Planchon), 13–14, 18, 25–26, 33,
35, 40, 86, 269n38
Phylloxera Exclusion Zones, 190, 277n4
phylloxera infestations, 4; Europe outside
France, 163–83; southern hemisphere,
184, 199–200; spread through Europe,
163–64; viewed as inevitable, 186–87,
189–90; wind as vector, 188, 195,
213–14, 277n7. See also California
phylloxera infestations; French
phylloxera infestation; other specific
locations
Phylloxera Infested Zones, 277n4
Le Phylloxéra de la vigne: son organisation,
ses moeurs (Girard), 49
Phylloxera quercus, 33
phylloxera resistance, 45, 269n39, 282n36;
Bazille's argument for, 102, 105–6; in
California research programs, 222, 223,
224, 225, 226–29, 244; current state of
knowledge, 236–37, 270n41; French

doubts and concerns about, 102, 105–7,
268n31; heritability of, 83–84, 91–92,
114–15, 138, 141, 142, 263n29,
270nn1,45; hybridization and, 93,
114–15, 268n24; labrusca-based
varieties, 86–90, 92, 106, 114–15, 127,
267n21, 269n39; Lafitte on, 105–7, 114,
270n45; Laliman's observations and
advice, 38–39, 82, 86–87; Millardet's
work, 91–93, 106, 137, 141,
268nn24,25, 270n40; Old Americans,
203, 219; Planchon's work and
positions, 83–84, 86–90, 127,
267nn16,21. See also American vine
controversy; resistant rootstocks;
individual species and varieties
phylloxera responses, 4–5, 199–200,
247–50; successful responses outside
France, 183. See also denial; French
expertise abroad; government responses;
specific locations
Phylloxera vastatrix, 259n1. See also
phylloxera
Phylloxéra vastatrix cause prétendue de la
maladie actuelle de la vigne (Signoret),
33, 42
phylloxeristes. See phylloxera-cause
theory
physiological theories of disease, 18–19,
24–26, 27, 29. See also phylloxera-effect
theory
Pickering, Andrew, 11, 260n16
Pierce's disease, 3
Piernavieja, Herminie, 208–9
Pieyre, M. le Baron, 130
Pinch, Trevor, 11, 260n16
pineau de Noah, 206–7
Pinney, Thomas, 220, 243
Pinot Noir, 265n18
Pinot Noir Précoce, 272n15
Piola, M., 74–75
Placer County, 213
Planchon, J.-É., 13, 17fig., 44fig., 155,
156–62, 249, 268n29; on American vine
restrictions, 56; on California wines,
266–67n14; on cash prize for effective
control, 46–47; and cause controversy,
18, 31, 40; contributions to Bushberg
catalogue, 129; credited with
discovery of phylloxera, 159, 162,
275n35; death and funeral of, 156–57;
on foxiness, 91; grafting lectures, 124;
Montpellier monument to, 157–60,
158fig., 274n33, 275n34; on National

Phylloxera Commission, 6, 41, 55–56; phylloxera identification and life cycle research, 33, 35, 37, 275n35; on phylloxera's appearance and spread, 13–14; as proponent of American vines, 90–91, 93, 95–96, 127; and Riley, 35, 42–43; and submersion control, 38, 262n19; teachers and intellectual influences, 27–28, 160–62; U.S. study tour (1873), 45–46, 84–86, 89, 250, 262–63n27, 266n12, 273n20; views on resistance of American vines, 83–84, 86–90, 127, 263n29, 267nn16,21; at Zaragoza congress, 171, 249. *See also* Hérault phylloxera commission

Planchon, J.-É, works: *Les cépages américaines dans le département de l'Hérault pendent l'année 1870*, 15, 89–90, 267n21; "Le phylloxéra en Europe et en Amérique, I; II," 13–14, 18, 25–26, 33, 35, 40, 86, 269n38; "La question du phylloxéra en 1876," 88–91, 269n38; *Les vignes américaines, leur culture, leur résistance au phylloxéra et leur avenir en Europe*, 46, 86, 94, 263n30, 267n15

PLM carbon disulfide research and distribution, 72–74, 99

pollination mechanisms, 136–37, 272n10

polyculture, 77, 123, 166–67, 170

Ponts et Chaussées, 69, 96, 264n13

port, 163, 171, 172

Portugal, 163, 171, 172–75

Portuguese workers, 123, 124

potassium, 65

potassium arsenate, 72

potassium sulfocarbonate (KSCS$_2$), 73, 74–75, 99

potato famine, 259n7

Pouget, R., 14–15, 45, 261n6, 263n3, 264–65n15; on American vine debate, 55, 265–66n3; on Millardet's work, 93; on the cause controversy, 261n13

pourri des racines, 32, 261n12

pourridié, 14–15

powdery mildew (oidium), 35, 138, 146, 203, 261nn8,11

Prades, M., 67–68

Prairie Farmer, 34, 35

press. *See* agricultural press; *Revue des deux mondes*; *other specific publications*

Provence, 77, 87

Prussian War, 6, 41

pucerons, 17, 261n9

Pulliat, Victor, 96–97, 268n29

purple loosestrife, 3

QUANGOs (quasi-autonomous nongovernmental organizations), 6, 260n11

quarantines, 5, 56, 165–66, 178, 283n45; Algeria, 280n11; Argentina, 197; Australia, 186, 187–88, 199, 259n8; California, 217, 218; South Africa, 193, 194, 195, 217

quasi-autonomous nongovernmental organizations, 6, 260n11

Quatre ans de luttes pour nos vignes et nos vins de France (Lafitte), 103

Queensland, 185, 188

La question des vignes américaines au point de vue théorique et pratique (Millardet), 91–93, 106

"La question du phylloxéra en 1876" (Planchon), 88–91, 269n38

"La question du phylloxéra et le rôle des vignes *américaines*" (Lafitte), 104–9

quinine, 72

railroads: role in pesticide research and distribution, 72–74, 99, 174, 180

Ramsey, 223

Ravaz, M., 142, 152, 155, 270–71n2, 271–72n9

reconstitution, 79–80, 133–34, 170, 282n34; Argentina, 198–99; Australia, 186–87, 188–89; Austria, 178, 276n9; Bazille on, 103; as birth of new viticulture, 164–65, 170–71, 196, 247–48; Bulgaria, 182, 183, 277n14; California, 218–22, 234–36; Croatia and Dalmatia, 179; Gervais' view, 164, 275–76n2; Hungary, 180–81; initial failures in France, 87–88, 121–22, 127, 131, 134; Italy, 169–71, 223; Portugal, 174–75; Romania, 181–82; scale and expense of, 124, 126–27, 127*table*; South Africa, 193, 194, 196; Spain, 172, 175–76, 223. *See also* American vine controversy; Old Americans; resistant rootstocks

reed mulches, 60

Regent, 279n10

Reiss, Stephen, 235–36

research funding, 6, 7, 240–41. *See also* control efforts and methods; pesticides

resistant direct producers. *See* hybrid direct producers; Old Americans

resistant rootstock research: Argentina, 199;
Australia, 188–89, 190–91; ease of
propagation as goal, 139, 141, 145, 147,
148, 152, 243–44; funding for, 240–41;
Italy, 169, 254–55; Montpellier research,
80, 121, 154, 245, 249; Montpellier's
focus on, 122, 148, 270n1, 274n28; one
stock fits all notion, 189, 221, 226, 227,
277n9; selection for local conditions,
169, 190–91, 199, 276n6, 278n15;
South Africa, 193. *See also* California
rootstock research; hybrid rootstocks;
seedling rootstocks
resistant rootstocks, 79, 80, 120–62, 248,
272n13; adaptation concerns, 120–21,
267n22; American species as, 93, 104,
114–17, 147–48, 149–50, 268n25;
Argentina, 198–99; Australia, 186,
188–89, 190–91; Bazille's experimentation, 82–83; Eastern Europe, 180–82;
Fitz-James's recommendations, 111*table*;
Italy, 169, 170–71, 223; Lafitte's
concerns and opposition, 104, 107–8,
114–17, 119; Laliman's championing of,
82, 93, 127, 268n29; Old Americans as,
93, 97, 122, 220, 221, 223, 265n2;
Planchon's views, 90–91, 93, 127;
propagation difficulties, 122, 124, 139,
150, 152, 243–44; rootstock adaptation
in California, 221, 222, 223, 227–28,
244; scion-stock affinity, 108, 121, 221,
244, 267n22, 276n7; significant species
for, 253–56; South Africa, 193; Spain,
223; supply problems, 122–23, 190, 193;
white soils problem, 127–34, 143–56.
See also American vine controversy;
chalky soils; grafting; hybrid rootstocks;
reconstitution; resistant rootstock
research; seedling rootstocks
Rességuier, Euryale, 150
Revue des Deux Mondes: Fitz-James's 1881
article, 19, 109–13, 269n39; Lafitte's
1881 article, 104–9; Planchon's 1874
article, 13–14, 18, 25–26, 33, 35, 40, 86,
269n38, 275n35; Planchon's 1877
article, 88–91, 269n38. *See also* "Le
phylloxéra en Europe et en Amérique"
Revue des hybrides américains, 274n29
Revue de viticulture: Gervais articles, 164
Rhône-Gironde canal, 70
Ribera del Duero, 171
Richter rootstocks: 99R, 224, 226,
280–81n14, 281n20; 110R, 282n33
Riebeeck, Jan van, 191

Riley, C.V., 34–35, 34*fig.,* 262n25; Bushberg
catalogue contributions, 94–95;
Darwinian views, 261n14, 271n4;
phylloxera research, 20*fig.,* 35,
42–43, 82, 250; and Planchon's U.S.
visit, 46, 84–85, 250, 262–63n27;
at the USDA, 262–63n27; and views on
phylloxera resistance, 82, 87, 282n36;
visits to France, 43, 44*fig.,* 250,
262–63n27
Rioja, 175
riots, 166–67, 276n4
Riparia Gloire (de Montpellier), 131, 151,
221, 223, 271n6
Ripert, M., 31
riverbank grape. See *Vitis riparia*
Robinson, Jancis, 278n16
Rogers, Edward, 271n8
Rohart, M.F., 71, 72
Romania, 181–82, 183, 210, 249
root rot, 14–15, 32, 261n12
rootstocks. See resistant rootstocks; *specific
varieties*
Rouville, M., 156
Royal Botanical Gardens, Kew, 161–62
Royal Society of Medicine (France), 25
Ruggeri rootstocks, 169
Rulander, 181, 276n13
Rupestris du Lot (Rupestris St. George),
244, 256, 281n19; in California, 221,
223, 224, 226, 227, 230, 281n21,
282n29; in France, 151; in Portugal,
174–75
Rupestris Martin, 221

Sabatier, M., 160, 275n35
Sabran, Duc de, 69
Sahut, Felix, 122, 124, 156; on carbon
disulfide, 72; on failures and problems
of American vines, 87–88, 134, 146,
271n7; and phylloxera discovery, 16, 18,
275n35; on soil adaptation, 128,
133–34, 271n7
Saint-Clément, 88
Saint-Émilion, 74–75
Saintpierre, Camille, 27–28, 124, 155, 171
Saint-Pierre, Mme., 131
salt buildup, sand planting and, 60
sand planting, 10, 39, 58–62, 99, 248,
264n9, 278n17; Argentina, 198;
California, 218; Hungary, 180; Italy,
167; Portugal, 173; wine quality,
262nn21,22
sand treatment, 62–63, 264n8

sandy soils, 64, 152–53, 179–80. *See also* sand planting
San Juan, 197, 198
Santa Clara Valley, 215
Saône-et-Loire, 52
Sardinia, 166, 169
Sauvignon Vert, 281n26
SCAH (Société Centrale d'Agriculture de l'Hérault), 6, 51, 249; and first phylloxera commission, 16, 261n8; and Planchon, 46, 84, 156; and Planchon monument, 157, 158; and Viala's mission to the U.S., 144; as proponent of American vines, 100, 122, 124; educational activities, 96, 121, 270–71n2; founding and activities, 6, 261n8; officers of, 28–29, 82; survey of American vine plantings, 15, 89–90, 267n21
scientific cooperation, 10, 33–34, 43, 94, 250. *See also* French expertise abroad
scientific disagreements, 21–24, 28, 40
scientific practice: argument in, 7–8; impact of beliefs on, 21–24, 28, 40; as social construction, 11–12, 260n16. *See also* theory-practice interactions
scientific viticulture, phylloxera responses as birth of, 164–65, 170–71, 196, 247–48
seedling rootstocks, 147, 268n25; in California, 220; genetic variation in, 115, 150–51, 254; grafting technique and, 124–25; Lafitte's views, 104, 114–17; Millardet's research, 92–93, 114, 220, 254; in South Africa, 193
Seibel, Albert, 141, 219
Seigle, Dr., 264n10
Semichon, M., 154
Sequin, M., 32
Serres, M. de, 40
Sète, 59
Sétubal, 173
Seyval, 280n12
Seyve-Villard hybrids, 280n12
Shaw, Henry, 266n13
sherry, 171, 175, 279n7
shrimp shells, 276n12
Sicily, 163, 166, 169, 223
Sigean, 69
Signoret, Victor, 19, 21, 22*fig.*, 28, 48; cooperation in phylloxera research, 34, 42, 262n25; *Phylloxéra vastatrix cause prétendue de la maladie actuelle de la vigne,* 33, 42; Planchon's opinion of, 40; and Riley, 42

silkworms, silk industry, 19
Smith, Rhonda, 234–35
Smith-Lever Act, 260n13
social construction, scientific practice as, 11–12, 260n16
social issues: demographic impacts of phylloxera, 58, 179, 247, 264n5; tensions between large and small landholders, 69–70, 71, 74, 75, 76–77, 265n18
Société Centrale d'Agriculture de l'Hérault. *See* SCAH
Société d'agriculture de Vaucluse, 16, 261n8
Société des agriculteurs de France, 164
Société d'horticulture et d'histoire naturelle, 156
Société vigneronne de Beaune, 77
soils and soil composition, 9, 25, 60, 68, 198; California exceptionalism and, 242; chlorosis and, 130–31, 133–34, 153, 154–55, 271n5; Hungary, 179–80; Italy, 168–69; phylloxera-effect view and, 29; phylloxera resistance and, 193, 227–28, 242, 280–81n14; Portugal, 174–75; sand planting and, 59, 60–61; submersion and, 64, 67–68; sulfide treatment effectiveness and, 75. *See also* adaptation; chalky soils; sand planting
soil sterilization (extinction treatments), 57–58, 165–67, 177–78, 217–18
Solano County, 230
Solaris, 279n10
Solonis, 104, 115, 135, 271–72n9; chalk tolerance, 155*table;* Hungarian trials, 181; varietal characteristics, 258
Sonoma, 211–13, 214, 230, 234–36
Sonoma Vinicultural Association, 215
Soubeiran, M., 156
South African viticulture, 191–92, 196, 280–81n14; phylloxera, 164, 191, 192–96, 199, 237
South America. *See* Argentina; Chile
South Australia, 186, 187, 189–90
Southern Corn Leaf Blight, 4, 259n6
southern hemisphere, phylloxera in, 184–200; Argentina, 164, 196–99; Australia, 4, 164, 184–91, 259n8, 277n2; overviews, 184, 199–200; South Africa, 191–96
Southey, J.M., 278n14
Spain: AxR1 in, 228, 241, 245; phylloxera in, 171–72, 175–76, 223, 249; as source of *vinifera* in the Americas, 196, 197, 211, 280n1

Spanish vineyard workers, 123, 124
State Viticultural Commission (California), 214, 216
Station Viticole de Cognac, 152
Stel, Simon van der, 191–92
Stellenbosch, 191–92
St. George (Rupestris du Lot), 244, 256, 281n19; in California, 221, 223, 224, 226, 227, 230, 281n21, 282n29; in France, 151; in Portugal, 174–75
St-Martin-de-Crau, 14
strawberry wines, 219; Italian Fragolino, 203–4, 209–10; Uhudler, 204, 209, 276n9, 279nn9,20
submersion, 10, 37–38, 40, 63–70, 66fig., 248, 278n17; Argentina, 198; California, 218; Charentais vineyards, 152; duration and continuity, 66–67, 264n12; effectiveness of, 10, 63, 65–67, 68, 70, 97, 99, 264n12; estate size and, 69–70; expense of, 69, 104; Faucon and, 9, 37–38, 40, 63, 65–66, 249, 262nn17,19; Hungary, 180; Italy, 167; local syndicates, 63–64, 69; official/ government role, 56, 63, 69–70; Portugal, 173; problems and limitations, 64–65, 67–68, 167, 262n18; summer inundation, 66, 264n10; water requirements, 64–65, 68, 180, 262n18
Le Sud Est, 97, 98, 100
sulfiding, 57–58, 72–78, 73fig.; application equipment, 73–74, 73fig., 74fig.; Austria, 178; California, 218; expense of, 74, 76–77, 102, 104, 265n18; Hungary, 180; Italy, 165–67, 167–68; Lafitte as proponent of, 71–72, 103, 104, 109, 118; official/government support and oversight, 70–71, 75–76; opposition to, 71–72, 76, 166–67; Portugal, 174; with potassium sulfocarbonate, 73, 74–75, 99; problems and limitations, 5, 71–72, 74–75, 264n14; required by law, 75–76, 99; research and experimentation, 39, 46, 72–74; Swiss method, 57, 165–67; ultimate failure of, 77–78. See also carbon disulfide
sulfocarbonates, 70, 72, 264n14; potassium sulfocarbonate (KSCS₂), 73, 74–75, 99
sulfur, 62, 261n11
sulfuristes, 70–71, 95–103. See also American vine controversy; La Défense; Lafitte, Prosper de; sulfiding

Sullivan, Charles, 280n4
sweet summer grape. See Vitis aestivalis
Swiss sulfiding method, 57, 165–67
Switzerland, 57, 163, 165
Sylvaner, 281n26
symbolic logic, 7–8
syndicats de défense, 6–7; American vines and, 76, 77, 265n17; chemical treatments and, 75–76; submersion and, 63–64, 69
Syrah, 197

Talabot, M.P., 72–73
Tarn-et-Garonne, 132–33
Taylor, 89, 93, 107, 267n17; foxiness in, 91; Hungarian trials, 181; as rootstock, 90–91, 97, 155table; varietal characteristics, 258
technological progress, 9–11
Tennessee: Viala in, 146
Texas, 112, 116; Viala in, 146, 147–48
Thénard, Baron, 31–32, 39, 46, 72, 73, 96
theory-practice interactions, 7, 8–12, 260n16; impact of scientists' beliefs, 21–24, 28, 40
Third Annual Report on the Noxious, Beneficial and Other Insects of the State of Missouri (Riley), 42, 262n26
Thurber, George, 86
tobacco, 31, 72
Tocqueville, Alexis de, 283n48
Torrontés, 197
traitement d'extinction (extinction treatments), 57–58, 165–67, 177–78, 217–18
transport restrictions, 5, 6, 171, 194, 283n45. See also quarantines
Trimoulet, A.-H., 19, 21, 40, 48
Trouchaud-Verdier, L., 262n17
Type B phylloxera, 233–35, 283n44

Uhudler, 204, 209, 276n9, 279nn9,20
United States. See America; specific states
U.S. Geological Survey, 145, 262–63n27, 273n20
University of Bordeaux, 80, 114, 270n45. See also Millardet, Alexis
University of California, 220, 221, 224, 266–67n14, 283n51; Bushberg catalogue at, 268n27; Granett's 1980s research, 232–35; and 1980s–1990s phylloxera infestation, 232, 234–35, 240–41, 243; viticulture department

establishment, 214–15. *See also*
California rootstock research
University of Montpellier, 6, 7, 28, 46–47,
156; phylloxera research at, 33, 42;
Planchon at, 28, 160–62. *See also*
Planchon, J.-É.
University of Nancy, 162
USDA, 221–22, 240, 241, 262–63n27

Vallejo, Mariano, 211
Valmadonna riots, 166–67
Valmadrera, 167
Vannuccini, V., 61
Var, 57
Var agricultural society, 100
Vassal, M. le Comte de, 67
Vaucluse, 15, 264n10
Vaucluse agricultural society, 100
Vendée, 206–7
Vène, A., 60–61
Verein der Freunde des Uhudler, 209,
279n20
Vernette, M., 60
Verneuil, A., 152, 155
Viala, Pierre, 142, 144*fig.*, 152, 155,
268n29, 270–71n2; background,
145; as *berlandieri* proponent, 147–48,
149–50, 151; on Solonis' origins,
271–72n9; visit to Bulgaria, 182, 249,
277n14; visit to the U.S. (1883), 139,
144–49, 250, 254, 262–63n27, 267n22,
273nn19,20
Vialla, Louis, 38, 121–22, 124, 127,
128, 130; *Les cépages américaines
dans le département de l'Hérault
pendent l'année 1870*, 15, 89–90,
267n21
Vialla commission, 38, 58
Vialla grape, 108, 155*table*
Victoria (Australia), 185–86, 190,
277n2
Viger, M., 157, 159–60
La Vigne américaine, 96, 268n29
La vigne française, 52, 97–98, 140, 178,
268n31; Ganzin's 1888 article, 135,
138, 139, 140; Millardet's 1888 article,
135, 138–40
*Les Vignes américaines: Catalogue illustré et
descriptif*, 93–95, 116, 268nn27,28
*Les vignes américaines, leur culture, leur
résistance au phylloxéra et leur avenir en
Europe* (Planchon), 46, 86, 94, 263n30,
267n15
vignoble, defined, 260n1

Villard blanc, 255, 280n12
Vimont, Georges, 53
Vincendon-Dumoulin, M., 98
vineyard inspections, 56, 57, 75;
California, 218; South Africa, 193,
194, 195
vineyard workers: foreign workers in
France, 54, 74, 123, 124; grafting
and workforce issues, 123, 124,
126; grafting education, 121–24, 190,
193
Les vins mythiques (Couderc), 208
viral infections, 244, 276n7
Virginia creeper, 120, 266n9
Le Viticulteur submersionniste, 63
viticultural education: grafting education,
121–24, 190, 193; Montpellier/SCAH
programs, 80, 95–96, 121–24, 245, 249,
270–71n2; pesticide application, 71–72,
73; South Africa, 193; University of
California, 214–15
viticultural practices, 10; Australia, 185;
fertilization, 10, 60, 62, 65, 74, 99; for
Old American varieties, 203; phylloxera
and, 18, 25; phylloxera responses as
birth of scientific viticulture, 164–65,
170–71, 196, 247–48; polyculture, 77,
123, 166–67, 170; sand treatment,
62–63, 264n8; sulfuring, 62, 261n11;
traditional, preservation of, 54, 119. *See
also* grafting; sand planting; submersion;
sulfiding
viticultural research stations: Australia,
188–89; California, 221; Cognac, 152;
Hungary, 179, 180, 276n11. See also
specific institutions
Vitis: American species' range and diversity,
80, 238, 253, 266n5, 283n40
Vitis acerifolia, 258. *See also* Solonis
Vitis aestivalis and hybrids, 80, 265n1;
Fitz-James as proponent of, 111;
French importation and planting,
267n17; Millardet's trials, 92–93,
115; natural hybrids, 115; Old
American varieties listed, 257–58;
phylloxera resistance, 86–87, 89, 93,
141, 254, 267n21; Planchon on
potential as direct producers, 91;
promise for chalk soils, 145–46;
propagation difficulties, 10, 145–46,
254; seedling graft stock, 92–93; species
as rootstock, 124, 139; species
description, 254; in Texas, 116. See also
specific varieties

Vitis aestivalis Lincecumii, 124, 136, 146, 254; Jaeger #70 hybrid, 136, 254, 273n21

Vitis berlandieri and hybrids, 8, 149–56; chalk tolerance, 155*table,* 221; Cognac research program, 152; Couderc and Millardet's crosses, 142, 151–52, 154–56, 274n30; hybrid direct producers, 255, 280n12; phylloxera resistance, 141, 151–52, 254; propagation, 10, 124, 149–50, 152, 153, 254, 274nn30,32; rootstock hybrids, 142, 151–56, 169, 221, 254–55, 274n30; scion compatibility problems, 276n7; seedling graft stock, 124, 254; species description, 254–55; Viala and, 147–48, 149–50, 151, 153, 254. See also *specific varieties*

Vitis bourquiniana, 278–79n5. See also Herbemont; Jacquez

Vitis californica, 220, 280n13

Vitis candicans (V. mustangensis), 137, 147, 256, 273n22

Vitis cinerea, 147–48, 256; chalk tolerance, 155*table*

Vitis cordifolia, 111*table,* 256; chalk tolerance, 155*table,* 267–68n23; Jaeger's hybrids, 146; Millardet's work, 137, 139; Planchon's observations, 89, 90, 267n21; Viala's observations, 146, 147–48

Vitis labrusca and hybrids, 80, 104; adaptation to French conditions, 87–88, 106, 110, 111, 127; as direct producers, 85, 86–90, 266n7; failures of, 87–88, 121–22, 127, 272n11; flavor, 81, 91, 202–3, 255, 278n2; in Italy, 168; Millardet's trials, 92, 93; Old American varieties listed, 257–58; phylloxera resistance, 86–90, 92, 106, 114–15, 127, 267n21, 269n39; as rootstock, 90–91, 92, 114–16; species characteristics, 201–2, 255. See also Old Americans; *specific varieties*

Vitis lincecumii. See *Vitis aestivalis Lincecumii*

Vitis longii, 258, 271–72n9. See also Solonis

Vitis monticola, 133, 139, 145–46, 147, 155*table,* 256

Vitis mustangensis. See *Vitis candicans*

Vitis novo-Mexicana, 147

Vitis riparia and hybrids, 80, 255–56; chalk tolerance, 139, 153, 155*table;* as direct producers, 256; failure of, 121, 127, 131, 134, 272n11; French seed imports, 93, 220; Hungarian trials, 181; hybrid rootstocks, 138, 153, 174–75, 221, 228–29, 272nn13,15; Jaeger's hybrids, 146; Millardet's work, 92–93, 114, 115, 137, 138; natural hybrids, 115; Old American varieties listed, 257–58; Planchon's observations, 85–86; in Portugal, 174–75; soil adaptation, 221; species and seedling graft stock, 92–93, 114, 150–51, 220–21, 254, 265n2, 271n6; virus-infected scions and, 276n7. See also *specific varieties*

Vitis rotundifolia (muscadines), 86, 267nn15,16,21

Vitis rupestris and hybrids, 256; chalk tolerance, 153, 155*table;* as direct producers, 256; hybrid rootstocks, 153, 169, 174–75, 191, 221, 256, 272nn13,15, 278n11; in Italy, 169; Jaeger's hybrids, 136, 137, 146, 254, 273n21; in Portugal, 174–75; species and seedling graft stock, 93, 151, 220–21; virus-infected scions and, 276n7. See also AxR1; Couderc hybrids; *other specific varieties*

Vitis vinifera, 9, 143; in the Americas, 81, 110–11, 196, 202, 211, 253; in Old American varieties, 202; viral infections and rootstock compatibility, 244, 276n7. See also American-European hybrids; *specific varieties and locations*

Vorwerk, S., 238–39

Walker, M.A., 213, 234, 236

water requirements and supplies, 64–65, 68, 70, 74–75, 180

weather. See climate

Weber, Ed, 234–35

Wetmore, Charles A., 215, 216, 220

whip-and-tongue graft, 125*fig.,* 126

white soils problem, 149–56. See also chalky soils

Whiting, John, 188, 277n3

wind: as phylloxera vector, 188, 195, 213–14, 277n7; wind pollination, 137, 272n10

wine prices: control costs and, 76, 265n18; Languedoc riots, 276n4

wine production, phylloxera impacts, 16
Wines & Vines, 235
Wisconsin, 280n2
Wittgenstein, Ludwig, 261n10
Wolpert, J.A., 214, 220, 223–24, 231, 241, 242, 280n8
wood ashes, 62
Woodward, James, 260n16

yellow fever, 26
York's Madeira, 115, 181

Zaragoza, 176
Zaragoza congress (1880), 171–72
zebra mussels, 3
Zinfandel, 230–31, 280n4, 281n26
Zyl, D.J. van, 192–93

TEXT
10/13 Sabon

DISPLAY
Sabon

COMPOSITOR
Westchester Book Group

INDEXER
Thérèsè Shere

PRINTER AND BINDER
Sheridan Books, Inc.